# ADVANCES IN SEED BIOLOGY

Edited by **Jose C. Jimenez-Lopez**

**Advances in Seed Biology**
http://dx.doi.org/10.5772/intechopen.68178
Edited by Jose C. Jimenez-Lopez

**Contributors**
Timothy Grey, Charles Chen, Russel Nuti, Walter Monfort, George Cutts, Iii, Isaac Oludayo Daniel, Mustafa Yildiz, Yusuf Koç, Ramazan Beyaz, Murat Aycan, Mehtap Gursoy, Mustafa Kayan, Alberto Aníbal Galussi, María Esther Moya, Yang Liu, Balkrishna Ghimire, GoEun Choi, Hayan Lee, Kweon Heo, MiJin Jeong, Diego F. Villanueva-Mejía, Javier Correa Alvarez, Carlos Zapata, Javier Bernal, María López-Pedrouso, Daniel Franco, Lucio Garcia, Susana Bravo, Awatif Ali, Fadime Eryılmaz Pehlivan, Rodolfo López-Gómez, Fernando Sánchez-Albarrán, Huber León-Corona, Luis María Suárez-Rodríguez, Pinky Agarwal, Iny Mathew, Arunima Mahto, Michal Petrů, Ivan Mašín, Aviv Dombrovsky, Chandra Singh, Edward Bormashenko, Yekaterina Shapira, Gene Whyman, Bat-Chen Lubin, Elyashiv Drori

**Published by InTech**
Janeza Trdine 9, 51000 Rijeka, Croatia



**Publishing Process Manager** Maja Bozicevic
**Technical Editor** SPi Global
**Cover** InTech Design team

First published November, 2017
Printed in Croatia
Legal deposit, Croatia: National and University Library in Zagreb

Additional hard copies can be obtained from orders@intechopen.com

Advances in Seed Biology, Edited by Jose C. Jimenez-Lopez
p. cm.
Print ISBN 978-953-51-3621-7
Online ISBN 978-953-51-3622-4

For EU product safety concerns:
IN TECH d.o.o., Prolaz Marije Krucifikse Kozulić 3, 51000 Rijeka, Croatia,
info@intechopen.com or visit our website at intechopen.com.

# Meet the editor

Dr. Jose C. Jimenez-Lopez finished his BS degree in Biochemistry and Molecular Biology (1998) and Biological Sciences (2001); MS degree in Agricultural Sciences (2004) at the University of Granada and the Spanish National Research Council (CSIC). He received his PhD degree in Plant Cell Biology (April 2008) from CSIC. He was a postdoctoral research associate in Plant Biology at Purdue University (USA) until 2011. He was awarded with a prestigious European grant (2012–2015) from the Marie Curie program (EU-FP7) to work at the University of Western Australia in nutraceutical aspects and health benefits of the legume *Lupinus angustifolius* and at CSIC in lupin molecular allergy. Currently, he holds a senior research fellow position at CSIC Ramon y Cajal Research Program, working in the functional characterization of proteins from plant reproductive (pollen and seed) tissues in plant species of agro-industrial interest as legumes and olive tree.

# Contents

# Preface

The seed is the most important plant reproductive element, since it has conferred to plants a key advantage for the dispersal and distribution among land. The seed also represents the most economically important agricultural product worldwide, providing energy, nutrients, and raw materials for human nutrition, livestock feed, and innumerable manufactured goods.

Interestingly, investigation in seed has been primarily focused on understanding the biology of the embryo, the storage nutrient compartment, and the seed coat. A deeper understanding of seed biology has been significantly advanced in the last 10 years. The progress made in particular in research fields such as seed development and maturation, and seed dormancy and germination is remarkable. In addition, characterization of novel posttranslational modifications, including NO end products and S-nitrosylation/nitration, is of crucial importance for protein turnover and new functional interplays between signaling sensing and transduction, which also expanded our knowledge about the complexity of developmental physiology in seeds. These progresses made possible a better understanding of how seeds translate soil environmental signals into their internal biology to make decisions for breaking the dormancy and initiating with germination.

In the current era of metadata information, which is being obtained thanks to the advance of the high-throughput techniques and methodologies, many progresses have been made in multiple "omics" families—from functional genomics to phenomics, with the central goal of our current biology to establishing complete functional links between the genome and phenome, the so-called genotype-phenotype map. This will allow uncovering the connection of the molecular networks that control and integrate the physiological processes of seed development and germination, making seed research applicable to future challenges in agriculture.

Agriculture is charged with ensuring sufficient grain production in the face of pressing environmental constraints. A deeply integrated understanding of seed biology (genetics, development, and physiology) will play a key role in sustaining grain yield for the world's growing population.

The proposed research project aims to focus and integrate an updated vision of the recent advances in several different fields of seed biology that include, but not limited to, the evolution of seeds; genetic regulation during seed development and maturation; transcriptomics and proteomics; metabolic processes,ROS generation and seed physiological regulation; seed dormancy; seed vigor and germination; seed anatomy and morphology; nutritional and nutraceutical aspects of seeds; and seed biotechnological applications.

I would like to deeply thank the contributors of this project. I am also grateful to the editorial team of InTechOpen for their very good cooperation and constant help.

**Jose C. Jimenez-Lopez, PhD**
Spanish National Research Council (CSIC)
Granada, Spain

Section 1

# Seed Development

Chapter 1

# Roles of the Environment in Plant Life-History Trade-offs

Yang Liu, Jeffrey L. Walck and Yousry A. El-Kassaby

Additional information is available at the end of the chapter

http://dx.doi.org/10.5772/intechopen.70312

**Abstract**

Variation in plant life-history and functional traits at between- and within-species levels has key ecological consequences, in which environmental settings impose strong selective pressures and play a vital role throughout life cycles. Our general notion for plant life-history strategies may be that, relative to tall, long-lived plants, short-lived species have features of small stature, small-seededness, rapid growth, and low seedling survival (*k*- versus *r*-selection). Rate of evolution may be an important agent of selection and annals evolve more rapidly than perennial congeners. These empirical observations prompt a suite of enticing questions, such as how do life-history traits interplay with functional trait at late stages of regeneration? what are the primary trade-offs in a cohort of key life-history traits that may have undergone stabilizing selection? and how do environmental filters differently affect adaptive trait variation in annuals and perennials? In this chapter, we intend to address aforementioned questions via assembling our updated knowledge with emphasis on seed mass and temporal and spatial dimensions of seed dispersal. Through such synthesis, we wish to raise awareness about life-history trade-offs and provide a holistic understanding of the extent to which climate change is likely to impact plant adaptation and eco-evolutionary trajectories of life-history phenotypes.

**Keywords:** plant life histories, life-history trade-offs, seed mass, seed number, seed dormancy, seed dispersal syndrome, seed emergence, seed persistence, soil seed bank, life-cycle transitions, climate change

## 1. Introduction

Life-history traits, known as fitness components due to their predictable monotonic relationship with fitness, are related to the timing and success of development, reproduction, and senescence throughout the life cycle [1]. The environment has appreciable influences on plant life histories and in the life cycle, the timing of life-history traits (e.g., flowering, seed set, seed mass, seed number, seed dormancy intensity [i.e., delayed onset of germination], seed emergence, etc.) are covaried and thus probably coevolved. Examples for the interplay between the environment and life-history traits at ecological and/or evolutionary levels are instantiated as follows.

- At a global scale, seed dormancy tends to decrease and seed size to increase toward the equator [2–4].
- Life cycles with early flowering, small seeds, deep dormancy, and slow germination are associated with habitats exposed to high temperature, low rainfall, and high radiation [5]. (Note: warming selects for higher optimal photosynthetic temperatures.)
- Lower temperatures with as little as 1°C differences ($T_{critical}$ = 15°C) to the maternal plant in Arabidopsis, on the contrary, tend to enhance final seed dormancy levels [6–9].
- Under natural conditions, a given plant may produce seeds with different levels of dormancy in association with a particular temperature it has experienced during seed development [10].
- Variations in seed dormancy and mass often have a concomitant effect (reviewed by [4]) and are correlated in a negative manner [5, 11–15].
- Species showing very fast germination behavior have (very) small seeds and little or no endosperm, and there is a clear relationship between the phenomenon of very fast germination and high stress habitats (e.g., arid, saline, or in active floodplains), where seeds can rapidly exploit temporarily favorable conditions for germination [16].
- There is a positive correlation of relative embryo length with germination speed and negative correlations with the amount of habitat shade, longevity and precipitation [note that small embryo sizes are typical of primitive taxa] [17].
- Climate change is accelerating plant developmental transitions in temperate environments and advanced flower timing increases dormancy intensities [5, 8, 18].
- Early germination increases seed fecundity due to prolonged vegetative growth and nutrient accumulation but may also bring about high seedling mortality [5].
- There is a strong relatedness between seed mass and the depth of burial from which seedlings emerge [19, 20] and germination of large seeds is strongly facilitated by temperature fluctuations, ensuring germination after deep burial or in litter layers [21–23].
- There exists a negative correlation between seed dormancy and longevity [24] with small seeds persisting longer in soil seed banks than large seeds [25].
- There are strong correlations between seed mass and dispersal syndromes and their correlations hinge on dispersal vectors [26, 27].

Some life-history traits may have reciprocal effects with functional traits at late life stages of regeneration. Environmental challenges, mostly to the maternal plant, influence the resources

http://dx.doi.org/10.5772/intechopen.70312

| Trait one | Trait two | Correlation | References | Notes |
|---|---|---|---|---|
| Seed dormancy | Seed mass | – | [5, 11–15] | (1) |
| Seed dormancy | Seed persistence | No? | [143] | (2) |
| Seed dormancy | Seedling survival | – | [5, 14] | |
| Seed dormancy | Seed longevity | – | [24] | (3) |
| Seed mass | Time to seedling emergence given favorable germination conditions | + | [144] | |
| Seed mass | Seedling growth rate | – | [145–147] | |
| Seed mass | Maternal plant size | Neutral or + | [28, 148, 149] | (4) |
| Seed mass | Seed number per lifetime | No | [150] | |
| Seed mass | Seedling survival | No or weak + | [122, 123] | |
| Plant size | Time to reproductive maturity | + | [150] | |

Notes: (1) In general, species that produce light seeds are more likely to possess some type of seed dormancy [69, 142]. Other correlations were also documented and these inconsistencies may be explained by an incomplete consideration of other covarying factors (e.g., dispersal, fire, and predation) [13] or by phylogenetic constraints [151]. Variation in seed size and dormancy often results from a seed position effect within an inflorescence and within a dispersal unit [35, 50] and this also contributes to uncertainties of their associations. At molecular levels, the parent-of-origin effects on seed traits (e.g., dormancy and size), which are regulated by chromatin remodeling, have been documented for crossing between plants in different ploidy and mutants defective at reproduction [152].

(2) Such correlation remains in the question, as the cited study did not measure whether "the degree of dormancy" was related to persistence. Contrasted with seed release at maturity, canopy seed storage (i.e., serotiny) is reviewed in [153] and we do not extensively discuss canopy-stored seed banks in this chapter. Global warming is expected to reduce seedling emergence for some species [154, 155]. Moreover, the evolution of seed dormancy is favored by high seed persistence in the soil seed bank to alleviate the cost of delayed germination [156]. Both Cohen and Ellner's models suggested that an increase in seed survivorship selects for a low seed germination [88–90]. Soil temperature is the dominant environmental factor controlling the depth of seed dormancy during cycling in imbibed seeds [157]. Climate change engenders long-term exposure to high soil temperatures, which may reduce seed survival, thus selecting for decreased levels of seed dormancy [158]. Taken together, climate change may increase seed numbers in the life cycle and decrease dormancy levels due to increased seed mortality.

(3) Dormancy cycling coupled to seed longevity represents a bet-hedging strategy through persistence in the soil seed bank [159, 160]. As a consequence, seed persistence may be not simply associated with either seed dormancy or longevity.

(4) The mother plant has a significant influence over seed traits and instances have been documented for more than 10 decades. Factors such as age of the mother plant, position of the seed in the fruit, inflorescence, or canopy can affect seed properties, often accompanied by a dimorphism either of the seeds themselves or the fruits in which they arise [161].

**Table 1.** A summary of correlations of adaptive traits.

that are packaged into seeds (seed size) and may be critical for germination and initial seedling growth. For instance, small-seeded species have small plant size (e.g., a positive correlation between seed mass and plant height [28]). Larger plant size, in turn, has higher annual photosynthetic incomes, giving the plant more energy to allocate to seed yield (increased number, heavier mass, or a trade-off between the two in life cycles). Rather, there are studies reporting that seed mass, nature plant height, and leaf mass per unit area have little intercorrelations [29] and that seed dormancy strategy is largely independent of vegetative functional traits and range characteristics [30]. Those inconsistent reports supply us with clues to studying the interaction between adaptive traits in a broader scope (e.g., controlling for phylogeny, more species from different taxa, and/or more traits at different stages of life cycles).

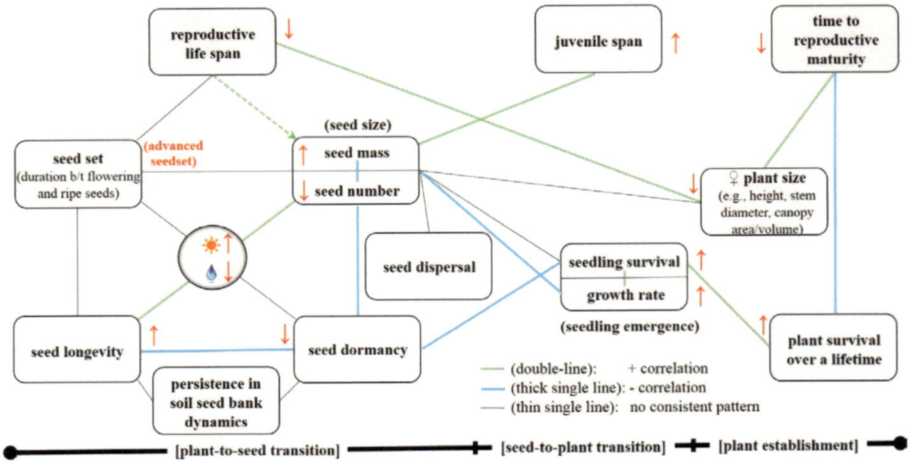


**Figure 1.** Relatedness among life-history and functional traits and the impact of climate change on the variation (and evolution) of these traits. Note: Lines give interactions between traits (boxes) in a positive or negative manner and the change of direction (↑ or ↓) depends on another trait(s). The arrow linking two traits prompts which trait affects the other. Sun and water drop symbols stand for temperature and rainfall, respectively.

| Altered environment (♀) | Affected traits in offspring | Test species | References |
|---|---|---|---|
| Temperature (high or low) | Seed production, seed mass, flowering time | *Arabidopsis thaliana, Plantago lanceolata* | [162–164] |
| Light (shade or over-exposure) | Seed provisioning, seed mass, germination, seedling survival, biomass, life-history schedule | *Polygonum hydropiper, Campanulastrum americanum* | [165, 166] |
| Rainfall (drought) | Flowering (select for early flowering and short life cycles), seed provisioning, germination | *Brassica rapa, Polygonum persicaria* | [167–170] |
| Seasonal environments | Germination timing, life-history schedule | *Arabidopsis thaliana* | [74] |
| Salinity (high) | Germination, seedling growth | *Arabidopsis thaliana* | [171] |
| Nutrient (deficiency) | Plant height, biomass | *Oryza sativa* | [172] |
| Herbivory | Seed mass, germination, seedling growth, flowering, plant height, biomass | *Raphanus raphanistrum, Impatiens capensis* | [137, 138, 173] |

Note: When maternal plants are deprived of resources, seed provisioning may be reduced (e.g., [145]), maintained or even increased (e.g., [167, 168, 174]), as trade-offs may exist between increased seed provisioning and decreased persistence in the soil seed bank [145, 167, 174] and thus the benefits may depend on a specific ecological setting.

Progressive global warming leads to widespread shifts toward earlier initiation of flowering in many plant populations, which contributes to an increase in the length of the flowering season in regions where flowering is temperature-dependent [175–177]. Note that longer growing seasons select for later flowering and thus warming and growing season may comprise a trade-off. As the detection of the relatedness between flowering locus and ambient temperature [72, 178], the flowering time diversity is associated with *cis*-regulatory variation [179] and further, flowering time loci restrict potential range size and niche breadth [180].

**Table 2.** Examples for the effect of different parental environments on offspring adaptive traits.

http://dx.doi.org/10.5772/intechopen.70312

Last, we provided a summary of interplay patterns between traits (life-history, functional traits, and a combination thereof) as influenced by climatic factors in **Table 1** and **Figure 1**. We also listed examples on adaptive traits with transgenerational plasticity as responses to altered maternal environmental conditions in **Table 2**.

## 2. Life-history traits and their trade-offs

### 2.1. Seed size

The reproductive output of an organism is a critical life-history trait defining its fitness and is the result of both offspring number and quality. Seeds are time capsules and receptacles of life and seed mass is a crucial life-history trait that links the ecology of reproduction and seedling establishment with that of vegetative growth. Seed mass commonly varies over 11.5 orders of magnitude among coexisting plant species [31], while within-species variation in seed mass is typically in the range of two- to fourfold [32]. Extant flowering plants exhibit a wider range of seed sizes than nonflowering seed plants, and in particular, some of them can bear very small seeds [31]. Seed mass variation is a type of heteromorphism, which represents a classic trade-off. Production of dimorphic or heteromorphic seeds by a single plant allows plants to decrease temporal variance in offspring success through bet-hedging [i.e., a strategy that reduced temporal variance in fitness at the expense of a lowered arithmetic mean fitness] [33–35], or rather, a blend of plasticity [i.e., a capacity of a genotype to produce different phenotypes when exposed to different environmental conditions] and bet-hedging [36]. Heteromorphism enables a fraction of propagules to adapt to any given environment and may increase long-term reproductive success by reducing the risk of extinction, but it comes at the cost of decreasing immediate fitness [37].

Seed mass is closely correlated with changes in plant form and vegetative type, followed by spatial dispersal syndrome and net primary productivity [27, 38, 39]. Besides, latitude, genome size, forest structure, and life history all have been linked to seed size variation [40–43]. Effects of temperature on seed mass are not consistent, as both increased [44, 45] or decreased [46] seed masses have been documented. Seed mass variation within species is largely related to seed position within pods and fruits [47–49] and upper grains in the spikelet tend to be larger than bottom ones [50]. This is possibly due to physiological or morphological constraints on optimum resource allocation to seeds. The diversity of seed mass may be maintained by tolerance-fecundity trade-offs (i.e., more tolerant (fecund) species gain more (less) stressful regeneration sites, respectively) [51].

Empirical evidence favors the notion that seed production during mast years (i.e., good-seed years) is tightly related to high temperature in the previous spring and summer, late spring frost and summer rainfall of the last two years. The difference in temperature from one growing season to the next effectively predicted the occurrence of mast years [52–54]. Considering yearly climatic variability, Kelly et al. [55] developed a model based on temperature differentials over multiple seasons to predict seed yield [55] and this model was further validated by

Pearse et al. [56]. The robustness of these models is emanated from the hypothesized correlation between seed mass and the environment and, in turn, lends support to the crucial role of climate in seed mass modulations.

From a genetic perspective, seed mass and number can evolve independently. Both traits are affected by a large number of mostly non-overlapping quantitative trait loci (QTLs) in their genetic architecture, which has been manifested by using mapping lines created by two [57–60] and multiple parents [61]. This indicates that the relatedness between seed mass and number may not be invariably direct.

### 2.2. Seed dispersal in time and space

While seed dormancy is a means to disperse in time, seed dispersal is an important way to disperse in space (hereafter simply referred to as dispersal) and also a risk-spreading strategy [62, 63]. Selection in heterogeneous or unpredictable environments may favor plants to synchronize seed dispersal with environmental conditions allowing or deferring germination until suitable conditions occur [64]. As risk-reducing strategies, can seed dormancy and dispersal substitute for one another so that selection for one may weaken selection for the other? Theoretical expectations support such a trade-off or a negative covariation (reviewed in [62]), but empirical evidence is inconsistent (there are reports on complicated patterns or no relationship, e.g., [65]).

#### *2.2.1. Seed dormancy*

Seed dormancy is an innate constraint on germination timing under conditions that would otherwise promote germination in nondormant seeds [66] and prevent germination during periods that are ephemerally favorable [67]. Dormancy is significantly higher in annuals than in perennials (note that perenniality is an alternative risk-reducing strategy; [68]) and dormant seed banks are thus better associated with annuals than perennials [13, 25, 69]. Dormancy is an important adaptive trait that links plant life-history to seasonal change. Dormancy exists as a continuum with multiple layers (blocks to germination completion) that are successively taken off by appropriate environmental signals. These signals inform the seed whether it is in an appropriate habitat and time of the year suitable for the resulting plant to survive and reproduce. Seed germination timing is the earliest trait in plant life-history, which allows plants to regulate when and where they grow. It affects the evolution of other life-history traits that follow in the life cycle, such as fecundity and survival [70]. As such, seed dormancy may be construed as an adaptive strategy for survival during bad seasons and can exert cascading selective pressures on subsequent life stages.

Seeds temporarily block germination through adaptation to the prevailing environments so that germination is timed to avoid unfavorable environmental conditions for subsequent plant establishment and growth and therefore sets the context for the traits that follow [71]. Dormancy levels are maternally manipulated [72, 73] and determined by maternal responses to day length and temperature in many species [6, 74, 75]. Notably, induction of primary dormancy was greatly influenced by the effect of maternal environments on embryo/endosperm

[74, 76, 77] and/or on seed coat properties [78]. Dormancy intensity can be manipulated via controlling the daily circadian clock at reproduction [79]. Such effects can be passed down for multiple generations [80, 81] and have been observed even in long-lived perennials, such as conifers [82]. Dormancy manipulation involves dormancy-specific genomic imprinting programs that mainly occur in mature endosperm [83]. Owing to the similar functionality between plant seed endosperm and mammalian placenta, the kinship or parental conflict theory is often proposed to account for the evolutionary origin of imprinting [84].

Plants distribute their offspring across time, hedging their bets against unpredictable environments [37, 85]. This increases the likelihood that some seeds will survive regardless of environmental perturbations. Seed dormancy variability among individuals is associated with environmental heterogeneity [86] and heterogeneous environments may select for bet-hedging strategies, as population growth is an inherently multiplicative process that is very sensitive to occasionally extreme values [87]. Cohen indicated that low germination probabilities can be expected in harsh environments as individuals can germinate in improved conditions and decrease their average mortality [88]. However, Ellner predicted that increasing the frequency of favorable years may also lead to lower germination rates due to increased density-dependent effects imposed by competitive interactions [89, 90].

Based on mathematical modeling, large nondormant seeds would be selected for under constant unfavorable environments, whereas in temporally unpredictable environments, dormant seeds would be selected for, and their size would rely on the likelihood of predation of large seeds [91]. This alludes to the adaptive value of dormancy that may depend on the variability of the environment and the probability of survival of large seeds in the soil seed bank. Nondormant seeds in legumes can only evolve in climates with long growing seasons and/or in lineages that produce larger seeds; conversely, dormancy should be evolutionarily stable in temperate lineages with small seeds [92]. In light of this, it may be reasonable to extrapolate that, within a given lineage, taxa producing larger, nondormant seeds necessarily predominate in aseasonal environments, while plants bearing small, dormant seeds are dominant under short growing seasons [92].

#### 2.2.2. *Seed dispersal syndrome*

Spatial aspects are especially important in the global climate change context, as temperature shifts depend on latitude and altitude gradients and species dispersal to higher altitudes and latitudes is thought to be a major constraint to their future survival. The evolution of local adaptation requires low dispersal (and selection against genotypes adapted to other localities) [93]. Also, spatial context influences gene flow and evolutionary dynamics with, again, important consequences for species competition and survival [94]. Differences in dispersal syndrome are likely to affect the shape of the dispersal kernel and the type of environments to which seeds are dispersed [95, 96]. On the one hand, they may evolve as phenotypic plasticity (e.g., bet-hedging) [97–99] in responses to selective factors, including reducing parent-offspring conflict or kin competition, the temporal heterogeneity of the environment, such as local population extinction [100–102] and avoiding inbreeding depression due to mating between related individuals (for dispersal only; [103]). Hence, seed dispersal promotes adaptation, stability, and persistence [104]. On the other hand, various costs of dispersal have been postulated in theoretical models [such as fleshy fruits dispersed by animals [105], getting lost during displacements, dispersing in fragmented habitats [106], etc.], which end up concluding that increasing the cost of dispersal (certain selective forces) selects for lower dispersal [107]. Collectively, selection acts on trade-offs in temporal and spatial dispersal and eventually maximizes fitness [62]. These trade-offs can, in turn, introduce patterns of covariation among functional and life-history traits that correlate with dispersal (reviewed in [108]).

Further, dispersal syndrome is a consistent predictor of seed size especially in nonflowering seed plant groups [109]. In conifers (serotiny), seeds are generally larger in animal- than wind-dispersed species [110, 111]. This is in alignment with our intuition, because animals should be attracted to large, copious seeds and can be better than wind to disperse large propagules [112–114]. Moreover, spatial differences at the levels of environmental stochasticity, restricted dispersal, increased fragmentation, and intermediate survival during dormancy favor the adaptive diversification of bet-hedging dormancy strategies [115] and spatial heterogeneity and restricted dispersal are essential for evolutionary branching of germination strategies [116]. This suggests that dispersal and above-ground environmental fluctuations have a significant effect on the development of dormancy or germination polymorphisms.

### 2.3. Trade-offs among multiple life-history traits via stabilizing selection: a compromised strategy to recoup the disadvantage of respective traits

The Smith-Fretwell theory (1974) modeled the optimal seed mass and formalized the concept of a trade-off between producing a few, well provisioned offspring versus producing many poorly nourished individuals. It considers the offspring's lifetime fitness, given the population is stable during their life time. The role of differential seed mass in promoting species coexistence has been stressed in previous theoretical studies [117–119]. Evolution of seed mass results in the fixation of a given strategy and evolved seed mass decreases when seed dormancy is lowered [120]. Large seed mass, on the one hand, confers direct advantages to many fitness-related plant characteristics, including recruitment and survivorship [121, 122] and establishment [32, 122], because large seeds accumulate copious nourishing substances for germination and have better tolerance in face of disturbances (e.g., abiotic stresses) [119, 123]. Furthermore, for a given reproductive investment, seed mass is negatively correlated with seed number [124–126] and large seeds are less dispersible due to their great mass [127]. However, can the survival advantage of large-seeded species really counterbalance the greater seed yield of small-seeded species?

The advantage of large-seededness is generally temporary, probably expiring when all maternal reserves have been deployed [123]. This means large-seeded species have a survival advantage over small-seeded species solely during early seedling establishment [122]. Actually, there is no or even weak negative correlation between seed mass and overall seedling survival [122, 123]. Because slow growth rate (due to slow metabolic rate) increases the capability of persistence under stress and the duration of exposure to juvenile mortality. Greater survival per unit time associated with large-seeded species may be canceled out by the longer time to maturity.

Here, we speculate possible impacts of elements of plant strategy systems (e.g., vegetative functional traits) on seed ecology. Larger plant size has higher annual photosynthetic incomes, giving the plant more energy to allocate to seed production (increased seed number, heavier mass, or a trade-off between the two). A species with a large adult stature will necessarily have a lengthy juvenile period to produce large, well-provisioned offspring, which makes up for a high rate of juvenile mortality. By contrast, smaller plant size of small-seeded species may have less photosynthate to allocate to reproduction. Further, they also tend to have shorter reproductive life span (i.e., less total investment to offspring and lower mortality). Less input from maternal plants further erodes the apparent advantage that small-seeded species harbor during seed production.

In addition to seed dormancy and dispersal, there are other risk-spreading strategies. For instance, there is a negative trade-off between dormancy and longevity (a survival strategy) to reduce risk of reproductive failures in time [69, 128, 129]. Iteroparous (i.e., reproducing more than once in a lifetime) perennial plants increase their probability of encountering favorable conditions for reproduction in time [130–132]. Therefore, the need for seed dormancy in perennials may be negated and perenniality may select against seed dormancy [128, 129].

Additionally, it is reported that dormancy is also dependent on the likelihood of seed predation and nondormancy may evolve as part of a predator avoidance strategy [133, 134] and also with the involvement of environmental pressures [135]. The risk of predation is thought to be proportional to seed size due to its detectability and nutritious contents. However, such risk is lower in large than small seeds, possibly because size is positively correlated with defense mechanisms [136]. Furthermore, maternal herbivory has a major impact on seed size [137, 138] (also see **Table 2**; mediated by phytohormone signaling pathways that affect seed filling [e.g., gibberellins, auxin, brassinosteriod] [139, 140]) and dormancy [141] (mediated by jasmonate signaling pathway and resultant changes in the sensitivity of seed germination to ABA). Thus, whether dormancy or nondormancy is favored by selection ultimately depends on seed size [92].

In a nutshell, there is no prevalent relationship between seed mass and number, in the sense that the advantages that small-seeded species gain during seed production must be counterbalanced somewhere else in the life cycle and seed number is just one of the possibilities to make up the loss of small seededness. Seed mass is determined via a process of stabilizing selection, which may operate through selections on trade-offs among seedling survival, dormancy, dispersal, seed number, and predation [142], and may play a central role among a correlated suite of traits that covaries across a spectrum of life-history strategies.

## 3. Conclusions

This chapter intends to direct readers' attention to consider multiple life-history traits in the life cycle when studying plant life-history evolution, as selection for one strategy (not necessarily limited to one single trait) may constrain another. Disentangling the trade-offs of how disadvantages in one studied trait may be made up by being an advantage in another trait(s) helps essentially understand the evolution of a particular trait. Moreover, these trade-offs are essential for understanding the complex response of species to climate change.

## Author details

Yang Liu[1*], Jeffrey L. Walck[2*] and Yousry A. El-Kassaby[1*]

*Address all correspondence to: yang.liu@ubc.ca, Jeffrey.Walck@mtsu.edu and y.el-kassaby@ubc.ca

1 Department of Forest and Conservation Sciences, University of British Columbia, Vancouver, BC, Canada

2 Department of Biology, Middle Tennessee State University, Murfreesboro, TN, USA

## References

[1] Calow P. The Encyclopedia of Ecology and Environmental Management. New York: Blackwell Science; 1998

[2] Moles AT, Westoby M. Latitude, seed predation and seed mass. Journal of Biogeography. 2003;**30**(1):105-128

[3] Morin X, Chuine I. Niche breadth, competitive strength and range size of tree species: A trade-off based framework to understand species distribution. Ecology Letters. 2006;**9**(2):185-195

[4] Baskin CC, Baskin JM. Seeds: Ecology, Biogeography, and Evolution of Dormancy and Germination. 2nd ed. San Diego, California: Academic Press; 2014. p. 666

[5] Vidigal DS, Marques AC, Willems LA, Buijs G, Méndez-Vigo B, Hilhorst HW, et al. Altitudinal and climatic associations of seed dormancy and flowering traits evidence adaptation of annual life cycle timing in *Arabidopsis thaliana*. Plant, Cell & Environment. 2016;**39**(8):1737-1748

[6] Fenner M. The effects of the parent environment on seed germinability. Seed Science Research. 1991;**1**:75-84

[7] Huang ZY, Footitt S, Finch-Savage WE. The effect of temperature on reproduction in the summer and winter annual *Arabidopsis thaliana* ecotypes Bur and Cvi. Annals of Botany. 2014;**113**(6):921-929

[8] Springthorpe V, Penfield S. Flowering time and seed dormancy control use external coincidence to generate life history strategy. eLIFE. 2015;**4**:e05557

[9] Schmuths H, Bachmann K, Weber WE, Horres R, Hoffmann MH. Effects of preconditioning and temperature during germination of 73 natural accessions of *Arabidopsis thaliana*. Annals of Botany. 2006;**97**(4):623-634

[10] Burghardt LT, Edwards BR, Donohue K. Multiple paths to similar germination behavior in *Arabidopsis thaliana*. New Phytologist. 2016;**209**(3):1301-1312

[11] Thompson K, Grime JP. Seasonal variation in the seed banks of herbaceous species in ten contrasting habitats. Journal of Ecology. 1979;**67**(3):893-921

[12] Grime JP, Mason G, Curtis AV, Rodman J, Band SR, Mowforth MAG, et al. A comparative study of germination characteristics in a local flora. Journal of Ecology. 1981;**69**(3):1017-1059

[13] Rees M. Evolutionary ecology of seed dormancy and seed size. Philosophical Transactions of the Royal Society B. 1996;**351**(1345):1299-1308

[14] Kiviniemi K. Evolution of recruitment features in plants: A comparative study of species in the Rosaceae. Oikos. 2001;**94**(2):250-262

[15] Larios E, Búrquez A, Becerra JX, Venable DL. Natural selection on seed size through the life cycle of a desert annual plant. Ecology. 2014;**95**(11):3213-3220

[16] Parsons RF. Incidence and ecology of very fast germination. Seed Science Research. 2012;**22**(3):161-167

[17] Vandelook F, Janssens SB, Probert RJ. Relative embryo length as an adaptation to habitat and life cycle in Apiaceae. New Phytologist. 2012;**195**(2):479-487

[18] Debieu M, Tang C, Stich B, Sikosek T, Effgen S, Josephs E, et al. Co-variation between seed dormancy, growth rate and flowering time changes with latitude in *Arabidopsis thaliana*. PloS One. 2013;**8**(5):e61075

[19] Jurik TW, Wang SC, Vandervalk AG. Effects of sediment load on seedling emergence from wetland seed banks. Wetlands. 1994;**14**(3):159-165

[20] Maun MA, Lapierre J. Effects of burial by sand on seed germination and seedling emergence of four dune species. American Journal of Botany. 1986;**73**(3):450-455

[21] Pearson TRH, Burslem DFRP, Mullins CE, Dalling JW. Germination ecology of neotropical pioneers: Interacting effects of environmental conditions and seed size. Ecology. 2002;**83**(10):2798-2807

[22] Ghersa CM, Arnold RLB, Martinezghersa MA. The role of fluctuating temperatures in germination and establishment of *Sorghum halepense* – Regulation of germination at increasing depths. Functional Ecology. 1992;**6**(4):460-468

[23] Xia Q, Ando M, Seiwa K. Interaction of seed size with light quality and temperature regimes as germination cues in 10 temperate pioneer tree species. Functional Ecology. 2016;**30**(6):866-874

[24] Nguyen TP, Keizer P, van Eeuwijk F, Smeekens S, Bentsink L. Natural variation for seed longevity and seed dormancy are negatively correlated in Arabidopsis. Plant Physiology. 2012;**160**(4):2083-2092

[25] Thompson K, Bakker JP, Bekker RM, Hodgson JG. Ecological correlates of seed persistence in soil in the north-west European flora. Journal of Ecology. 1998;**86**:163-169

[26] Leishman MR, Westoby M, Jurado E. Correlates of seed size variation: A comparison among five temperate floras. Journal of Ecology. 1995;**83**:517-529

[27] Moles AT, Ackerly DD, Webb CO, Tweddle JC, Dickie JB, Pitman AJ, et al. Factors that shape seed mass evolution. Proceedings of the National Academy of Sciences of the United States of America. 2005;**102**(30):10540-10544

[28] Moles AT, Ackerly DD, Webb CO, Tweddle JC, Dickie JB, Westoby M. Response to comment on "A brief history of seed size". Science. 2005;**310**(5749):783

[29] Westoby M. A leaf-height-seed (LHS) plant ecology strategy scheme. Plant and Soil. 1998;**199**(2):213-227

[30] Hoyle GL, Steadman KJ, Good RB, McIntosh EJ, Galea LME, Nicotra AB. Seed germination strategies: An evolutionary trajectory independent of vegetative functional traits. Frontiers in Plant Science. 2015;**6**:731

[31] Moles AT, Ackerly DD, Webb CO, Tweddle JC, Dickie JB, Westoby M. A brief history of seed size. Science. 2005;**307**(5709):576-580

[32] Leishman MR, Wright IJ, Moles AT, Westoby M. Seeds: The ecology of regeneration in plant communities. In: Fenner M, editor. The Evolutionary Ecology of Seed Size. New York: Oxford University Press; 2000

[33] Venable DL, Búrquez A, Corral G, Morales E, Espinosa F. The ecology of seed heteromorphism in *Heterosperma pinnatum* in central Mexico. Ecology. 1987;**68**:65-76

[34] Baskin JM, Lu JJ, Baskin CC, Tan DY, Wang L. Diaspore dispersal ability and degree of dormancy in heteromorphic species of cold deserts of northwest China: A review. Perspectives in Plant Ecology Evolution and Systematics. 2014;**16**(2):93-99

[35] Imbert E. Ecological consequences and ontogeny of seed heteromorphism. Perspectives in Plant Ecology Evolution and Systematics. 2002;**5**(1):13-36

[36] Lenser T, Graeber K, Cevik ÖS, Adigüzel N, Dönmez AA, Grosche C, et al. Developmental control and plasticity of fruit and seed dimorphism in *Aethionema arabicum*. Plant Physiology. 2016;**172**(3):1691-1707

[37] Venable DL. Bet hedging in a guild of desert annuals. Ecology. 2007;**88**(5):1086-1090

[38] Moles AT, Ackerly DD, Tweddle JC, Dickie JB, Smith R, Leishman MR, et al. Global patterns in seed size. Global Ecology and Biogeography. 2007;**16**(1):109-116

[39] Díaz S, Kattge J, Cornelissen JH, Wright IJ, Lavorel S, Dray S, et al. The global spectrum of plant form and function. Nature. 2016;**529**(7585):167-171

[40] Salisbury E. Seed size and mass in relation to environment. Proceedings of the Royal Society Series B-Biological Sciences. 1974;**186**(1083):83-88

[41] Lord J, Egan J, Clifford T, Jurado E, Leishman M, Williams D, et al. Larger seeds in tropical floras: Consistent patterns independent of growth form and dispersal mode. Journal of Biogeography. 1997;**24**(2):205-211

[42] Beaulieu JM, Moles AT, Leitch IJ, Bennett MD, Dickie JB, Knight CA. Correlated evolution of genome size and seed mass. New Phytologist. 2007;**173**(2):422-437

[43] Rueda M, Godoy O, Hawkins BA. Spatial and evolutionary parallelism between shade and drought tolerance explains the distributions of conifers in the conterminous United States. Global Ecology and Biogeography. 2017;**26**(1):31-42

[44] Murray BR, Brown AHD, Dickman CR, Crowther MS. Geographical gradients in seed mass in relation to climate. Journal of Biogeography. 2004;**31**(3):379-388

[45] Liu Y, Wang T, El-Kassaby YA. Contributions of dynamic environmental signals during life-cycle transitions to early life-history traits in lodgepole pine (*Pinus contorta* Dougl.). Biogeosciences. 2016;**13**:2945-2958

[46] Hovenden MJ, Wills KE, Chaplin RE, Vander Schoor JK, Williams AL, Osanai YUI, et al. Warming and elevated $CO_2$ affect the relationship between seed mass, germinability and seedling growth in *Austrodanthonia caespitosa*, a dominant Australian grass. Global Change Biology. 2008;**14**(7):1633-1641

[47] Wulff RD. Seed size variation in *Desmodium Paniculatum*. I. Factors affecting seed size. Journal of Ecology. 1986;**74**(1):87-97

[48] Michaels HJ, Benner B, Hartgerink AP, Lee TD, Rice S, Willson MF, et al. Seed size variation: Magnitude, distribution, and ecological correlates. Evolutionary Ecology. 1988;**2**(2):157-166

[49] Obeso JR. Seed mass variation in the perennial herb *Asphodelus albus*: Sources of variation and position effect. Oecologia. 1993;**93**(4):571-575

[50] Volis S. Seed heteromorphism in *Triticum dicoccoides*: Association between seed positions within a dispersal unit and dormancy. Oecologia. 2016;**181**(2):401-412

[51] Muller-Landau HC. The tolerance-fecundity trade-off and the maintenance of diversity in seed size. Proceedings of the National Academy of Sciences of the United States of America. 2010;**107**(9):4242-4247

[52] Schauber EM, Kelly D, Turchin P, Simon C, Lee WG, Allen RB, et al. Masting by eighteen New Zealand plant species: The role of temperature as a synchronizing cue. Ecology. 2002;**83**(5):1214-1225

[53] Smaill SJ, Clinton PW, Allen RB, Davis MR. Climate cues and resources interact to determine seed production by a masting species. Journal of Ecology. 2011;**99**(3):870-877

[54] Krebs CJ, LaMontagne JM, Kenney AJ, Boutin S. Climatic determinants of white spruce cone crops in the boreal forest of southwestern Yukon. Botany. 2012;**90**(2):113-119

[55] Kelly D, Geldenhuis A, James A, Penelope Holland E, Plank MJ, Brockie RE, et al. Of mast and mean: Differential-temperature cue makes mast seeding insensitive to climate change. Ecology Letters. 2013;**16**(1):90-98

[56] Pearse IS, Koenig WD, Knops JMH. Cues versus proximate drivers: Testing the mechanism behind masting behavior. Oikos. 2014;**123**(2):179-184

[57] Alonso-Blanco C, Blankestijn-de Vries H, Hanhart CJ, Koornneef M. Natural allelic variation at seed size loci in relation to other life history traits of *Arabidopsis thaliana*. Proceedings of the National Academy of Sciences of the United States of America. 1999;**96**(8):4710-4717

[58] Moore CR, Gronwall DS, Miller ND, Spalding EP. Mapping quantitative trait loci affecting *Arabidopsis thaliana* seed morphology features extracted computationally from images. G3-Genes Genomes Genetics. 2013;**3**(1):109-118

[59] Herridge RP, Day RC, Baldwin S, Macknight RC. Rapid analysis of seed size in Arabidopsis for mutant and QTL discovery. Plant Methods. 2011;**7**(1):3

[60] Van Daele I, Gonzalez N, Vercauteren I, de Smet L, Inzé D, Roldán-Ruiz I, et al. A comparative study of seed yield parameters in Arabidopsis thaliana mutants and transgenics. Plant Biotechnology Journal. 2012;**10**(4):488-500

[61] Gnan S, Priest A, Kover PX. The genetic basis of natural variation in seed size and seed number and their trade-off using *Arabidopsis thaliana* MAGIC Lines. Genetics. 2014;**198**(4):1751-1758

[62] Buoro M, Carlson SM. Life-history syndromes: Integrating dispersal through space and time. Ecology Letters. 2014;**17**(6):756-767

[63] Cohen D, Levin SA. The interaction between dispersal and dormancy strategies in varying and heterogeneous environments. In: Teramoto E, Yamaguti M, editors. Mathematical Topics in Population Biology, Morphogenesis and Neurosciences. Vol. 71. Heidelberg: Springer-Verlag; 1987. pp. 110-122

[64] Freas KM, Kemp PR. Some relationships between environmental reliability and seed dormancy in desert annual plants. Journal of Ecology. 1983;**71**:211-217

[65] Siewert W, Tielbörger K. Dispersal-dormancy relationships in annual plants: Putting model predictions to the test. American Naturalist. 2010;**176**(4):490-500

[66] Simpson G. Seed Dormancy in Grasses. Cambridge, UK: Cambridge University Press; 1990

[67] Bewley JD. Seed germination and dormancy. Plant Cell. 1997;**9**(7):1055-1066

[68] de Waal C, Anderson B, Ellis AG. Dispersal, dormancy and life-history tradeoffs at the individual, population and species levels in southern African Asteraceae. New Phytologist. 2016;**210**(1):356-365

[69] Rees M. Trade-offs among dispersal strategies in British plants. Nature. 1993;**366**(6451): 150-152

[70] Hamilton WD. The moulding of senescence by natural selection. Journal of Theoretical Biology. 1966;**12**:12-45

[71] Donohue K, de Casas RR, Burghardt L, Kovach K, Willis CG. Germination, postgermination adaptation, and species ecological ranges. Annual Review of Ecology, Evolution, and Systematics. 2010;**41**:293-319

[72] Chen M, MacGregor DR, Dave A, Florance H, Moore K, Paszkiewicz K, et al. Maternal temperature history activates Flowering Locus T in fruits to control progeny dormancy according to time of year. Proceedings of the National Academy of Sciences of the United States of America. 2014;**111**:18787-18792

[73] Chiang GCK, Barua D, Kramer EM, Amasino RM, Donohue K. Major flowering time gene, *FLOWERING LOCUS C*, regulates seed germination in Arabidopsis thaliana. Proceedings of the National Academy of Sciences of the United States of America. 2009;**106**(28):11661-11666

[74] Donohue K. Completing the cycle: Maternal effects as the missing link in plant life histories. Philosophical Transactions of the Royal Society B. 2009;**364**(1520):1059-1074

[75] Kendall S, Penfield S. Maternal and zygotic temperature signalling in the control of seed dormancy and germination. Seed Science Research. 2012;**22**:S23–S29

[76] Schmitt J, Niles J, Wulff R. Norms of reaction of seed traits to maternal environments in *Plantago lanceolata*. American Naturalist. 1992;**139**(3):451-466

[77] Postma FM, Ågren J. Maternal environment affects the genetic basis of seed dormancy in *Arabidopsis thaliana*. Molecular Ecology. 2015;**24**(4):785-797

[78] MacGregor DR, Kendall SL, Florance H, Fedi F, Moore K, Paszkiewicz K, et al. Seed production temperature regulation of primary dormancy occurs through control of seed coat phenylpropanoid metabolism. New Phytologist. 2015;**205**(2):642-652

[79] Footitt S, Ölcer-Footitt H, Hambidge AJ, Finch-Savage WE. A laboratory simulation of Arabidopsis seed dormancy cycling provides new insight into its regulation by clock genes and the dormancy-related genes *DOG1, MFT, CIPK23* and *PHYA*. Plant, Cell & Environment. 2017 (in press)

[80] Lu JJ, Tan DY, Baskin CC, Baskin JM. Effects of germination season on life history traits and on transgenerational plasticity in seed dormancy in a cold desert annual. Scientific Reports. 2016;**6**:25076

[81] Galloway LF, Etterson JR. Transgenerational plasticity is adaptive in the wild. Science. 2007;**318**(5853):1134-1136

[82] Stoehr MU, L'Hirondelle SJ, Binder WD, Webber JE. Parental environment after effects on germination, growth, and adaptive traits in selected white spruce families. Canadian Journal of Forest Research. 1998;**28**(3):418-426

[83] Piskurewicz U, Iwasaki M, Susaki D, Megies C, Kinoshita T, Lopez-Molina L. Dormancy-specific imprinting underlies maternal inheritance of seed dormancy in *Arabidopsis thaliana*. eLIFE. 2016;**5**:e19573

[84] Haig D, Westoby M. Parent-specific gene expression and the triploid endosperm. American Naturalist. 1989;**134**(1):147-155

[85] Poisot T, Bever JD, Nemri A, Thrall PH, Hochberg ME. A conceptual framework for the evolution of ecological specialisation. Ecology Letters. 2011;**14**(9):841-851

[86] Angevine MW, Chabot BF. Seed germination syndromes in higher plants. In: Solbrig OT, Jain S, Johnson GB, Raven PH, editors. Topics in plant population biology. New York, NY: Columbia University Press; 1979

[87] Dempster ER. Maintenance of genetic heterogeneity. Cold Spring Harbor Symposia on Quantitative Biology. 1955;**20**:25-32

[88] Cohen D. Optimizing reproduction in a randomly varying environment. Journal of Theoretical Biology. 1966;**12**(1):119-129

[89] Ellner S. ESS germination strategies in randomly varying environments. I. Logistic-type models. Theoretical Population Biology. 1985;**28**(1):50-79

[90] Ellner S. ESS germination strategies in randomly varying environments. II. Reciprocal yield-law models. Theoretical Population Biology. 1985;**28**(1):80-116

[91] Volis S, Bohrer G. Joint evolution of seed traits along an aridity gradient: Seed size and dormancy are not two substitutable evolutionary traits in temporally heterogeneous environment. New Phytologist. 2013;**197**(2):655-667

[92] Rubio de Casas R, Willis CG, Pearse WD, Baskin CC, Baskin JM, Cavender-Bares J. Global biogeography of seed dormancy is determined by seasonality and seed size: A case study in the legumes. New Phytologist. 2017 (in press)

[93] Kawecki TJ, Ebert D. Conceptual issues in local adaptation. Ecology Letters. 2004;**7**(12): 1225-1241

[94] Norberg J, Urban MC, Vellend M, Klausmeier CA, Loeuille N. Eco-evolutionary responses of biodiversity to climate change. Nature Climate Change. 2012;**2**(10):747-751

[95] Garcia-Castaño JL, Kollmann J, Jordano P. Spatial variation of post-dispersal seed removal by rodents in highland microhabitats of Spain and Switzerland. Seed Science and Technology. 2006;**16**(3):213-222

[96] Russo SE, Portnoy S, Augspurger CK. Incorporating animal behavior into seed dispersal models: Implications for seed shadows. Ecology. 2006;**87**(12):3160-3174

[97] Slatkin M. Hedging ones evolutionary bets. Nature. 1974;**250**:704-705

[98] Gomez-Mestre I, Jovani R. A heuristic model on the role of plasticity in adaptive evolution: Plasticity increases adaptation, population viability and genetic variation. Proceedings of the Royal Society B: Biological Sciences. 2013;**280**(1771):20131869

[99] Philippi T, Seger J. Hedging one's evolutionary bets, revisited. Trends in Ecology & Evolution. 1989;**4**(2):41-44

[100] Ellner S. Germination dimorphisms and parent offspring conflict in seed germination. Journal of Theoretical Biology. 1986;**123**(2):173-185

[101] Vitalis R, Rousset F, Kobayashi Y, Olivieri I, Gandon S. The joint evolution of dispersal and dormancy in a metapopulation with local extinctions and kin competition. Evolution. 2013;**67**(6):1676-1691

[102] Gremer JR, Venable DL. Bet hedging in desert winter annual plants: Optimal germination strategies in a variable environment. Ecology Letters. 2014;**17**(3):380-387

[103] Bengtsson BO. Avoiding inbreeding: At what cost? Journal of Theoretical Biology. 1978;**73**(3):439-444

[104] Kovach-Orr C, Fussmann GF. Evolutionary and plastic rescue in multitrophic model communities. Philosophical Transactions of the Royal Society B. 2013;**368**(1610):20120084

[105] Herrera CM. Plant-vertebrate seed dispersal systems in the Mediterranean: Ecological, evolutionary, and historical determinants. Annual Review of Ecology and Systematics. 1995;**26**:705-727

[106] Cheptou PO, Carrue O, Rouifed S, Cantarel A. Rapid evolution of seed dispersal in an urban environment in the weed *Crepis sancta*. Proceedings of the National Academy of Sciences of the United States of America. 2008;**105**(10):3796-3799

[107] Gandon S, Michalakis Y. Multiple causes of the evolution of dispersal. In: Clobert J, Danchin E, Dhondt AA, Nichols JD, editors. Dispersal. Oxford, New York: Oxford University Press; 2001

[108] Ronce O, Clobert J. Dispersal syndromes. In: Clobert J, Baguette M, Benton TG, Bullock JM, Ducatez S, editors. Dispersal ecology and evolution. Oxford, UK: Oxford University Press; 2012

[109] Leslie AB, Beaulieu JM, Mathews S. Variation in seed size is structured by dispersal syndrome and cone morphology in conifers and other nonflowering seed plants. New Phytologist. 2017 (in press)

[110] Tomback DF, Linhart YB. The evolution of bird-dispersed pines. Evolutionary Ecology. 1990;**4**(3):185-219

[111] Lanner RM. Seed dispersal in *Pinus*. In: Richardson DM, editor. Ecology and Biogeography of *Pinus*. Cambridge, UK: Cambridge University Press; 1998. pp. 281-295

[112] Leishman MR, Westoby M. The role of seed size in seedling establishment in dry soil conditions – Experimental evidence from semiarid species. Journal of Ecology. 1994;**82**(2):249-258

[113] Primack RB. Relationships among fowers, fruits, and seeds. Annual Review of Ecology and Systematics. 1987;**18**:409-430

[114] Hughes L, Dunlop M, French K, Leishman MR, Rice B, Rodgerson L, et al. Predicting dispersal spectra – A minimal set of hypotheses based on plant attributes. Journal of Ecology. 1994;**82**(4):933-950

[115] Rajon E, Venner S, Menu F. Spatially heterogeneous stochasticity and the adaptive diversification of dormancy. Journal of Evolutionary Biology. 2009;**22**(10):2094-2103

[116] Mathias A, Kisdi E. Adaptive diversification of germination strategies. Proceedings of the Royal Society B: Biological Sciences. 2002;**269**(1487):151-155

[117] Geritz SAH. Evolutionarily stable seed polymorphism and small-scale spatial variation in seedling density. American Naturalist. 1995;**146**(5):685-707

[118] Rees M, Westoby M. Game-theoretical evolution of seed mass in multi-species ecological models. Oikos. 1997;**78**(1):116-126

[119] Geritz SA, van der Meijden E, Metz JA. Evolutionary dynamics of seed size and seedling competitive ability. Theoretical Population Biology. 1999;**55**(3):324-343

[120] Liu Y, Barot S, El-Kassaby YA, Loeuille N. Impact of temperature shifts on the joint evolution of seed dormancy and size. Ecology and Evolution. 2017;**7**(1):26-37

[121] McGinley MA, Temme DH, Geber MA. Parental investment in offspring in variable environments – Theoretical and empirical considerations. American Naturalist. 1987;**130**(3):370-398

[122] Moles AT, Westoby M. Seedling survival and seed size: A synthesis of the literature. Journal of Ecology. 2004;**92**(3):372-383

[123] Westoby M, Falster D, Moles A, Vesk P, Wright I. Plant ecological strategies: Some leading dimensions of variation between species. Annual Review of Ecology Evolution and Systematics. 2002;**33**:125-159

[124] McGinley MA, Charnov EL. Multiple resources and the optimal balance between size and number of offspring. Evolutionary Ecology. 1988;**2**(1):77-84

[125] Jakobsson A, Eriksson O. A comparative study of seed number, seed size, seedling size and recruitment in grassland plants. Oikos. 2000;**88**:494-502

[126] Harper JL, Lovell PH, Moore KG. The shapes and sizes of seeds. Annual Review of Ecology and Systematics. 1970;**1**:327-356

[127] Salisbury E. The survival value of modes of dispersal. Proceedings of the Royal Society B: Biological Sciences. 1975;**188**:183-188

[128] Rees M. Delayed germination of seeds: A look at the effects of adult longevity, the timing of reproduction, and population age/stage structure. American Naturalist. 1994;**144**(1):43-64

[129] Tuljapurkar S. Delayed reproduction and fitness in variable environments. Proceedings of the National Academy of Sciences of the United States of America. 1990;**87**(3):1139-1143

[130] Bulmer MG. Selection for iteroparity in a variable environment. American Naturalist. 1985;**126**(1):63-71

[131] Ehrlén J, van Groenendael JM. The trade-off between dispersability and longevity: An important aspect of plant species diversity. Applied Vegetation Science. 1998;**1**(1):29-36

[132] Zeineddine M, Jansen VAA. To age, to die: Parity, evolutionary tracking and Cole's paradox. Evolution. 2009;**63**(6):1498-1507

[133] Dalling JW, Davis AS, Schutte BJ, Arnold AE. Seed survival in soil: Interacting effects of predation, dormancy and the soil microbial community. Journal of Ecology. 2011;**99**(1):89-95

[134] Paulsen TR, Colville L, Kranner I, Daws MI, Högstedt G, Vandvik V, et al. Physical dormancy in seeds: A game of hide and seek? New Phytologist. 2013;**198**(2):496-503

[135] Jayasuriya KMGG, Athugala YS, Wijayasinghe MM, Baskin JM, Baskin CC, Mahadevan N. The crypsis hypothesis: A stenopic view of the selective factors in the evolution of physical dormancy in seeds. Seed Science Research. 2015;**25**(2):127-137

[136] Schutte BJ, Davis AS, Peinado SA, Ashigh J. Seed-coat thickness data clarify seed size-seed-bank persistence trade-offs in *Abutilon theophrasti* (Malvaceae). Seed Science Research. 2014;**24**(2):119-131

[137] Agrawal AA. Transgenerational consequences of plant responses to herbivory: An adaptive maternal effect? American Naturalist. 2001;**157**(5):555-569

[138] Agrawal AA, Laforsch C, Tollrian R. Transgenerational induction of defences in animals and plants. Nature. 1999;**401**(6748):60-63

[139] Hu J, Mitchum MG, Barnaby N, Ayele BT, Ogawa M, Nam E, et al. Potential sites of bioactive gibberellin production during reproductive growth in *Arabidopsis*. Plant Cell. 2008;**20**(2):320-336

[140] Li N, Li YH. Signaling pathways of seed size control in plants. Current Opinion in Plant Biology. 2016;**33**:23-32

[141] Singh P, Dave A, Vaistij FE, Worrall D, Holroyd GH, Wells JG, et al. Jasmonic acid-dependent regulation of seed dormancy following maternal herbivory in Arabidopsis. New Phytologist. 2017 (in press)

[142] Venable DL, Brown JS. The selective interactions of dispersal, dormancy, and seed size as adaptations for reducing risk in variable environments. American Naturalist. 1988;**131**:360-384

[143] Thompson K, Ceriani RM, Bakker JP, Bekker RM. Are seed dormancy and persistence in soil related? Seed Science Research. 2003;**13**(2):97-100

[144] Moles AT, Westoby M. Seed mass and seedling establishment after fire in Ku-ring-gai Chase National Park, Sydney, Australia. Austral Ecology. 2004;**29**(4):383-390

[145] Fenner M, Thompson K. The Ecology of Seeds. Cambridge, UK: Cambridge University Press; 2005

[146] Poorter L, Rose S. Light-dependent changes in the relationship between seed mass and seedling traits: A meta-analysis for rain forest tree species. Oecologia. 2005;**142**(3):378-387

[147] Maranon T, Grubb PJ. Physiological basis and ecological significance of the seed size and relative growth rate relationship in Mediterranean annuals. Functional Ecology. 1993;**7**(5):591-599

[148] Hendrix SD, Sun IF. Inter-and intraspecific variation in seed mass in seven species of umbellifer. New Phytologist. 1989;**112**:445-452

[149] Stocklin J, Favre P. Effects of plant size and morphological constraints on variation in reproductive components in two related species of *Epilobium*. Journal of Ecology. 1994;**82**(4):735-746

[150] Moles AT, Falster DS, Leishman MR, Westoby M. Small-seeded species produce more seeds per square metre of canopy per year, but not per individual per lifetime. Journal of Ecology. 2004;**92**(3):384-396

[151] Willis CG, Baskin CC, Baskin JM, Auld JR, Venable DL, Cavender-Bares J, et al. The evolution of seed dormancy: Environmental cues, evolutionary hubs, and diversification of the seed plants. New Phytologist. 2014;**203**(1):300-309

[152] Scott RJ, Spielman M, Bailey J, Dickinson HG. Parent-of-origin effects on seed development in *Arabidopsis thaliana*. Development. 1998;**125**(17):3329-3341

[153] Lamont BB, Lemaitre DC, Cowling RM, Enright NJ. Canopy seed storage in woody plants. Botanical Review. 1991;**57**(4):277-317

[154] Hoyle GL, Venn SE, Steadman KJ, Good RB, McAuliffe EJ, Williams ER, et al. Soil warming increases plant species richness but decreases germination from the alpine soil seed bank. Global Change Biology. 2013;**19**(5):1549-1561

[155] Cochrane A, Holye G, Yates C, Wood J, Nicotra A. Climate warming delays and decreases seedling emergence in a Mediterranean ecosystem. Oikos. 2015;**124**(2):150-160

[156] Childs DZ, Metcalf CJE, Rees M. Evolutionary bet-hedging in the real world: Empirical evidence and challenges revealed by plants. Proceedings of the Royal Society B: Biological Sciences. 2010;**277**(1697):3055-3064

[157] Probert RJ. The role of temperature in the regulation of seed dormancy and germination. In: Fenner M, editor. Seeds: The Ecology of Regeneration in Plant Communities. Wallingford, Oxon: CABI; 2000. pp. 261-292

[158] Ooi MKJ, Auld TD, Denham AJ. Climate change and bet-hedging: Interactions between increased soil temperatures and seed bank persistence. Global Change Biology. 2009;**15**:2375-2386

[159] Evans ME, Dennehy JJ. Germ banking: Bet-hedging and variable release from egg and seed dormancy. The Quarterly Review of Biology. 2005;**80**(4):431-451

[160] Walck JL, Hidayati SN, Dixon KW, Thompson K, Poschlod P. Climate change and plant regeneration from seed. Global Change Biology. 2011;**17**(6):2145-2161

[161] Roach DA, Wulff RD. Maternal effects in plants. Annual Review of Ecology. 1987;**18**: 209-235

[162] Whittle CA, Otto SP, Johnston MO, Krochko JE. Adaptive epigenetic memory of ancestral temperature regime in *Arabidopsis thaliana*. Botany. 2009;**87**(6):650-657

[163] Case AL, Lacey EP, Hopkins RG. Parental effects in *Plantago lanceolata* L.II. Manipulation of grandparental temperature and parental flowering time. Heredity. 1996;**76**:287-295

[164] Lacey EP. Parental effects in *Plantago lanceolata* L. I. A growth chamber experiment to examine pre- and postzygotic temperature effects. Evolution. 1996;**50**(2):865-878

[165] Lundgren MR, Sultan SE. Seedling expression of cross-generational plasticity depends on reproductive architecture. American Journal of Botany. 2005;**92**(2):377-381

[166] Galloway LF, Etterson JR. Plasticity to canopy shade in a monocarpic herb: Within- and between-generation effects. New Phytologist. 2009;**182**(4):1003-1012

[167] Sultan SE. Phenotypic plasticity for offspring traits in *Polygonum persicaria*. Ecology. 1996;**77**(6):1791-1807

[168] Sultan SE. Phenotypic plasticity for fitness components in *Polygonum* species of contrasting ecological breadth. Ecology. 2001;**82**(2):328-343

[169] Sultan SE, Barton K, Wilczek AM. Contrasting patterns of transgenerational plasticity in ecologically distinct congeners. Ecology. 2009;**90**(7):1831-1839

[170] Franks SJ, Sim S, Weis AE. Rapid evolution of flowering time by an annual plant in response to a climate fluctuation. Proceedings of the National Academy of Sciences of the United States of America. 2007;**104**(4):1278-1282

[171] Boyko A, Blevins T, Yao YL, Golubov A, Bilichak A, Ilnytskyy Y, et al. Transgenerational adaptation of *Arabidopsis* to stress requires DNA methylation and the function of Dicer-Like proteins. PloS One. 2010;**5**(3):e9514

[172] Kou HP, Li Y, Song XX, Ou XF, Xing SC, Ma J, et al. Heritable alteration in DNA methylation induced by nitrogen-deficiency stress accompanies enhanced tolerance by progenies to the stress in rice (*Oryza sativa* L.). Journal of Plant Physiology. 2011;**168**(14):1685-1693

[173] Steets JA, Ashman TL. Maternal effects of herbivory in *Impatiens Capensis*. International Journal of Plant Sciences. 2010;**171**(5):509-518

[174] Donohue K, Schmitt J. Maternal environmental effects in plants: Adaptive plasticity? In: Mousseau T, Fox CW, editors. Maternal effects as adaptations. New York: Oxford University Press; 1998

[175] Fitter AH, Fitter RSR. Rapid changes in flowering time in British plants. Science. 2002;**296**(5573):1689-1691

[176] Parmesan C, Yohe G. A globally coherent fingerprint of climate change impacts across natural systems. Nature. 2003;**421**(6918):37-42

[177] Miller-Rushing AJ, Primack RB. Global warming and flowering times in Thoreau's concord: A community perspective. Ecology. 2008;**89**(2):332-341

[178] Verhage L, Angenent GC, Immink RG. Research on floral timing by ambient temperature comes into blossom. Trends in Plant Science. 2014;**19**(9):583-591

[179] Rosas U, Mei Y, Xie Q, Banta JA, Zhou RW, Seufferheld G, et al. Variation in Arabidopsis flowering time associated with *cis*-regulatory variation in *CONSTANS*. Nature Communications. 2014;**5**:3651

[180] Banta JA, Ehrenreich IM, Gerard S, Chou L, Wilczek A, Schmitt J, et al. Climate envelope modelling reveals intraspecific relationships among flowering phenology, niche breadth and potential range size in *Arabidopsis thaliana*. Ecology Letters. 2012;**15**(8):769-777

# Chapter 2

# Decoding the Transcriptome of Rice Seed During Development

Arunima Mahto, Iny Elizebeth Mathew and
Pinky Agarwal

Additional information is available at the end of the chapter

http://dx.doi.org/10.5772/intechopen.70574

**Abstract**

Rice seed development is a continuous process wherein it undergoes complex molecular and tissue reprogramming. It is a collective effect of embryo and endosperm development, each of which undertakes its own developmental paths, with endosperm development significantly affecting embryo. Understanding the mechanistics of the regulatory networks administrating this process is the building block for any future research on grain yield and quality. High-throughput transcript profiling and small RNA profiling studies have proved useful in providing information about the molecular changes occurring in various tissues associated with seed development. Transcriptome sequencing studies have highlighted the significant genes and pathways that are operating during seed development. The involvement of TFs and hormones has also been implicated in regulating key aspects of seed development, including embryo patterning and seed maturation. This chapter will review the information provided by high-throughput sequencing studies on various aspects of rice seed development, highlighting the developmental complexities of embryo and endosperm.

**Keywords:** cell cycle, embryo, endosperm, hormones, rice, seed development, transcription factors, transcriptome

## 1. Introduction

Seed development is a unique attribute of plants providing them the privilege of perpetuating genetic information over generations by safeguarding against environmental atrocities. Physiologically, it is a combined effect of two complex developmental processes, embryo and endosperm development. In case of dicots, majority of the seed volume is formed by the embryo at maturity and the endosperm is consumed by the embryo during the course of seed development. The structure of monocot seed, such as rice, is different from a dicot seed by the presence of a starchy endosperm which occupies most of the space inside the seed coat and the embryo is positioned at the ventral side. Furthermore, the seed is covered entirely by the husk, which is formed by drying of the lemma and the palea. Seeds serve as the storage factories for synthesizing carbohydrates, proteins and lipid molecules, hence act as nutrition suppliers to the germinating seedling as well as to animals and humans. Rice seeds, in particular, are the major calorie providers constituting about 20% of the human nutrition worldwide [1, 2]. Therefore, it becomes imperative to understand seed development in rice to produce varieties with improved nutritional content and yield.

Seed development in rice incorporates development of the embryo and the endosperm and occurs in a systematic and sequential manner followed by desiccation and seed dormancy. The entire process of seed development in rice has been summated into five different stages from S1 to S5, categorized as 0–2, 3–4, 5–10, 11–20 and 21–29 days after pollination (DAP) seeds, respectively. Developmental period of the seed consisting of post-fertilization to middle globular embryo constitutes the first stage followed by embryo patterning and endosperm cellularization in second stage. The third stage is concerned with embryo morphogenesis, formation of a milky endosperm and initiation of endoreduplication. In the maturation phase, the milky endosperm transits from soft dough and hard dough stages in S4, and the seeds progress towards dormancy and desiccation in S5 stage [3, 4]. These developmental changes are channelized impeccably through the skillful operation of several genes and complex regulatory networks upon perception of internal and external stimuli [4–6]. Recent technological advances have facilitated the identification of genes responsible for guiding various steps of seed development. High-throughput mRNA profiling studies or transcriptomics is one such technology that has helped in deriving vital information about a myriad of molecular events that orchestrate seed development [7, 8]. Transcriptome profiling of a wide range of rice tissues, including vegetative and reproductive tissues, have proved beneficial in providing primary information about the genes expressed during seed development including their levels, patterns and molecular functions [6, 9–11]. With the aid of advanced bioinformatics platforms, transcriptome data is now being processed to derive more complex interpretations including pathways and regulatory networks that provide more complete picture of the molecular changes regulating seed development [12–15].

## 2. Expression atlases capture the dynamicity of transcriptome during seed development

Seed development is a continuous process and is effected by the participation of many tissues that undergo various developmental changes over a course of time. Such dynamic alterations would be difficult to be depicted entirely by studying tissues in isolation. Gene expression atlases incorporate transcriptome profiles of a wide range of cell types and/ or developmental stages. Such global profiling studies become important when tracing the complex transcriptional changes associated with transition of tissues from one phase of development to another. Several high-throughput studies of this nature have been conducted in rice, employing MPSS, microarray and transcriptome sequencing, which span both vegetative and reproductive tissues [6, 9–11, 16, 17]. The primary information obtained from expression atlases is about the transcriptional status of the tissues/organs with respect to one another. For instance, the steady increase in the number of down regulated genes in seed tissues with respect to vegetative tissues indicates gradual decrease in transcriptional activity with the progression of seed development [6]. The similarity and disparity in the transcriptome profiles has also been used to assess the relatedness of tissues [16]. Through microarray analysis of 39 vegetative and reproductive tissue types in rice, it has been shown that rice endosperm forms a separate cluster and exhibits relatively lesser gene expression than other tissues including panicle stages. However, the expression levels of these genes are significantly high and a large portion of these are endosperm-specific. Also, most of the development related genes show variation in expression levels among different tissues revealing the fluctuations at the molecular level that these genes experience as they pass from one stage of development to another. These observations provide several important inferences regarding seed development, such as transcriptional dynamics and tissue-specificity. Also, an inevitable inference from here is that the transcriptome undergoes intense spatiotemporal reprogramming during transition from vegetative to seed development in order to manifest the expression of seed-specific/preferential genes [11]. Atlases also provide information about the contributions of various tissues in seed development. Transcriptome study encompassing various vegetative and reproductive tissues of rice indicate that out of the total seed-specific genes obtained, the proportion of endosperm-specific genes is higher than that of embryo-specific genes. This might imply that the endosperm has more disparate transcript profile shifting the balance towards the role of endosperm in seed development in comparison to embryo [10]. Nevertheless, conclusions from such studies can be subjective and will vary according to the tissues and developmental stages under investigation and the methods of data analysis. For instance, the discovery of seed-specific genes will be influenced by the variety of vegetative tissues that are being considered for assessing specificity or the parameters set to call a gene expressed or differentially expressed. In a study published by our group, encompassing 19 vegetative and reproductive tissues, the number of seed-specific genes has been found to vary when expression is considered against different vegetative controls such as root and mature leaf [6]. This indicates that a single transcriptome data can emanate various circumstantial biological interpretations. However, to minimize erroneous accounting it is necessary to exercise caution while sampling and have precise knowledge of the query that is being pursued.

## 3. Functional intricacies of seed development revealed by atlases explain the conundrum of seed development

Functional annotations of the enormous data generated by atlases further elaborate on the predominant activities occurring during seed development including their mode of regulation. The enrichment of genes involved in embryonic development in the transcriptome of endosperm suggests a possible communication between the embryo and the endosperm [11]. This exchange of information between the embryo and the endosperm highlights that a certain level of cross-talk might be necessary for their growth. Induction of expression of seed-specific genes after 5 days of flowering, when the endosperm development is accelerated and starch accumulation is initiated, suggests that majority of the seed-specific genes are associated with later stages of seed development and are required for seed filling and maturation [10]. Information about the key processes, pathways and genes involved in seed development can also be identified by comparative transcript profiling of tissues from various stages of development. Seed-specific proteins, seed allergens, genes involved in starch biosynthesis/degradation and ubiquitin-mediated protein degradation pathway show specific expression throughout the seed stages in comparison to four vegetative stages [6]. Transcription factors (TFs) have been reported to be enriched in both early and later stages of seed development suggesting their involvement throughout the process. In a study involving 48 organs of a *japonica* rice variety, it has been found that out of 41 tissue-specific TFs obtained, 29 are seed-specific. These include several members of MADS, NAC, AP2-EREBP and CCAAT families that are expressed in a seed-specific manner with predominant expression either in the endosperm or the embryo and this expression is also driven by the stage of the tissues (**Figure 1**). TFs such as *VP1* and *LEC1*, which have been reported to be active during seed maturation, have been found to be present in the seed-specific category whereas AP2-EREBP TFs are found to express mostly in the embryo through its entire development [10]. In our study on an *indica* rice variety, it has been seen that 27 TFs families have higher number of members expressing in the seed stages, which include MYB, NAM, HSF, MADS, POZ, and bZIP. About 47 TFs are found to be specific to seed stages, of which most have specificity to S2 stage [6]. Such preferential expression patterns observed in various studies would imply that these TFs are administrating seed development by regulating downstream genes and pathways required for individual growth and development of the embryo and the endosperm at specific time points and stages.

Hormonal regulation is another indispensable component of the multi-faceted regulation of seed development. Rhythmic fluctuations in levels of hormones such as auxins, gibberellic acid (GA) and abscisic acid (ABA) have been observed during the course of seed development suggesting a complex interplay between these [18]. Different hormones have been known to be controlling different modules of seed development such as organ patterning, cell enlargement, desiccation and dormancy [19–22]. Seed-specific differential regulation of various hormones has been encountered in several reports. Auxin biosynthesis genes have been found to be induced during early stages of seed development [23]. This emphasizes on the role of auxins in the initial seed development processes that are mainly associated with active cell division in the embryo and the endosperm and organ initiation [4]. Along with auxins, implications for the role of gibberellin and ethylene have also been proposed. Genes associated with ent-kaurene biosynthesis, a precursor of gibberellin biosynthesis, were found to be up regulated in the S1 stage of seed development indicating the role of gibberellin in early phases of seed development [23]. A negative regulator of gibberellin signaling, *SLRL2* and a putative ethylene receptor that negatively regulates ethylene signaling, *OsETR2;2*, have been seen to be showing seed-specific expression [10]. These findings enumerate the significance of differential

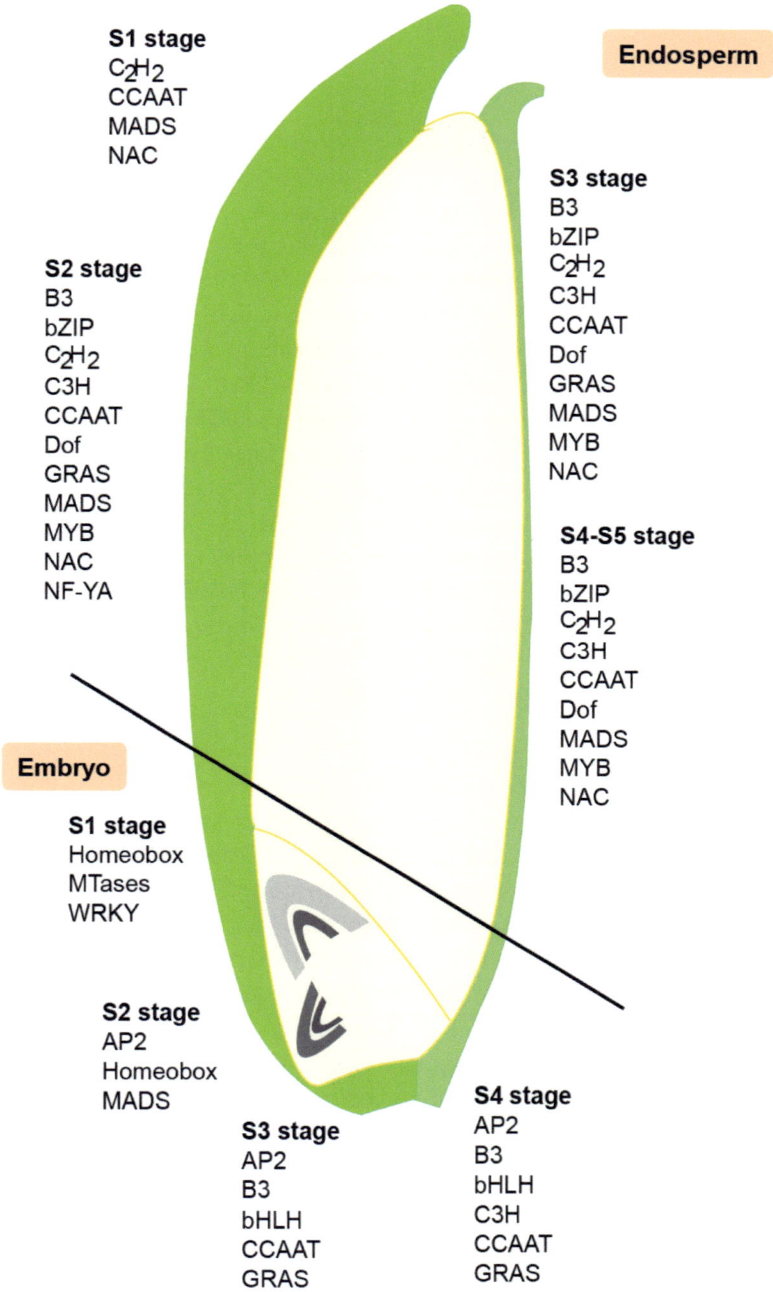


**Figure 1.** Schematic representation of TFs involved in rice seed development. TF families whose members express during the five stages of seed development in the endosperm and embryo have been mentioned above and below the solid line, respectively.

regulation of these hormones in materializing seed development. The localized expression of different phytohormone-related genes during embryo development as well as their critical role in endosperm development and grain filling has been discussed in further sections.

## 4. Genes involved in polarity establishment and organ initiation are expressed in the embryo

Embryogenesis is an important aspect of seed development, which involves cellular division and establishment of the embryo body plan. The process of embryo development is classified into ten major stages in rice. The first stage is the zygotic stage formed at 0 days after fertilization (DAP). The initial stages of post-zygotic development are characterized by the repeated cellular division without morphogenesis. This leads to the formation of a globular embryo, which can be observed from 1 to 3 DAP. It is the stage at which embryo specifies its body axis. Subsequently, formation and proper positioning of the shoot apical meristem (SAM), coleoptile primordium and the radicle primordium occurs at 4 DAP. This is followed by the origin of leaf primordium, which initiates at 5 DAP and is completed by 8 DAP, by the formation of all the three leaf primordia. With the enlargement of various organs, embryo morphogenesis is completed in rice by 10 DAP. This is followed by a stage of maturation which lasts till about 20 DAP, and thereafter, it enters into a dormant state [4, 5]. Profiling the RNA expression levels has greatly broadened the understanding of rice embryo development. Unlike animal embryos, both maternal and paternal genomes contribute equally in the process of embryogenesis and the plant embryonic development is majorly under zygotic control. Maternal-to-zygotic transition (MZT) initiates 50 hours after fertilization [24–26]. Thus, zygotic genome gets "switched on" almost immediately after fertilization.

Microarrays coupled with laser microdissection (LMD) have made available the expression profiles of developing female gametophyte from the pre-meiotic to the mature embryo-sac stages in rice [27]. Similarly, expression profiles of stigma as well as embryo sac are also available from high-throughput RNA-Seq technology [28]. These cell-type specific expression analyses have been compared with various developmental stages of embryo to identify genes that are differentially expressed during the post fertilization period. Identification of such genes expands our idea of embryogenesis as many of them can be candidates for maintaining various stages or aspects of embryo morphogenesis. Microarray-based comparison of gametic and zygotic tissues has identified a total of 325 genes up regulated in zygote in comparison with the egg cell. Majority of these up regulated genes are involved in DNA/chromatin organization and their assembly is probably involved in the induction of genes participating in zygotic development. Further, methyltransferase 1 (MET1) and different TFs belonging to the homeobox proteins are highly up regulated in the zygote, apparently affecting polarity or asymmetric division in zygote. Specific METs show higher expression during early rice seed development and are essential for cytosine methylation, regulating the genes involved in various developmental processes [29]. Also, microarray analysis shows a significant transcript accumulation of one such member, *OsMET1–2* during the early seed stages [23]. This is an indication of the role played by various DNA methyl transferases (MTases) in gene regulation during early seed development (**Figure 1**). Microarray analysis has also shown that a total of 94 genes are down regulated in the zygotic stage. Subsequent gene ontology (GO) and pathway analyses suggest the involvement of these genes in metabolic pathways possibly associated with suppression of the maternal genes [26, 29]. Thus, various transcript analyses indicate the existence of an active zygotic genome during early seed development.

A systematic expression profiling at three developmental stages of the embryo categorized as early (3–5 DAP), middle (7 DAP) and late (14 DAP) have shown the expression of about 20,856 common genes, suggesting their role in housekeeping functions. However, many genes show specific expression in each category suggesting their involvement in imparting functions unique to the stage. These genes belong to different functional categories as metabolic processes, binding and cell part and cellular processes. About 1131 genes show specific expression at 3–5 DAP, possibly involved in determining the embryo axis. Polarized expression of different TF and transcription regulator (TR) genes has been identified at the apical-basal and dorsal-ventral axis in the globular embryo. This spatiotemporal expression of specific TFs and TRs might be involved in the establishment of early embryo patterning in rice [30]. Different phytohormone-related genes including GA biosynthetic genes, auxin efflux *PIN* genes, cytokinin A-type response regulators, and brassinosteroid (BR)-perception genes also show an embryonic axis-dependent expression. The repressor of GA-signaling *OSSLR1* shows a preferential expression in the basal region of the embryo. On the other hand, a GA biosynthetic gene, *OsGA20ox1* is expressed in the apical-dorsal region. Apical to basal auxin transport is initiated at the early globular stage by the auxin transport proteins, thus, regulating various aspects of embryonic pattern formation. Transcript accumulation for cytokinin response-regulator occurs in the apical-ventral region whereas BR and ethylene biosynthesis occurs in the basal region [31]. Among TFs, homeobox gene family members show differential expression during different phases of embryo development suggesting their inevitable roles during the process [4, 30]. Different MADS-box transcripts show seed-preferential expression with about 12 of them showing a specific expression in the seed, including Arabidopsis ABCDE class gene orthologs, suggesting their involvement in early embryo development [30, 32].

A gradual transition in the transcript profile from early to late stages of embryo has been observed. A large number of genes are shared between the early and middle stages of embryo development although, unique expression of different genes is also observed [30, 31]. The genes up regulated in the early and middle stages of embryo development are majorly involved in amino acid, lipid and energy metabolism, nucleic acid replication/processing, signal transduction and transcriptional regulation. The enrichment of these pathway genes provides the energy required for the early developing embryo. As the embryo progresses towards the middle stage, additional genes as ribosomal protein components, translational machinery components are up regulated. Thus, the protein biosynthesis genes show a greater expression during the middle phase. There occurs a significant enhancement in the expression of genes belonging to different categories as the embryo progresses towards the maturation phase. Many of these genes show differential expression between 7 and 14 DAP embryos. Pathway and gene ontology studies suggest significant differences in the physiological processes that occur during early and late stages of rice embryogenesis [30]. Further, the maturation phase is characterized by the accumulation of protein modification and starch biosynthesis genes. Auxin related *Aux/IAA, OsIAA18,* shows a significant up regulation during the middle and late stages of embryo development. Auxin-biosynthetic genes have been shown to be induced during different stages of embryo development [6, 14]. Also, the biosynthesis of ethylene is down regulated during embryo development by the enhanced level of ABA. Thus, the two phytohormones ABA and ethylene function antagonistically during embryo development in rice [14]. Additionally, GA also functions in the seed development process during maturation. Seed maturation process is majorly determined by the GA/ABA ratio [22]. To add, profiling studies have also identified the accumulation of long-lived mRNAs between 10 and 20 DAF within the embryo. Long-lived mRNAs present in the mature dry seeds are required for proper seed germination. These majorly code for proteins related to the signaling of ABA, calcium ions and phospholipids as well as a heat shock protein HSP DNA J, essential for rice seed germination [33].

## 5. Rice endosperm shows structural and developmental complexities

The endosperm of rice occupies a major portion of the seed and defines the shape of the grain. It is the storehouse of nutrients including carbohydrates, lipids and storage proteins and serves as an important source of nutrition for the developing embryo. The formation of the endosperm starts with triple fusion, wherein the male nucleus fertilizes the bi-nucleate central cell to produce a triploid cell. Thereafter, it sequentially undergoes events of cell division, cell fate determination, tissue differentiation and programmed cell death (PCD) to produce the mature endosperm. Structurally, the endosperm features four major types of cells, the starchy endosperm, the aleurone cells, the transfer cells and the cells in the vicinity of the embryo [34]. The cells in the peripheral region, except those near the vascular tissue, form the aleurone layer which varies in thickness from one to five cell layers [35]. Cells immediately above the vascular bundles form the transfer cells. Cells enclosed within these two cell layers form the starchy endosperm. The embryo surrounding cells create the cavity in which the embryo is housed. These differentiated cell layers perform specific functions which are required by the embryo during its growth and afterwards for seed germination. Photosynthate (sucrose) produced in the leaves (source) is transported into the endosperm (sink) via the transfer cells [36, 37]. Endosperm surrounding region separates the embryo and the endosperm and might also be involved in providing nutrition to the embryo by apoplastic transport [38, 39]. Starchy endosperm stores starch and proteins that start accumulating soon after cellularization is complete [39]. The aleurone layer is composed of terminally differentiated cells that produce proteolytic, hydrolytic and cell wall degrading enzymes that digest the starch and proteins stored in the endosperm into sugars and amino acids for utilization by the growing embryo during seed germination [40, 41]. Hence, the development of endosperm is complex and singular owing to the modifications in its structure occurring through a short span of time and the accumulation of reserve materials and cell cycle activities that are switched on and off at precise time points.

The functional uniqueness of the endosperm is reflected in its transcriptome which has been found to be quite distinct from several other tissue types including reproductive stages and embryo [10]. Transcriptome analysis of three developmental stages of endosperm, spanning from during this time, has shown an overall decline in gene expression during this time. Even more down regulation has been observed in the later stages [12]. Studies also indicate that in the young endosperm stages (0–4 DAP), the number of specific genes increases with age suggesting that the complexity of molecular changes rapidly increases with progression of endosperm development [9]. In another study involving 7, 14 and 21 DAP endosperm tissue, it has been observed that the expression of specific genes can be clustered into distinct patterns. About 79 genes are expressed in all the three stages suggesting that they are constitutively required throughout endosperm development. A set of 32 genes express highly in 14 and 21 DAP indicating their role in nutrient accumulation and PCD. About 22 genes and 15 genes show higher expression in 7 DAP alone and in both 7 and 14 DAP, respectively. These genes can be presumed to be regulating cell proliferation and cellularization during initial development and synthesis and accumulation of storage compounds [42]. Thus, amalgamation of such information from transcriptome data with knowledge from previous developmental studies can be useful in generating knowledge about the functions performed by various genes in different stages.

Endosperm involves the precise operation of several transcription factors throughout its course of development. About 1118 transcription factors belonging to 55 families have been reported to be expressing in early stages of endosperm development [12]. TFs have emerged as a major functional category in later stages (7–21 DAP) of endosperm development [42]. Expression pattern of TFs has also been indicated to be subjected to temporal regulation. Members of the transcription factor families such as, MADS, NAC, AP2-EREBP, MYB and CCAAT, have been observed to show higher expression in the endosperm (**Figure 1**). Out of these, *MADS* genes are expressed through the early stages (1–14 DAP), *AP2-EREBP* and *MYB* are expressed during early through middle stages (7–21 DAP), whereas, *NAC* and *CCAAT* are expressed through all stages (2–42 DAP) of endosperm development [10, 42]. MADS TFs have been shown to regulate endosperm development by a mechanism affecting the cytokinin level. Overexpression of *MADS29* activates the genes involved in starch biosynthesis and promotes the differentiation of proplastids to starch-containing amyloplasts [43]. In our study encompassing five different rice varieties, three NAC TFs exhibit seed-specific/preferential expression with significantly higher expression in S3-S5 stages suggesting their role in accumulation of storage reserves. They also show significant association with seed traits emphasizing their role in regulation of seed development [44]. Similarly, genome-wide analysis of 14 vegetative and reproductive tissues has indicated the expression of 21 $C_2H_2$ proteins in seeds of which 12 are specific to seed tissue. The expression of these genes shows variable pattern among the five stages. Some of them are expressed from S1–S5, while most of them show higher expression in the later stages of seed development implying their function in both initial seed development and seed maturation [3]. In another report including three endosperm stages covering 3–10 DAP, different types of expression patterns of the transcription factors have been observed. Six TFs families including Dof are up regulated through 3–10 DAP. Three transcription factor families including GRAS are down regulated from 3 to 6 DAP then up regulated till 10 DAP. NF-YA family members on the other hand are up regulated from 3 to 6 DAP then down regulated until 10 DAP [12]. In summary, TFs are expressed throughout the development of the endosperm and their expression is highly preferential. The heterogeneity in the expression patterns of TFs is an indicator of the intricate molecular regulation of endosperm transcriptome probably required for proper completion of a stage and subsequent transition to another.

As mentioned previously, hormones are known to be regulators of embryo development and this raises the possibility of them being key ingredients in the regulatory network of endosperm development. In this context, it has been observed that several hormone response *cis*-elements are present in the promoters of endosperm-specific genes that are expressed from 7 DAP to 21 DAP. The most abundant *cis*-elements belong to abscisic acid responses, including ABADESI1 and ABREMOTIFAOSOSEM. Since abscisic acid is a well-known hormone for desiccation and dormancy, which is associated with seed maturation, the aforesaid observation would imply that these processes are very eminent in endosperm and are initiated from the middle stages of development [45]. Along with ABA, *cis*-elements for gibberellic acid, such as GARE1OSREP1 and PYRIMIDINEBOXOSRAMY1A, auxins, ARFAT, and ethylene-responsive element, such as ERELEE4 have also been observed [42]. The simultaneous expression of the genes regulated by hormones unambiguously indicates the significance of hormonal interplay in the growth of endosperm. Although, the specific effects of these hormones can only be understood from detailed functional characterization studies.

One eccentric yet indispensable feature of endosperm development is the occurrence of two types of cell cycles at different stages of development. First is the free nuclear division without cellularization leading to syncytium formation in the initial stages of development (0–5 DAP), and second is endoreduplication that occurs in the later stages (8–10 DAP) and is associated with increasing cell size and endosperm volume [4]. In coherence with this information, two CDKs, *CDKB;1* and *CDKB;2*, have been found to be showing higher expression in early stages of endosperm (1–2 DAP). Also, one A-type cyclin and four B-type cyclins exhibit patterns of expression overlapping with these CDKs. It is noteworthy that cell cycle defects associated with the endosperm can influence the growth of the embryo. Knockdown of a rice cyclin gene, *CycB1;1*, results in the formation of a large embryo and abortive endosperm suggesting that normal mitotic activity of the endosperm is imperative for the development of the embryo as well [46].

One of the primary objectives of the endosperm is stocking of nutrients which will eventually be assigned to various metabolic pathways required for seed development process. Bulk of the endosperm is constituted by starch and prolamin storage proteins [47–49]. Transcriptome profiling studies advocate that genes associated with accumulation of starch and sugars are significantly up regulated in the endosperm. Genes related to starch metabolism and storage protein biosynthesis have been found to be among the highly up regulated genes as development progresses from 3 to 10 DAP. Also, 11 members of Dof TFs have been found to be up regulated in endosperm [12]. Dof TFs are known to be associated with synthesis of storage proteins in the endosperm [50]. Pathway studies have also indicated that in the endosperm, starch and sugar metabolism are highly up regulated followed by amino sugar and nucleotide sugar metabolism and carbon fixation by photosynthesis. It has also been observed that as endosperm moves from 3 DAP to 10 DAP most of the genes and pathways are down regulated except for those related to accumulation of storage materials [12]. Functional annotation of endosperm-specific genes from 7 DAP to 21 DAP have shown that seed storage protein, carbohydrate and energy metabolism, seed maturation, protein metabolism, lipid metabolism and transport emerge as the major categories. Seed storage proteins, including prolamins, glutelins and globulins have been reported to constitute the third largest category of endosperm-specific genes after transcription factors and stress responsive genes. Apart from this, overrepresentation of CATGCA motif or the RY element has been seen in the promoters of the endosperm-specific genes expressed from 7 DAP to 21 DAP. These genes show varied molecular function, including hydrolase activity, nutrient reservoir activity and transcription factor activity [42]. From these findings, it can be concluded that the endosperm starts gathering storage material quite early in its development which continues till maturation.

This continuous process is controlled by the collaborative efforts of several genes, pathways and regulatory networks, which are primarily associated with synthesis and accumulation of starch and proteins.

Towards the end of its development, after the complete size has been attained and storage materials are being accumulated, the endosperm undergoes programmed cell death (PCD) which is initiated from 16 DAP in cereal seeds [4, 51, 52]. This results in degeneration of the storage cells of the endosperm surrounded by living cells of the aleurone layer. Although PCD has been less explored in rice endosperm, some reports of PCD genes from pollen tissues and rice protoplasts are available [11, 53, 54]. Transcriptome studies of endosperm have detected several PCD related genes. *AIP5*, a positive regulator of PCD in the tapetum, has been found to be up regulated, whereas, *hsp70*, a negative regulator of PCD in rice protoplasts, is down regulated in the endosperm. Apart from these, 11 PCD related genes have also been found to express in the endosperm [12]. PCD has also been implicated to be influenced by hormones. Ethylene and gibberellic acid have been suggested to be positive regulators and abscisic acid has been shown to be a negative regulator of cell death [51, 55]. Up regulation of six gibberellic acid pathway genes and down regulation of 20 abscisic acid pathway genes has been observed in the 6 and 10 DAP endosperm tissue [12]. These results provide additional support to the earlier reports and emphasize on the active involvement of these hormones in the regulation of PCD in the endosperm.

## 6. Expression profiling unravels the complex molecular machinery involved in grain filling

Quality of rice grains, the major human calorie provider is very significant in the present scenario of ensuring global food security. Quality and quantity of grain production is majorly dependent on the synthesis and storage of various macromolecules and minerals during the grain filling stage. In rice, grain filling happens in the endosperm tissue and is regulated by highly coordinated and synchronous pathways [4]. Endosperm acts as the nutrient reservoir for the developing embryo initially and to the germinating embryo over the course of time. Endosperm functions in the supply of nutrients to the growing embryo right from its syncytial state. Growth and expansion of the endosperm cells are limited by programmed cell death in mature seeds. Thus, the accumulation of storage reserve is dependent on the life span of endosperm cells [56]. Understanding the intricate machineries involved in grain filling is imperative in the identification and manipulation of the key regulatory pathways aimed at improving the quality and productivity of the crop varieties available. Expression analyses serve as a promising tool facilitating the identification of candidate genes regulating grain filling process in rice.

Major reserves accumulating in seeds include carbohydrates, storage proteins and lipid compounds. Biosynthesis of these storage macromolecules are coordinately controlled by different TFs and other TRs. Expression profiling has shown the co-expression of different TF genes including *bZIP*, *Dof* and *MYB* with many grain filling genes in rice [57]. Members belonging to these protein families have shown to play significant roles in the regulation of storage protein and starch biosynthesis [57–59]. Additionally, genes involved in the biosynthesis of macromolecules, various transporters for amino acid, sugar, phosphate, peptide, nitrate and ABC transporters show enrichment in the grain filling stage [57]. Transporter genes are essential for the uptake of nutrient and precursor molecules from the source tissues. Expression profiling also gives information regarding the genes involved in specific pathways. Furthermore, a detailed analysis can be useful in the identification of *cis*-elements enriched in specific process. Many of the grain filling genes contain a conserved *cis*-element 'AACA' in their promoters, suggesting its importance in the process. AACA element is essential for conferring the expression in rice seed [57, 60]. Transcript profiling has also shown that the milling yield and eating quality of rice grains depends on the proportion of starch and proteins in the grain. High quality rice grains contain a high composition of starch and protein biosynthetic transcripts. Massively parallel signature sequencing (MPSS) and sequencing by synthesis (SBS) have shown a higher level of alternative splicing and antisense transcripts for different metabolic genes in the high milling yield and eating quality varieties. These transcripts belong to starch, aspartate amino acid, storage protein and allergenic protein metabolism genes, indicating the complex transcriptional cascade involved in the regulation of rice grain quality [61]. Thus, expression profiling has not only improved our understanding in the grain filling process but also identified different transcripts essential for the process. Many of these genes can also serve as potential markers for the identification of superior rice varieties.

Grain chalkiness is another important agronomic trait influencing the market value of rice. It negatively affects the consumer preference and culinary quality. Analyses have shown the differential expression of a large repertoire of genes involved in signal transduction, cell rescue/defense, transcription, protein degradation, carbohydrate metabolism and redox homeostasis in a high chalky rice variety. Out of the different metabolic genes, starch metabolism genes can be considered as the major reason for grain chalkiness because of their opposite expression pattern in varieties showing varying levels of chalkiness. The sucrose and starch biosynthetic genes show up regulation in the chalky variety. Moreover, the non-starchy polysaccharide transcripts show significant down regulation. Thus, the expression profiling suggests a positive correlation of the starchy polysaccharide transcripts with the chalky phenotype in rice grains. Additionally, the genes involved in oxido-reductive homeostasis also show significant up regulation in the chalky rice variety [62]. Thus, transcript profiling has the potential for the identification of candidate genes underlying a phenotype, including grain chalkiness.

A delay in the expression of various genes involved in the transformation of sucrose to starch has been identified as the major reason for poor grain filling in the inferior spikelets located on the lower secondary panicle branches. RNA-Seq analysis shows the lower expression of these genes in the inferior spikelets at an early stage of grain filling in comparison with the superior spikelets. However, it was reversed during the later stages of grain filling process. Low capacity of the sink tissue and the associated limited carbohydrate supply at the later stages of grain filling has been proposed as the probable reason for the poor filling of grains [63]. Thus, profiling of seed transcripts has greatly deepened our understanding of the molecular machinery involved in seed filling and panicle branching. This has served to identify the cascade of TRs involved in the process. It will definitely pave way for the identification of candidate genes and their introgression for the production of improved variety with better consumer preference.

## 7. Dissecting out the effect of temperature stress on seed filling at the molecular level using expression profiling

Plants are sessile organisms and show ideal growth and development only when they are grown under optimal conditions. Temperature is one of the most important environmental factors which has a great influence on the growth and development of various plant species. It has got much attention because of the recent global warming. Rice grain weight and yield have been shown to be greatly reduced by the environmental temperature. Grain filling process is more affected by the increase in night temperature than by the day temperature, particularly. High temperature reduces grain yield by impairing the filling process. Transcriptome profiling has greatly helped in revealing the underlying mechanism of yield loss associated with high temperature stress.

An increase in the temperature affects the grain development process right from the opening of the glumes till the grain filling stage. At high temperature, the glumes remain unclosed which affects the germination rate as well as commercial quality of the seeds. Genes involved in cell wall metabolism and response to water and carbohydrate metabolism are up regulated following high temperatures stress [64]. A large number of genes show differential expression in heat sensitive and tolerant cultivars in the early milky stage of the rice seed. These genes are majorly involved in oxidation-reduction, metabolic, transport, transcript regulation, defense response and photosynthetic processes. It has been shown that high temperature disrupts the mitochondrial electron transport system. This further induces a higher concentration of hydrogen ions in the matrix, which affects the functional state of different enzymes involved in TCA cycle and other metabolic pathways [65]. Further, the elevated temperature has been shown to negatively affect the grain filling process. Impairment in grain filling has been shown to result from the shortage of storage material. RNA profiling shows the down regulation of different genes involved in sucrose import and starch biosynthesis. Concomitantly, many of the genes involved in starch degradation show up regulation in the heat sensitive variety. Further, high temperature also results in the inhibition of the respiratory chain. This leads to the inefficient production of energy/ATP, eventually leading to the poor filling in rice grains [66]. An elevation in temperature can negatively affect the development of rice grains in different ways as non-closure of glumes or poor grain filling. Sequence profiling show the down regulation of different genes involved in carbohydrate synthesis, sucrose transport and ETC genes, altogether resulting in poor energy production in the system. Thus, profiling studies have not only helped in understanding the mechanisms associated with poor grain filling and elevated temperature but also serve as a promising tool for the identification of candidates for improving the heat tolerant character in rice.

## 8. Summary and concluding remarks

Transcriptome studies have gained momentum in the last decade by the advancements in sequencing techniques, bioinformatics tools and functional genomics. This has strengthened our knowledge of the seed development program by boosting data generation and interpretation. The outlook from these high-throughput studies is predominantly determined by the tissues under investigation. Seed development process can be studied in entirety through expression atlases. They provide information about the dynamics and specificity of transcriptomes across a range of tissues/organs associated with the growth and development of rice seed. Larger proportion of endosperm-specific genes obtained in comparative studies between vegetative and seed stages indicate the amount of reprogramming happening at the molecular level. The extent of involvement of various tissues in seed development as well as their molecular relatedness can also be assessed by means of transcript profiling of tissues in groups. This helps in visualizing the molecular changes occurring during progression of development in rice seed through subsequent stages by highlighting the factors that undergo change or remain constant. Several TFs and hormone-related genes show specific or preferential expression in seed tissues indicating their active involvement with the process. Molecular interactions between tissues, such as cross talk between embryo and endosperm, can also be revealed by transcriptome atlases. However, for obtaining detailed information about the developmental changes related to a particular tissue, it will be beneficial to study that tissue in isolation, as it will be cost effective and less time consuming. Transcriptome studies of rice embryo and endosperm have identified the genes and pathways that control various phases of their development. Several MTases, including MET1, genes associated with DNA/chromatin remodeling and homeobox TFs are up regulated in the initial stages of embryo development indicating induction of genes associated with zygotic development and organ formation. Polarized expression of various TFs, TRs and phytohormones, including auxin, GA and cytokinin, suggests their role in establishing apical-basal polarity in the embryo. In the early and middle stages of embryo development, pathways related to amino acid metabolism, lipid and energy metabolism, nucleic acid replication/processing and signal transduction are up regulated, while in the later stages, pathways related to starch biosynthesis and protein modification are up regulated. Also, up regulation of ABA biosynthesis and down regulation of ethylene biosynthesis in later stages suggests their antagonistic role in embryo maturation. Endosperm development starts little later than embryo in seed development. Higher expression of cell cycle related genes in the initial stages suggest that early endosperm development is mostly concerned with cell division and expansion. Variable expression pattern of TFs, such as MADS, AP2-EREBP, NACs and CCAAT, throughout different stages of endosperm development emphasizes their specific roles throughout the process. Up regulation of PCD related genes, starch and storage protein synthesis genes and carbohydrate and energy metabolism genes in the middle and later stages suggests that these stages of endosperm development are active in accumulation of storage reserves and programmed cell death.

Expression profiling studies performed till now have provided information about important genes and pathways, interrelatedness of various processes and cross talks between the key players. Nevertheless, obtaining a fully functional knowledge of this complex development process would require the enormous task of consolidation of the data generated by various studies in a single platform uniting all the pathways and regulatory networks controlling different aspects of seed development and the functional validation of these genes.

## Acknowledgements

IEM and AM acknowledge the Junior and Senior Research Fellowship from University Grants Commission (UGC). PA thanks the Department of Biotechnology (DBT), India for grants supporting research and NIPGR core grant.

## Author details

Arunima Mahto†, Iny Elizebeth Mathew† and Pinky Agarwal*

*Address all correspondence to: pinky.agarwal@nipgr.ac.in

National Institute of Plant Genome Research, Aruna Asaf Ali Marg, New Delhi, India

† Both authors contributed equally to the manuscript.

## References

[1] Das P, Nutan KK, Singla-Pareek SL, Pareek A: Understanding salinity responses and adopting 'omics-based' approaches to generate salinity tolerant cultivars of rice. Frontiers in Plant Science. 2015;**6**:712. DOI: 10.3389/fpls

[2] Jacquemin J, Bhatia D, Singh K, Wing RA: The international Oryza map alignment project: Development of a genus-wide comparative genomics platform to help solve the 9 billion-people question. Current Opinion in Plant Biology. 2013;**16**:147-156. DOI: 10.1016/j.pbi

[3] Agarwal P, Arora R, Ray S, Singh AK, Singh VP, Takatsuji H, Kapoor S, Tyagi AK: Genome-wide identification of C2H2 zinc-finger gene family in rice and their phylogeny and expression analysis. Plant Molecular Biology. 2007;**65**:467-485. DOI: 10.1007/s11103-007-9199-y

[4] Agarwal P, Kapoor S, Tyagi AK. Transcription factors regulating the progression of monocot and dicot seed development. BioEssays. 2011;**33**:189-202. DOI: 10.1002/bies.201000107

[5] Itoh J, Nonomura K, Ikeda K, Yamaki S, Inukai Y, Yamagishi H, Kitano H, Nagato Y. Rice plant development: From zygote to spikelet. Plant & Cell Physiology. 2005;**46**:23-47. DOI: 10.1093/pcp/pci501

[6] Sharma R, Agarwal P, Ray S, Deveshwar P, Sharma P, Sharma N, Nijhawan A, Jain M, Singh AK, Singh VP, Khurana JP, Tyagi AK, Kapoor S. Expression dynamics of metabolic and regulatory components across stages of panicle and seed development in indica rice. Functional & Integrative Genomics. 2012;**12**:229-248. DOI: 10.1007/s10142-012-0274-3

[7] Agarwal P, Parida SK, Mahto A, Das S, Mathew IE, Malik N, Tyagi AK. Expanding frontiers in plant transcriptomics in aid of functional genomics and molecular breeding. Biotechnology Journal. 2014;**9**:1480-1492. DOI: 10.1002/biot.201400063

[8] Agarwal P, Parida SK, Raghuvanshi S, Kapoor S, Khurana P, Khurana JP, Tyagi AK. Rice improvement through genome-based functional analysis and molecular breeding in India. Rice (N Y). 2016;**9**:1. DOI: 10.1186/s12284-015-0073-2

[9] Fujita M, Horiuchi Y, Ueda Y, Mizuta Y, Kubo T, Yano K, Yamaki S, Tsuda K, Nagata T, Niihama M, Kato H, Kikuchi S, Hamada K, Mochizuki T, Ishimizu T, Iwai H, Tsutsumi N, Kurata N. Rice expression atlas in reproductive development. Plant & Cell Physiology. 2010;**51**:2060-2081. DOI: 10.1093/pcp/pcq165

[10] Sato Y, Antonio B, Namiki N, Motoyama R, Sugimoto K, Takehisa H, Minami H, Kamatsuki K, Kusaba M, Hirochika H, Nagamura Y. Field transcriptome revealed critical developmental and physiological transitions involved in the expression of growth potential in japonica rice. BMC Plant Biology. 2011;**11**:10. DOI: 10.1186/1471-2229-11-10

[11] Wang L, Xie W, Chen Y, Tang W, Yang J, Ye R, Liu L, Lin Y, Xu C, Xiao J, Zhang Q. A dynamic gene expression atlas covering the entire life cycle of rice. The Plant Journal. 2010;**61**:752-766. DOI: 10.1111/j.1365-313X.2009.04100.x

[12] Gao Y, Xu H, Shen Y, Wang J. Transcriptomic analysis of rice (*Oryza sativa*) endosperm using the RNA-Seq technique. Plant Molecular Biology. 2013;**81**:363-378. DOI: 10.1007/s11103-013-0009-4

[13] Wei T, He Z, Tan X, Liu X, Yuan X, Luo Y, Hu S: An integrated RNA-Seq and network study reveals a complex regulation process of rice embryo during seed germination. Biochem and Biophys Res Commun. 2015;**464**:176-181. DOI: 10.1016/j.bbrc.2015.06.110

[14] Xue LJ, Zhang JJ, Xue HW. Genome-wide analysis of the complex transcriptional networks of rice developing seeds. PloS One. 2012;**7**:e31081. DOI: 10.1371/journal.pone.0031081

[15] Yang Y, Zhong J, Ouyang Y-D, Yao J. The integrative expression and co-expression analysis of the AGO gene family in rice. Gene. 2013;**528**:221-235. DOI: 10.1016/j.gene.2013.07.002

[16] Jiao Y, Tausta SL, Gandotra N, Sun N, Liu T, Clay NK, Ceserani T, Chen M, Ma L, Holford M, Zhang HY, Zhao H, Deng XW, Nelson T. A transcriptome atlas of rice cell types uncovers cellular, functional and developmental hierarchies. Nature Genetics. 2009;**41**:258-263. DOI: 10.1038/ng.282

[17] Nobuta K, Venu RC, Lu C, Belo A, Vemaraju K, Kulkarni K, Wang W, Pillay M, Green PJ, Wang GL, Meyers BC. An expression atlas of rice mRNAs and small RNAs. Nature Biotechnology. 2007;**25**:473-477. DOI: 10.1038/nbt1291

[18] Locascio A, Roig-Villanova I, Bernardi J, Varotto S: Current perspectives on the hormonal control of seed development in Arabidopsis and maize: A focus on auxin. Frontiers in Plant Science. 2014;**5**:412. DOI: 10.3389/fpls.2014.00412

[19] Santos-Mendoza M, Dubreucq B, Baud S, Parcy F, Caboche M, Lepiniec L. Deciphering gene regulatory networks that control seed development and maturation in Arabidopsis. The Plant Journal. 2008;**54**:608-620. DOI: 10.1111/j.1365-313X.2008.03461.x

[20] Sun TP, Gubler F. Molecular mechanism of gibberellin signaling in plants. Annual Review of Plant Biology. 2004;**55**:197-223. DOI: 10.1146/annurev.arplant.55.031903.141753

[21] Uchiumi T, Okamoto T. Rice fruit development is associated with an increased IAA content in pollinated ovaries. Planta. 2010;**232**:579-592. DOI: 10.1007/s00425-010-1197-7

[22] Vicente-Carbajosa J, Carbonero P. Seed maturation: Developing an intrusive phase to accomplish a quiescent state. The International Journal of Developmental Biology. 2005;**49**:645-651. DOI: 10.1387/ijdb.052046jc

[23] Sharma R, Mohan Singh RK, Malik G, Deveshwar P, Tyagi AK, Kapoor S, Kapoor M. Rice cytosine DNA methyltransferases – gene expression profiling during reproductive development and abiotic stress. The FEBS Journal. 2009;**276**:6301-6311. DOI: 10.1111/j.1742-4658.2009.07338.x

[24] Nodine MD, Bartel DP. Maternal and paternal genomes contribute equally to the transcriptome of early plant embryos. Nature. 2012;**482**:94-97. DOI: 10.1038/nature10756

[25] Yuan J, Chen S, Jiao W, Wang L, Wang L, Ye W, Lu J, Hong D, You S, Cheng Z, Yang D-L, Chen ZJ. Both maternally and paternally imprinted genes regulate seed development in rice. The New Phytologist. 2017. DOI: 10.1111/nph.14510

[26] Zhao J, Xin H, Qu L, Ning J, Peng X, Yan T, Ma L, Li S, Sun MX: Dynamic changes of transcript profiles after fertilization are associated with de novo transcription and maternal elimination in tobacco zygote, and mark the onset of the maternal-to-zygotic transition. The Plant Journal. 2011;**65**:131-145. DOI: 10.1111/j.1365-313X.2010.04403.x

[27] Kubo T, Fujita M, Takahashi H, Nakazono M, Tsutsumi N, Kurata N: Transcriptome analysis of developing ovules in rice isolated by laser microdissection. Plant & Cell Physiology. 2013;**54**:750-765. DOI: 10.1093/pcp/pct029

[28] Yu L, Ma T, Zhang Y, Hu Y, Yu K, Chen Y, Ma H, Zhao J: Identification and analysis of the stigma and embryo sac-preferential/specific genes in rice pistils. BMC Plant Biology. 2017;**17**:60. DOI: 10.1186/s12870-017-1004-8

[29] Abiko M, Maeda H, Tamura K, Hara-Nishimura I, Okamoto T: Gene expression profiles in rice gametes and zygotes: Identification of gamete-enriched genes and up- or down-regulated genes in zygotes after fertilization. Journal of Experimental Botany. 2013;**64**:1927-1940. DOI: 10.1093/jxb/ert054

[30] Xu H, Gao Y, Wang J. Transcriptomic analysis of rice (Oryza Sativa) developing embryos using the RNA-Seq technique. PloS One. 2012;**7**:e30646. DOI: 10.1371/journal.pone.0030646

[31] Itoh J, Sato Y, Hibara K, Shimizu-Sato S, Kobayashi H, Takehisa H, Sanguinet KA, Namiki N, Nagamura Y. Genome-wide analysis of spatiotemporal gene expression patterns during early embryogenesis in rice. Development. 2016;**143**:1217-1227. DOI: 10.1242/dev.123661

[32] Arora R, Agarwal P, Ray S, Singh AK, Singh VP, Tyagi AK, Kapoor S. MADS-box gene family in rice: Genome-wide identification, organization and expression profiling during reproductive development and stress. BMC Genomics. 2007;**8**:242. DOI: 10.1186/1471-2164-8-242

[33] Sano N, Ono H, Murata K, Yamada T, Hirasawa T, Kanekatsu M. Accumulation of long-lived mRNAs associated with germination in embryos during seed development of rice. Journal of Experimental Botany. 2015;**66**:4035-4046. DOI: 10.1093/jxb/erv209

[34] Olsen OA. Endosperm development: cellularization and cell fate specification. Annual Review of Plant Physiology and Plant Molecular Biology. 2001;**52**:233-267. DOI: 10.1146/annurev.arplant.52.1.233

[35] Del Rosario AR, Briones VP, Vidal AJ, Juliano BO. Composition and endosperm structure of developing and mature rice kernel. Cereal Chemistry. 1968;**45**:225-235

[36] Cochrane MP, Duffus CM. The nucellar projection and modified aleurone in the crease region of developing caryopses of barley (*Hordeum vulgare* var. *distichum*). Protoplasma. 1980;**103**:361-376

[37] Wang HL, Offler CE, Patrick JW. The cellular pathway of photosynthate transfer in the developing wheat grain. II. A structural analysis and histochemical studies of the pathway from the crease phloem to the endosperm cavity. Plant, Cell & Environment. 1995;**18**:373-388

[38] Opsahl-Ferstad HG, Le Deunff E, Dumas C, Rogowsky PM. ZmEsr, a novel endosperm-specific gene expressed in a restricted region around the maize embryo. The Plant Journal. 1997;**12**:235-246

[39] Sabelli PA, Larkins BA. The development of endosperm in grasses. Plant Physiology. 2009;**149**:14-26. DOI: 10.1104/pp.108.129437

[40] Jones RL, Jacobsen JV. Regulation of synthesis and transport of secreted proteins in cereal aleurone. International Review of Cytology. 1991;**126**:49-88

[41] Wang M, Oppedijk BJ, Lu X, Van Duijn B, Schilperoort RA. Apoptosis in barley aleurone during germination and its inhibition by abscisic acid. Plant Molecular Biology. 1996;**32**:1125-1134

[42] Nie DM, Ouyang YD, Wang X, Zhou W, Hu CG, Yao J. Genome-wide analysis of endosperm-specific genes in rice. Gene. 2013;**530**:236-247. DOI: 10.1016/j.gene.2013.07.088

[43] Nayar S, Sharma R, Tyagi AK, Kapoor S. Functional delineation of rice MADS29 reveals its role in embryo and endosperm development by affecting hormone homeostasis. Journal of Experimental Botany. 2013;**64**:4239-4253. DOI: 10.1093/jxb/ert231

[44] Mathew IE, Das S, Mahto A, Agarwal P: Three rice NAC transcription factors heteromerize and are associated with seed size. Frontiers in Plant Science. 2016;**7**:1638. DOI:10.3389/fpls.2016.01638

[45] Nakashima K, Yamaguchi-Shinozaki K. ABA signaling in stress-response and seed development. Plant Cell Reports. 2013;**32**:959-970. DOI: 10.1007/s00299-013-1418-1

[46] Guo J, Wang F, Song J, Sun W, Zhang XS. The expression of Orysa;CycB1;1 is essential for endosperm formation and causes embryo enlargement in rice. Planta. 2010;**231**:293-303. DOI: 10.1007/s00425-009-1051-y

[47] Muntz K. Deposition of storage proteins. Plant Molecular Biology. 1998;**38**:77-99

[48] Shewry PR, Napier JA, Tatham AS. Seed storage proteins: Structures and biosynthesis. Plant Cell. 1995;**7**:945-956. DOI: 10.1105/tpc.7.7.945.

[49] Smith AM. Making starch. Current Opinion in Plant Biology. 1999;**2**:223-229. DOI: 10.1016/S1369-5266(99)80039-9

[50] Mena M, Vicente-Carbajosa J, Schmidt RJ, Carbonero P. An endosperm-specific DOF protein from barley, highly conserved in wheat, binds to and activates transcription from the prolamin-box of a native B-hordein promoter in barley endosperm. The Plant Journal. 1998;**16**:53-62

[51] Wei CX, ZX X, Lan SY. Ultrastructural features of nucleus degradation during programmed cell death of starchy endosperm cells in rice. Acta Botanica Sinica. 2002;**44**:1396-1402

[52] Young TE, Gallie DR, DeMason DA. Ethylene-mediated programmed cell death during maize endosperm development of wild-type and shrunken2 genotypes. Plant Physiology. 1997;**115**:737-751

[53] Li X, Gao X, Wei Y, Deng L, Ouyang Y, Chen G, Zhang Q, Wu C. Rice APOPTOSIS INHIBITOR5 coupled with two DEAD-box adenosine 5′-triphosphate-dependent RNA helicases regulates tapetum degeneration. Plant Cell. 2011;**23**:1416-1434. DOI: 10.1105/tpc.110.082636

[54] Qi Y, Wang H, Zou Y, Liu C, Liu Y, Wang Y, Zhang W. Over-expression of mitochondrial heat shock protein 70 suppresses programmed cell death in rice. FEBS Letters. 2011;**585**:231-239. DOI: 10.1016/j.febslet.2010.11.051

[55] Steffens B, Sauter M. Epidermal cell death in rice is regulated by ethylene, gibberellin, and abscisic acid. Plant Physiology. 2005;**139**:713-721. DOI: 10.1104/pp.105.064469

[56] Kobayashi H, Ikeda TM, Nagata K. Spatial and temporal progress of programmed cell death in the developing starchy endosperm of rice. Planta. 2013;**237**:1393-1400. DOI: 10.1007/s00425-013-1854-8

[57] Zhu T, Budworth P, Chen W, Provart N, Chang HS, Guimil S, Su W, Estes B, Zou G, Wang X. Transcriptional control of nutrient partitioning during rice grain filling. Plant Biotechnology Journal. 2003;**1**:59-70. DOI: 10.1046/j.1467-7652.2003.00006.x

[58] Onodera Y, Suzuki A, Wu CY, Washida H, Takaiwa F. A rice functional transcriptional activator, RISBZ1, responsible for endosperm-specific expression of storage protein genes through GCN4 motif. The Journal of Biological Chemistry. 2001;**276**:14139-14152. DOI: 10.1074/jbc.M007405200

[59] Vicente-Carbajosa J, Moose SP, Parsons RL, Schmidt RJ. A maize zinc-finger protein binds the prolamin box in zein gene promoters and interacts with the basic leucine zipper transcriptional activator Opaque2. Proceedings of the National Academy of Sciences of the United States of America. 1997;**94**:7685-7690

[60] Wu C, Washida H, Onodera Y, Harada K, Takaiwa F. Quantitative nature of the Prolamin-box, ACGT and AACA motifs in a rice glutelin gene promoter: Minimal cis-element requirements for endosperm-specific gene expression. The Plant Journal. 2000;**23**:415-421

[61] Venu R, Sreerekha M, Nobuta K, Belo A, Ning Y, An G, Meyers BC, Wang GL. Deep sequencing reveals the complex and coordinated transcriptional regulation of genes related to grain quality in rice cultivars. BMC Genomics. 2011;**12**:190. DOI: 10.1186/1471-2164-12-190

[62] Liu X, Guo T, Wan X, Wang H, Zhu M, Li A, Su N, Shen Y, Mao B, Zhai H, Mao L, Wan J. Transcriptome analysis of grain-filling caryopses reveals involvement of multiple regulatory pathways in chalky grain formation in rice. BMC Genomics. 2010;**11**:730. DOI: 10.1186/1471-2164-11-730

[63] Sun H, Peng T, Zhao Y, Du Y, Zhang J, Li J, Xin Z, Zhao Q. Dynamic analysis of gene expression in rice superior and inferior grains by RNA-Seq. PloS One. 2015;**10**:e0137168. DOI: 10.1371/journal.pone.0137168

[64] Fu C, Wang F, Liu W, Liu D, Li J, Zhu M, Liao Y, Liu Z, Huang H, Zeng X, Ma X: Transcriptomic analysis reveals new insights into high-temperature-dependent glume-unclosing in an elite rice male sterile line. Frontiers in Plant Science. 2017;**8**:112. DOI: 10.3389/fpls.2017.00112

[65] Liao JL, Zhou HW, Peng Q, Zhong PA, Zhang HY, He C, Huang YJ. Transcriptome changes in rice (*Oryza sativa* L.) in response to high night temperature stress at the early milky stage. BMC Genomics. 2015;**16**:18. DOI: 10.1186/s12864-015-1222-0

[66] Yamakawa H, Hakata M. Atlas of rice grain filling-related metabolism under high temperature: Joint analysis of metabolome and transcriptome demonstrated inhibition of starch accumulation and induction of amino acid accumulation. Plant & Cell Physiology. 2010;**51**:795-809. DOI: 10.1093/pcp/pcq034

Chapter 3

# Transcriptome (ESTs) of Avocado "Native" Mexicano Early Seed Development Shows Abundance of Regulatory, Antioxidant and Defense Genes

Luis María Suárez-Rodríguez,
Fernando Sánchez-Albarrán,
Huber León-Corona and Rodolfo López-Gómez

Additional information is available at the end of the chapter

http://dx.doi.org/10.5772/intechopen.70573

**Abstract**

Avocado native "Mexicano" (*Persea americana* var drymifolia) has been a really important species in agricultural and indigenous medicine. In the agricultural world, it has been the germplasm source for the generation of economically important cultivars like Hass and it is the main source of rootstocks for the world production of Hass avocado fruit. In spite of its importance, little is known about the molecular network of seed-fruit development. The aim of this work was to know the expressed genes (ESTs) during the early avocado native "Mexicano" seed development. Using total RNA we constructed cDNA libraries of fourth months seed development, sequencing, assembling and bioinformatic analysis was made. For validation, a semi-quantitative PCR experiments with the most abundant genes were made. About 5005 ESTs from the 5′ representing 1653 possible unigenes were isolated. After assembling process, we have 171 genes that are closely related to *Nelumbo nucifera* sequences. The transcriptome is dominating by one bHLH transcription factor, three metallothioneins and snakin, suggesting its main role in seed development. Until now, there are no molecular studies in avocado seed development.

**Keywords:** avocado fruit, transcriptome, transcription factor, stress, antimicrobials, cotyledon, alignment, domain, phylogenetic, metal homeostasis, cysteine, seed development, gene expression

## 1. Introduction

Avocado (*Persea americana* Mill.) is an oleaginous fruit produced by a tree belonging to the magnoliid clade, a basal linage of flowering plants. It belongs to the large plant family of Lauraceae, with approximately 2500–3000 species [1, 2]. Avocado has been rapidly incorporated as a component of human diet in many countries [3]. Due to low cost, vigor of seedling growth and easy propagation, most of the countries are still using seeds to produce rootstocks for grafted avocado trees despite their genetic variability [4]. Several Mexican varieties are derived from seeds that are resistant to attack by *Phytophthora cinnamomi* [5, 6] and are adapted to the soil and environmental conditions of the region. The Mexican state of Michoacán is the primary avocado-producing region in the world, and all the rootstocks used for the commercial production of cultivar Hass are obtained from *P. americana* var. drymifolia ("nativo mexicano") [7]. The principal consume form is as fresh fruit, but is really important in cosmetic industry [8]. Avocado plant has medicinal properties, including cancer prevention [9–11]. There is ethnopharmacological information on the use of avocado seeds for the treatment of health-related conditions, especially in America. Recent research has shown that the avocado seeds are rich in phenolic compounds and these maybe play a role in putative health effects [12]. The avocado fruit is a berry of one carpel containing a single seed. This large and very conspicuous seed is made up of two fleshy cotyledons and a central attached plumule, hypocotyl and radicle, the whole surrounded by two papery seed coats closely adherent to each other. There is no endosperm left in the seed at maturity. The cotyledons are formed of indifferentiated parenchyma tissue interspersed with occasional idioblasts. Starch is the main storage material of the cotyledons and is present in great abundance [13]. Despite its importance, avocado seed development remains uncharacterized. To date, little information is available regarding the molecular biology of the seed. Analysis of expressed sequence tags (ESTs) is a rapid and effective method to identify novel genes or to investigate gene expression in different tissues, organs and plants [14, 15]. Furthermore, EST libraries and databases could provide valuable resources for functional genomic studies [16].

In principle, the frequency with which the sequence of a given gene is read in ESTs sequencing projects should reflect the relative abundance of the corresponding mRNA. This approach uses EST counts to infer the relative level of expression of a gene [17–19].

In this work, we report the analysis of an ESTs collection from immature avocado nativo mexicano seeds and the analysis of expression of bHLH transcription factor, metallothioneins (MTs) and snakin like more abundantly expre3ssed.

## 2. Materials and methods

### 2.1. Biological material

Seeds from avocado nativo mexicano fruits of three stages of development (1, 4 and 8 months) were excised from fruits and frozen immediately in liquid nitrogen until use. The materials were collected in the avocado Germplasm Bank of the Instituto Nacional de Investigaciones Forestales y Agropecuarias (INIFAP; Uruapan, Michoacán México).

### 2.2. cDNA library construction and sequencing by Sanger method

Total RNA from frozen seed tissue of 4 months of development was extracted using López-Gómez et al.'s protocol with some modifications [20]. The cDNA complementary library was built from 1 μg of total RNA using SMART™ cDNA Library Construction Kit (Clontech). The obtained cDNA sequences were cloned into pTriplEx2 vector. Cleavage experiments were made using *E. coli* BM25.8 cells to obtain the plasmid pTriplEx2. Sequencing reactions were performed using ABI PRISM BigDye Terminators v3.0 kit (Applied Biosystems), by 5′ end of plasmids extracted from random clones. The sequences obtained were filtered by quality using PHRED [21]; vector masked and trimming of poly A/T were performed using LUCY2 software [22] resulting in 5002 high-quality reads.

### 2.3. Assembly and identification of full-length cDNA

Sanger sequences were assembled using default parameters of the 454 Newbler-Assembler v1.1.03.24 (454 Life Science, Branford, CT) using 16,526 generated by the University of Florida and the Washington University Genome Sequencing Center.

Unigene set was generated by combining all assembled contigs and non-assembled reads (singlets). The consensus sequences of the Unigenes were analyzed with EuGeneHom [23] to identify the Unigenes that contained the components of a complete cDNA (5′ UTR, ORF and 3′ UTR): http://genoweb.toulouse.inra.fr/eugene/EuGeneHom/cgi-bin/EuGeneHom.pl.

### 2.4. Functional annotation

Stand-alone BLAST software was obtained from the National Center for Biotechnology Information (NCBI, http://www.ncbi.nlm.nih.gov). The unigenes were compared by BLAST against nucleotides and proteins plant databases. The BLAST results from different databases were used for gene ontology (GO) mapping and annotation. Blast2Go software was used to perform GO functional classification.

### 2.5. Transcript characterization and homologous search

The transcripts were analyzed by UGENE V1.26.1 software for the identification of ORF, CDS and hypothetical protein sequence and physicochemical parameters, using the standard genetic code.

### 2.6. Alignment and phylogeny

Mega7.0.14 software was used for alignment (Clustal W algorithm with Blossom 62 matrix) and reconstruction of the phylogenetic pattern (Neighbor Joining model with JTT matrix-based model for distance computing and 1000 replicates as bootstrap test).

| GenBank Accession | Gene name | Primer sequence (5′–3′) | Tm (°C) |
|---|---|---|---|
| MF353071 | PabHLHS1 | **F:** TAGCGACAGGATTTGGCAGTTT | 65°C |
| | | **R:** TGCCCCGTCGTCTCTTTCTTCTACC | |
| MF353072 | PaMT2a | **F:** TGAGAGGGAGATTGGAGGAG | 54°C |
| | | **R:** TCCCAACCACAGCATAGTACC | |
| MF353073 | PaMT2b | **F:** CATGCACCTGCAAATGAGAG | 60°C |
| | | **R:** CACCCAGATACAGCAGGAGAA | |
| MF353074 | PaMT3 | **F:** TTGGTGTTGGTGTCCTATGC | 57°C |
| | | **R:** CATGACTCAACCACACACACC | |
| AGC92009 | PaSn | **F:** TCCTTGCTTTCCTTCTCTTCAC | 64°C |
| | | **R:** AGATGTCCCGCAATCTATTTGT | |
| | SUMO | **F:** GATAAGAAGCCCACGGATCA | 55°C |
| | | **R:** GACGGCCATCGAATAAGAAC | |

**Table 1.** Primers used in semi-quantitative PCR.

### 2.7. Semi-quantitative RT-PCR

The RNA from different seed stages (1, 4 and 8 months of development) digested with DNAse I amplification grade (Invitrogen) was used as a template in sqPCR reactions, and the synthesis of cDNAs was carried out with the First Strand cDNA Synthesis Kit (Thermo Scientific). In Veriti 96-well Thermal cycler (Applied Biosystems), in reactions of 10 μl (100 ng/μl cDNA,10X Reaction Buffer, 2 mM dNTPs, 50 mM $MgCl_2$, 1 U Taq DNA polymerase), sqPCR primers (**Table 1**) were designed by Primer3 webtool (http://bioinfo.ut.ee/primer3-0.4.0/); the primers for bHLH transcription factor, three metallothioneins, snakin and ubiquitin as reference gene were selected. The amplification procedures were 95°C - 10min., 30 cycles (95°C - 45 s, annealing 30 s, amplification 72°C - 45 s), 72°C - 7 min. Gene expression ratio between selected gene and endogenous control was calculated using band intensity measured with GelAnalyzer 2010. The semi-quantitative PCR was performed with three repeats.

## 3. Results

### 3.1. Sequencing and assembly of *Persea americana* var. drymifolia seed transcriptome

A total of 5005 sequences was assembled and obtained 3328 sequences (171 contigs of 3222 sequences + 106 singletons); these sequences were imported into BLAST analysis. In contig assembly for the avocado seed ESTs, a total number of contigs was 171, derived from 3222 sequences and unique transcripts represented was 106 sequences. Contig length ranged from 84 to 1149 bp (**Figure 1**), the peak EST length was 84–195 bp with 478 sequences in range. The shortest sequence examined was 84 bp to known genes like GTP-binding nuclear protein.

### 3.2. Functional annotation

Functional interpretation is an important step in the analysis of transcriptomics which cannot be done without the availability of functional annotation. The most widespread and probably most extensive functional annotation schema for gene and protein sequences is the Gene Ontology (GO) [24] as standard in all public databases. Automatic functional annotation methods basically rely on sequence, structure, phylogenetic or co-expression relationships between known and novel sequences [25]. A total of 277 uniESTs sequences were manually annotated for a closer understanding of gene expression in avocado seed. The annotation proceeds through three basic steps: homologs search, GO term mapping and actual annotation. At the first step, NCBI-BLASTX and BLASTN are typically used, and for this work, the e-value $1xE^{-5}$, cut-off:33 and the number of 20 retrieved BLAST hits are used. These uniESTs were classified into five functional categories, including antioxidative protection (677, 20.34%), transcription regulatory (1013, 30.44%), defense (507, 15.23%), cellular structure and organization (287, 8.62%) and unknown (844, 25.36%).

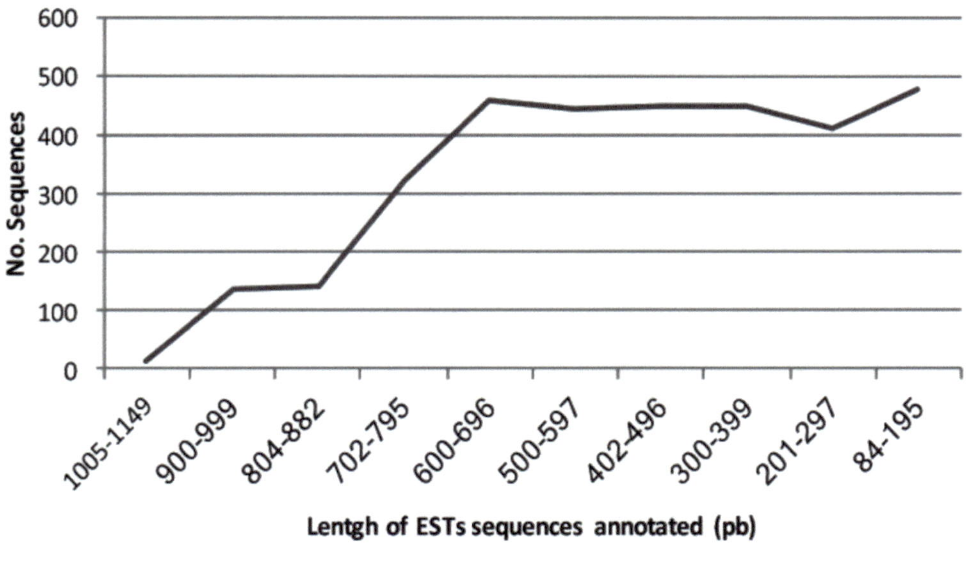


**Figure 1.** Sequence length distribution of ESTs from avocado native mexicano seed transcriptome. Total number of sequences analyzed were 3328. Data include both contigs and singletons.

Gene ontology annotations and functional analyses of avocado seed transcriptome were carried out with automated software Blast2GO. These were assigned into three standard classifications: biological processes, molecular functions and cellular components, and summarized according to GO criteria. The majority GO annotation was for biological process (65.34%), cellular component (19.34%) and molecular function (15.33%). In addition, the organisms closely related to the genetic load on avocado seed were reviewed within the databases, and the majority of analyzed sequences (23.47%) (**Figure 2**) were closely related to *Nelumbo nucifera* sequences (**Figure 3**). The 20 most abundant uniESTs and their annotations are shown in **Table 2**.

### 3.3. Avocado seed abundant genes and validation

The most abundant sequences match with metallothionein genes. This result suggests that metallothionein genes dominate the avocado seed transcriptome like avocado fruit [26]. Metallothioneins (MTs) were discovered by Margoshes and Vallee as cadmium-bound proteins isolated from the cortex of the equine kidney. These proteins were named for the high sulfur content and metals they are able to bind; depending on the metal species, these may possess more than 20% of its nature of metal ions [27, 28]. Mammalian metallothioneins are 60 amino acid peptides with 20 Cys residues and a molecular mass of about 6–7 kDa. Mammalian MTs are capable of binding up to 7 divalent metal ions via mercaptide bonds (sulfur-metal) with the Cys residues. By convention, any peptide or protein that resembles several characteristics of mammalian metallothioneins can be classified as metallothionein [28]. Plant metallothioneins have two (highly conserved) sequence similarity regions corresponding to the two Cys-rich terminal domains joined by a less conserved "spacer" (about 40 aa without Cys residues). In plants, the most distinctive feature is to have a large spacer, which differs from the MT of the

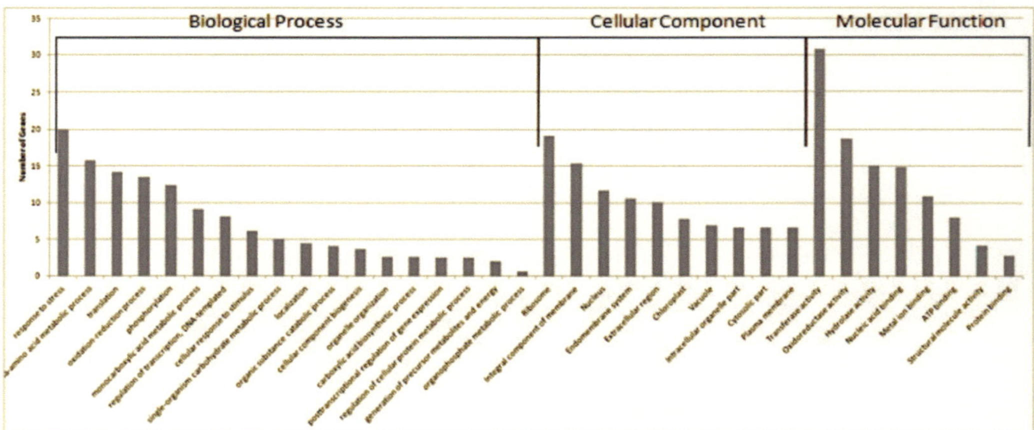


**Figure 2.** Gene ontology classification of *P. americana* var. drymifolia seed transcriptome. Unigenes with BLASTX matches were classified into three main GO categories: biological process, cellular components and molecular functions.

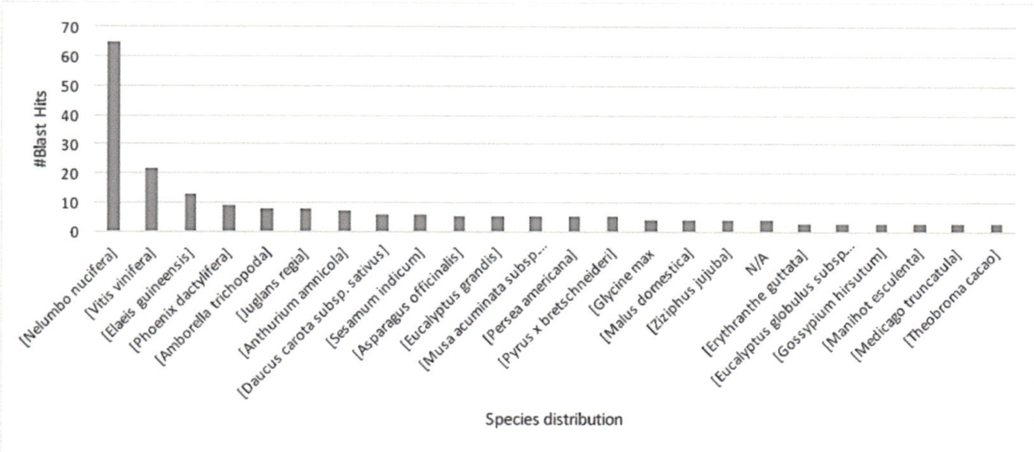


**Figure 3.** Main organisms with which the avocado seed transcriptome is closely related.

animals in which the Cys-rich domains are separated by a short spacer of less than 10 amino acids which do not include aromatic residues. The distribution of Cys residues as well as the length of the spacer region served to classify more MT of plants into four types, namely group 1, 2, 3 and 4 [29].

From the analysis of these abundant transcripts, we founded the existence of three MT's genes on avocado seed: PaMT2a, PaMT2b and PaMT3, which were registered in GenBank database with an accession code shown in **Table 3**. Characteristics predicted in silico for avocado metallothioneins, we have two sequences belonging to Methallotionein-2 superfamily due to the two highly conserved Cys-rich motifs and the long spacer in the middle of them. PaMT3 keeps the spacer but Cys-rich motifs are not so conserved grouping this as part of the third family of plant metallothioneins (MT3). Alignment shown that the most conserved amino acids are around the Cys residues for both MT2 and MT3 groups, which are associated with the "metal binding clusters" (**Figures 4** and **5**). The alignments performed identify the amino and terminal carboxyl regions as having the most conserved Cysteine sequences, which correspond to the metal binding clusters. The intermediate spacer of about 40 amino acids is much more variable but has no cysteines. In mammalian metallothioneins, this spacer is very small (8 amino acids) and has no aromatic amino acids. However, in the family 2 to which the plants belong, we can find conserved tyrosine residues in the spacer, as well as several less conserved phenylalanines present in the metallothionein sequences of types 2 and 3 (**Figures 4** and **5**). Seed metallothioneins correspond with the three reported: NnMT2a, NnMT2b and NnMT3 [30]; the avocado seed metallothioneins are closely related with *Nelumbo nucifera* metallothioneins like 79% for NnMT2a-PaMT2a; 66% NnMT2b-PaMT2b and 71% for NnMT3-PaMT3 **Figure 6**. NnMT2a and NnMT3 were associated with processes of seed germination, tolerance to accelerated aging and salt on Arabidopsis [30]. However, avocado seeds have a low tolerance

| ESTs | Description | Length | e-value | GO IDs | Abundance |
|---|---|---|---|---|---|
| arlgES1 | Transcription factor bHLH96 | 942 | 1.35E-103 | GO:0046983 | 123 |
| arlgES153 | Metallothionein type 3 | 195 | 4.52E-28 | GO:0043167 | 123 |
| arlgES154 | Stress response nst1 | 696 | 1.68E-34 | GO:0009507; GO:0010207 | 113 |
| arlgES2 | 40S ribosomal S29 | 171 | 1.77E-32 | GO:0003735; GO:0043167; GO:0005829; GO:0005840; GO:0006412 | 100 |
| arlgES3 | PREDICTED: uncharacterized protein LOC104611921 | 282 | 4.53E-37 | N/A | 96 |
| arlgES4 | Vesicle-associated membrane 726 | 579 | 5.71E-130 | GO:0016192; GO:0005575 | 90 |
| arlgES6 | ras-related RIC2 | 525 | 2.61E-118 | GO:0007165; GO:0043167; GO:0005622 | 78 |
| arlgES7 | Metallothionein 2 | 240 | 3.91E-30 | GO:0043167 | 77 |
| arlgES8 | Isocitrate dehydrogenase [NADP] | 588 | 1.02E-132 | GO:0044281; GO:0006091; GO:0016491; GO:0043167 | 74 |
| arlgES10 | kDa proline-rich | 483 | 4.35E-25 | N/A | 69 |
| arlgES155 | Early nodulin-93-like | 318 | 8.41E-43 | GO:0005575 | 69 |
| arlgES156 | PREDICTED: uncharacterized protein LOC103701850 isoform X3 | 552 | 2.83E-19 | GO:0005739 | 66 |
| arlgES157 | Potassium transporter 12 isoform X1 | 366 | 2.97E-55 | GO:0022857; GO:0009536 | 64 |
| arlgES12 | 60S ribosomal L24 | 492 | 5.32E-63 | GO:0003735; GO:0022618; GO:0005829; GO:0042254; GO:0005840; GO:0006412 | 58 |
| arlgES13 | 60S ribosomal L7–2-like | 738 | 1.65E-136 | GO:0003735; GO:0005829; GO:0042254; GO:0005840; GO:0006412 | 57 |
| arlgES15 | Ethanolamine utilization eutQ | 291 | 3.63E-61 | N/A | 54 |
| arlgES158 | Programmed cell death 4 | 636 | 1.00E-80 | N/A | 52 |
| arlgES16 | Type 2 metallothionein | 243 | 5.56E-24 | GO:0043167 | 51 |
| arlgES159 | Mitochondrial | 178 | 4.72E-35 | GO:0005739 | 50 |
| arlgES161 | Snakin | 318 | 4.00E-57 | GO:0009740 | 48 |

**Table 2.** Top 20 most abundant annotated ESTs of avocado nativo mexicano seed transcriptome.

to aging. Basic function of MTs is metal homeostasis and has been reported during biotic and abiotic stress conditions too [31]. Recently, it has been reported that one metallothionein interacts with a cytoskeleton protein in the nucleus of rice cells in response to salt stress [32].

| Name | GenBank | Length (aa) | Weight (kDa) | Isoelectric Point | Cys Residues | % Cys |
|---|---|---|---|---|---|---|
| PaMT2a | MF353072 | 80 | 7.9 | 5.11 | 14 | 17.5 |
| PaMT2b | MF353073 | 80 | 7.9 | 4.54 | 14 | 17.5 |
| PaMT3 | MF353074 | 64 | 67 | 4.5 | 10 | 15.6 |

**Table 3.** Characterization of Persea americana var. drymifolia metallothioneins.

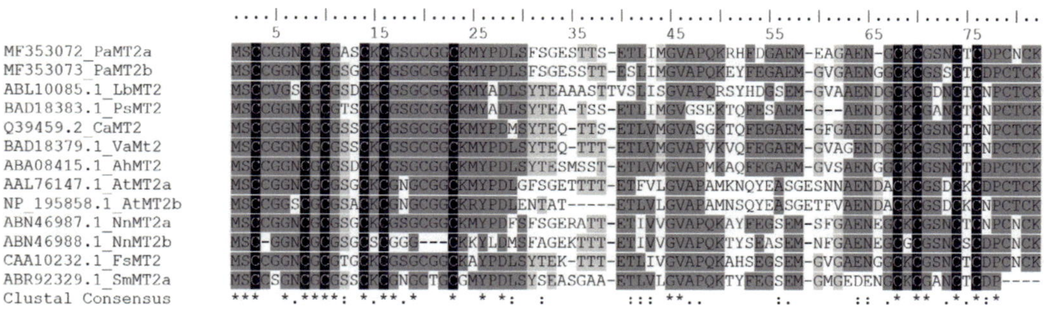


**Figure 4.** Alignment of plant metallothioneins of subfamily 2.

**Figure 5.** Alignment of plant metallothioneins of subfamily 3.

Sequences of related proteins were obtained from Zhou et al. [30] and downloaded from NCBI-GenBank: BAD18376.1 GmMT1 (*Glycine max*); BAD18378.1 VaMT1 (*Vigna angularis*); AAF04584.1 MsMT1 (*Medicago sativa*); BAD18382.1 PsMT1 (*Pisum sativum*); BAD18380.1 VfMT1 (*Vicia faba*); ABL10085.1 LbMT2 (*Limonium bicolor*); BAD18383.1 PsMT2 (*Pisum sativum*); Q39459.2 CaMT2 (*Cicer arietinum*); BAD18379.1 VaMT2 (*Vigna angularis*); ABA08415.1 AhMT2 (*Arachis hypogaea*); AAL76147.1 AtMT2a (*Arabidopsis thaliana*); NP_195858.1 AtMT2B (*Arabidopsis thaliana*); ABN46987.1 NnMT2a (*Nelumbo nucifera*); ABN46988.1 NnMT2b (*Nelumbo nucifera*); CAA10232.1 FsMT2 (*Fagus sylvatica*); ABR92329.1 SmMT2a (*Salvia miltiorrhiza*); NP_566509.1 AtMT3 (*Arabidopsis thaliana*); AAS99234.1 NcMT3 (*Noccaea caerulescens*); BAB85599.1 BjMT3 (*Brassica juncea*); ACB10219.2 EgMT3 (*Elaeis guineensis*); ACV51811.1 TaMT3 (*Typha angustifolia*); Q40256.1 MaMT3 (*Musa acuminata*); ABN46986.1 NnMT3 (*Nelumbo

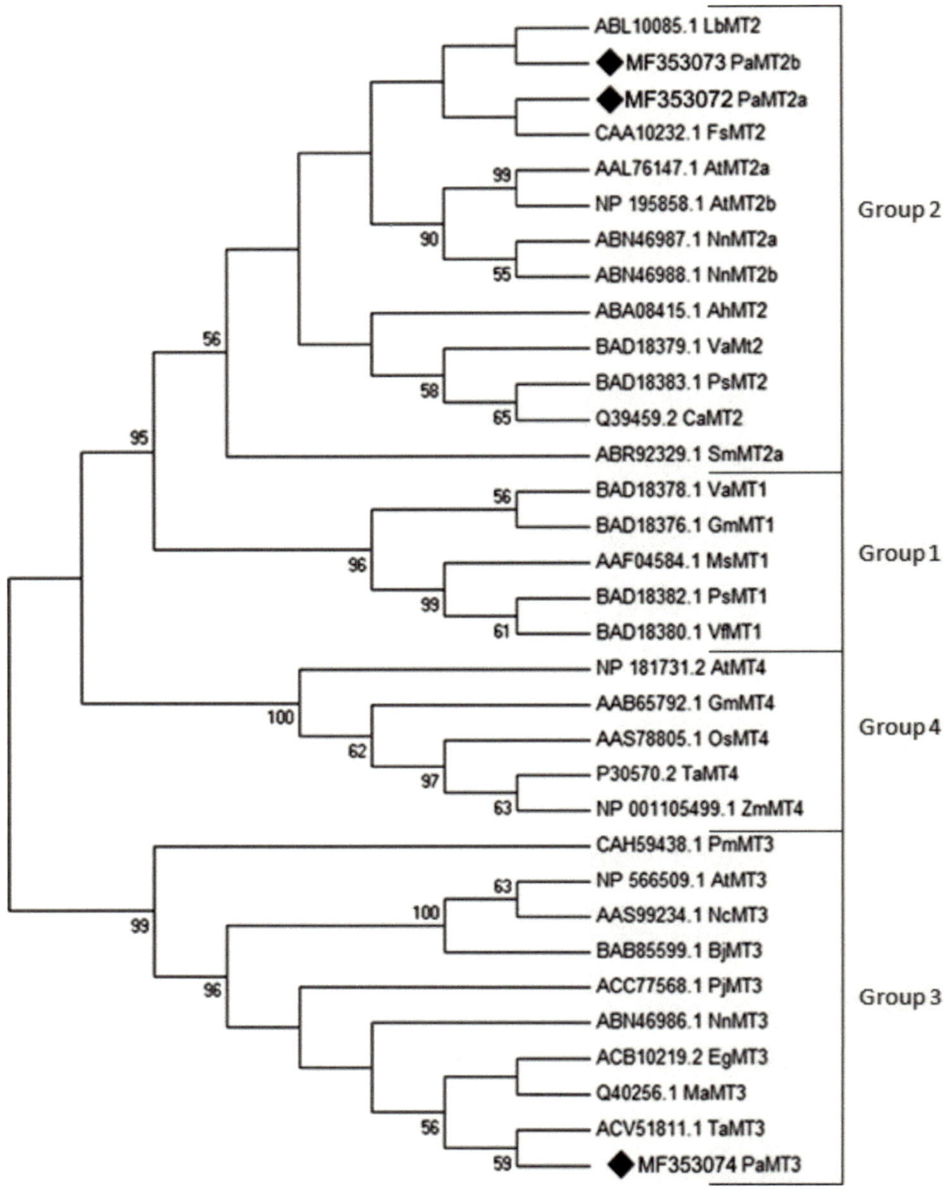


**Figure 6.** Phylogenetic analysis of avocado and plant metallothioneins. The evolutionary history was inferred using the neighbor-joining method. The bootstrap consensus tree inferred from 1000 replicates. The evolutionary distances were computed using the JTT matrix-based method. The analysis involved 33 amino acid sequences. All positions with less than 85% site coverage were eliminated.

*nucifera*); ACC77568.1 PjMT3 (*Prosopis juliflora*); CAH59438.1 PmMT3 (*Plantago major*); NP_181731.2 AtMT4 (*Arabidopsis thaliana*); AAB65792.1 GmMT4 (*Glycine max*); P30570.2 TaMT4 (*Triticum aestivum*); AAS78805.1 OsMT4 (*Oryza sativa* Japonica Group) and NP_001105499.1 ZmMT4 (*Zea mays*), [30].

Another abundant transcript is a messenger codified for a transcription factor type bHLH. The basic domain (bHLH) is a highly conserved amino acid motif that defines a group of transcription factors, which was initially described in animals and was soon discovered in all major eukaryotic lineages [33]. Proteins containing a bHLH domain (referred to as bHLH proteins) are involved in a variety of regulatory processes; their functions include the regulation of neurogenesis, myogenesis and the development of the heart in animals [34, 35], control of phosphate uptake and glycolysis in yeast [36] or modulation of secondary metabolism pathways, epidermal differentiation and environmental responses in plants [37]. The bHLH domain consists of two distinct segments composed by 50–60 amino acids, 10–15 mostly basic amino acids form the stretch (basic region) and approximately 40 amino acids form the two amphipathic helices separated by a loop (helix-loop-helix region). The analysis of the structure of bHLH proteins (yeast and mammalian) showed the basic region made in the DNA contact, while the two helices promote the formation of heterodimers between bHLH proteins [35]. These bHLH transcription factors are generally classified into six major groups (FAs) based on their ability to bind to DNA [35, 38, 39]. Most bHLH proteins are classified into group A or B; in group A, it is expected to bind to E-box consensus sequences (CACCTG or CAGCTG), in group B, it is specifically bind to G-Box consensus sequences (CACGTG or CATGTTG) and in group C, bHLH proteins share a PAS domain and bind to the recognized sequences without a need a E-box (ACGTG or GCGTG) sequences. The E group includes bHLH proteins containing a conserved Pro or Gly residue at a key position within the basic region, preferably bind to sequences referred to as N-boxes (CACGCG or CACGAG), and further share an additional WRPW motif. Groups D and F represent particularly proteins which were separated in the basic region. Some group D proteins have been described as being unable to bind to DNA and could form heterodimers that function as negative regulators of bHLH binding to DNA [40]. Group F includes so-called COE proteins. A phylogenetic study indicated that group A contained mammalian bHLH proteins and lacked bHLH plant proteins. The other groups had a mixture of different species and most of the bHLH proteins of plants belonged to group B [41, 42], It has been shown that the bHLH family of proteins in plants is monophyletic and subjected to significant radiation before the evolution of mosses; bHLH groups established in terrestrial plants during the first 400 million years were conserved during the later evolution of plants, although there were many duplications of genes. The transcription factors are very varied since it does not have many amino acids conserved throughout its sequence; nevertheless in the sites of union to the DNA like the case of the bHLH, the great majority of its amino acids is conserved within its main motive. Due to their propensity to form homodimers or heterodimers, bHLH proteins can participate in an extensive set of a combinatorial interactions leading to the regulation of multiple transcriptional programs. The development of fleshy fruits involves complex physiological and biochemical changes. Recent studies have described the involvement of bHLH proteins in the determination of plant organ size. The SPATULA protein was shown to control cotyledon, leaf and petal expansion by affecting cell proliferation in *Arabidopsis thaliana* [43]. Nicolas et al. [44] described a bHLH transcription factor preferentially expressed in grape berry fruit, but is weakly detected in seeds. This gene is involved in cell size determination. Three basic helix-loop-helix transcription factors (bHLH) were also found to be involved in *Arabidopsis* fruit dehiscence process: ALCATRAZ (*ALC*), SPATULA (*SPT*) and INDEHISCENT (IND); they form a regulatory network that orchestrates the differentiation of the valve margin, allowing seed dispersal [45]. A protein Blast was performed and the sequences that were selected for a multiple alignment were the ones that presented greater coverage and identity with the transcription factor bHLH of Mexican native avocado seed. High sequence variability was found for the nearest bHLH motif, which presenting a large number of conserved amino acids **(Figure 7)**.

However, since they do not have any additional information regarding the function and/or tissue in which the function is performed, we analyzed our sequence with bHLH sequences of which their function or organospecificity is known; two bHLH factors were chosen with these characteristics of *Arabidopsis thaliana*, bHLH Zoupi (GenBank Accession: OAP16519) involved in seed development and SPATULA (GenBank Accession: AT4G36930) involved in

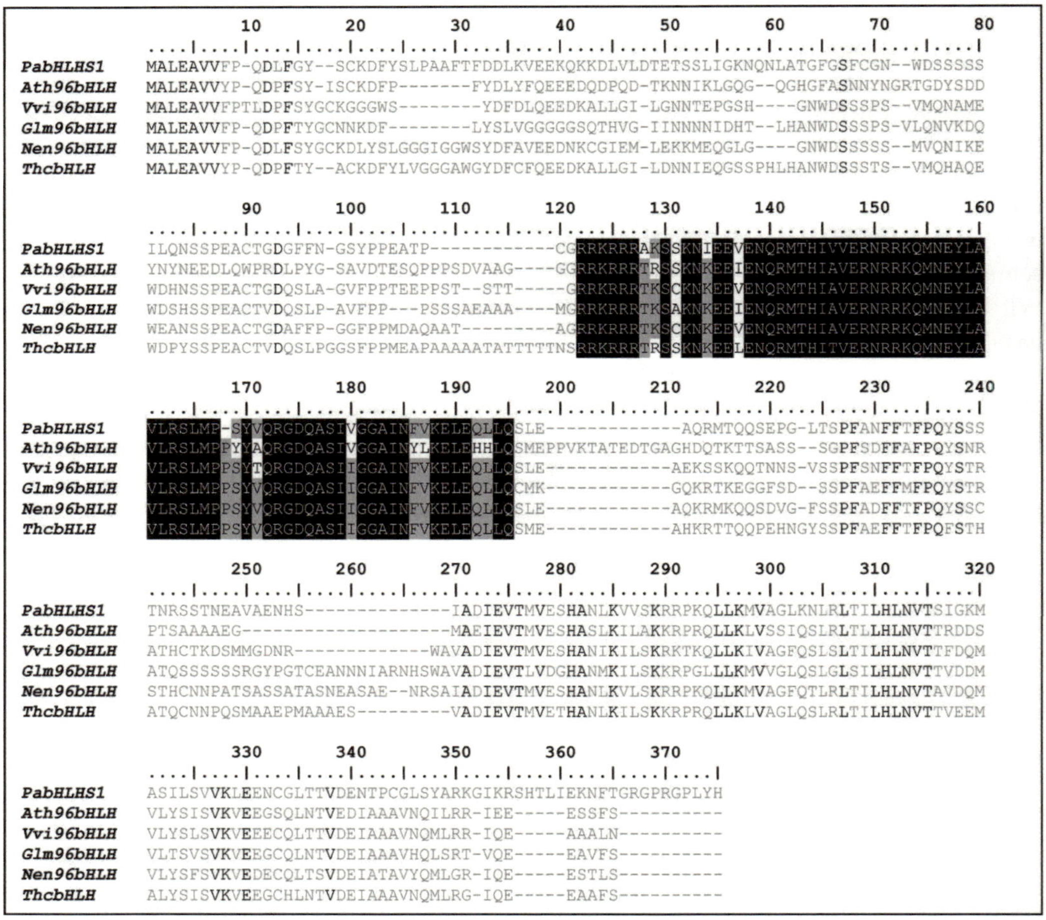


**Figure 7.** bHLH (PabHLHS1) avocado seed transcription factor alignment with bHLH transcription factors of some plant species. *Arabidopsis thaliana* (Ath96bHLH), *Vitis vinifera* (Vvi96bHLH), *Glicine max* (Glm96Bhlh), *Nelumbo nucifera* (96bHLH), *Theobroma cacao* (ThcbHLH), in shadow black are the bHLH motif represented in black letters the amino acids that are conserved in these proteins and gray letters the amino acids differs between them.

cell development; however, no conserved amino acids were found with the avocado bHLH (data not shown). To try to infer the function of this transcription factor highly expressed in avocado seed, it will be necessary to carry out research studies of the recognition boxes inside in DNA to infer the possible association to the group to which it belongs. It is probable that the bHLH transcription factor of the avocado seed has a principal paper in seed differentiation and development.

Multicellular organisms produce small cysteinerich antimicrobial peptides (AMPs) as an innate defense against pathogens. Native Mexican avocado seed abundantly express Snakin (PaSn) gene. These kind of AMPs were initially isolated from potato but were later found to be ubiquitous. Novel plant APs isolated include in *Arabidopsis* (family of 12-cysteine peptides).

We identified a single cDNA sequence for snakin/GASA (gibberellic acid-stimulated), which contains a coding sequence of 318 bp and encodes a predicted 106 amino acid peptide. This molecule comprises a 26 amino acid signal peptide (residues 1–26), identified by SignalP (http://www.cbs.dtu.dk/services/SignalP/) and a 79 amino acid mature peptide (**Figure 8**). An amino acid alignment of avocado snakin with other similar APs (**Figure 9**) showed that PaSn has the longest sequence compared with the previously reported StSN1 and StSN2 genes from potato. In addition, PaSn has the 12 characteristic Cys residues of this type of AP. In addition to the highly conserved 12 Cys residues, the other motifs in the PaSn protein consist of residues that are mostly polar, non-polar and basic (**Figure 9**). From these analyses, we hypothesized that the Mexican avocado snakin gene could be involved in plant defense in a similar way to that of the StSN1 and StSN2 genes in potato [7]. Until now, this is the first Snakin gene isolated from a seed.

```
  1 5'CGTGCCGGGGAAGCTAATATACGCTCCAATTGTCTATCTCTACAATACCCTTTCCTCATT
 21                               M  A  L  S  K  T  F  L  A  F  L
 61   TCCTCCATCTGTTGAGTTCTCCTTCTTGAATGGCTCTGTCAAAGACTTTCCTTGCTTTCC
 41     L  F  T  I  L  A  L  N  L  M  D  D  A  L  G  F  Q  S  H  S
121   TTCTCTTCACTATCCTTGCTCTCAATCTCATGGATGATGCTCTTGGATTTCAGTCTCATT
 61     L  P  T  P  A  D  E  R  N  L  L  Q  Q  I  D  C  G  T  S  C
181   CGCTGCCTACACCTGCGGATGAGAGAAATCTCCTTCAACAAATAGATTGCGGGACATCTT
 81     S  A  R  C  R  L  S  S  R  P  R  L  C  K  R  A  C  G  T  C
241   GTTCTGCGAGGTGTCGATTGTCATCAAGGCCTAGACTCTGCAAGAGGGCTTGTGGGACAT
101     C  A  R  C  N  C  V  P  S  G  T  A  G  N  L  D  E  C  P  C
301   GTTGTGCGAGGTGCAACTGCGTACCATCAGGAACTGCAGGCAACTTAGACGAGTGCCCTT
121     Y  A  N  M  T  T  H  G  N  K  R  K  C  P  *
361   GCTACGCCAACATGACTACTCATGGGAACAAACGCAAATGTCCTTAAAAGACCTTATCTC
421   TCTCTCTCTCTCTCTCTCTCTCTCTCTCTCTCTATGCTTGCTCTGTCCTTCTTTAGTA
481   TTCTGTGAGATATCTGTAAGAGAAATAAGGATGATTTGCTCTTGAGTGTGTGAATGATGA
541   AAACTGTGAGATGTGTGTAAACTTCTCTATCTGAGTAAGAGAAGTTATTCGAAAAAAAAA
601   AAAAAAAAAAAAAAAAAAAA3'
```

**Figure 8.** PaSn nucleotide sequence with amino acid frame translation (GenBank: KC012806). ORF in gray shaded; signal peptide (bold-underlined) is followed by the amino acid sequence of the mature peptide; the signal peptide prediction was performed using SignalP program (http://www.Cbs.Dtu.Dk/services/SignalP/).

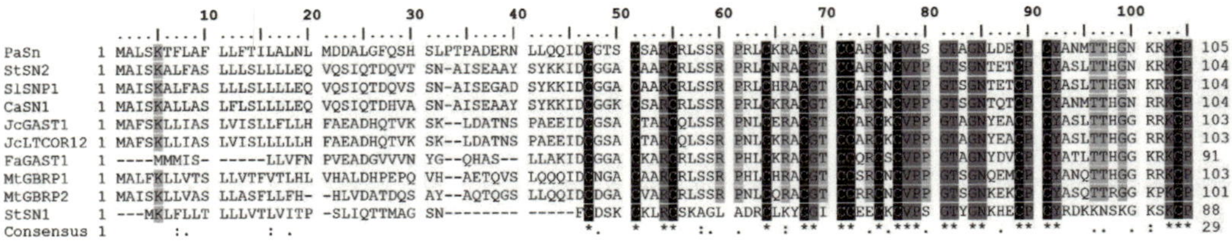


**Figure 9.** Plant snakin amino acid sequences alignment. Amino acids conserved are indicated by an asterisk in the consensus sequence. The black-shaded cysteine residues are present in all sequences. The sequences included are: Avocado PaSn (KC012806); potato: StSN1 (Q948Z4), StSN2 (Q93X17); *Medicago truncatula:* MtGBRP1 (XP_003603759), MtGBRP2 (XP_003589486); *Capsicum annum:* CaSN1 (ACC91329); *Jatropha curcas:* JcGAST1 (ACV70139), JcLTRCOR12 (ACU30848); *Solanum lycopersicum:* SlSNP1 (ADR32106) and *Fragaria x ananassa:* FaGAST1 (AAB97006).

## 3.4. Expression patterns of selected genes measured by sqPCR

Expression patterns of five genes from the seed library were studied by semi-quantitative PCR (**Figure 10**); bHLH transcription factor, three metallothioneins, antimicrobial peptide snakin and SUMO like reference gene during avocado seed development. These genes could be divided into three stages according to the time of growth of the seed in avocado fruit. The bHLH gene has an expression pattern comparable to the endogenous gene SUMO (Ubiquitin), suggesting a role throughout the formation and development of the avocado fruit seed possibly modulating the biogenesis of the seed or embryo; since from the first month of formation (E1) to ripening (E8), similar expression levels were present. For the Metallothionein gene group, PaMT3 presented a pattern of constant expression in the three stages of seed

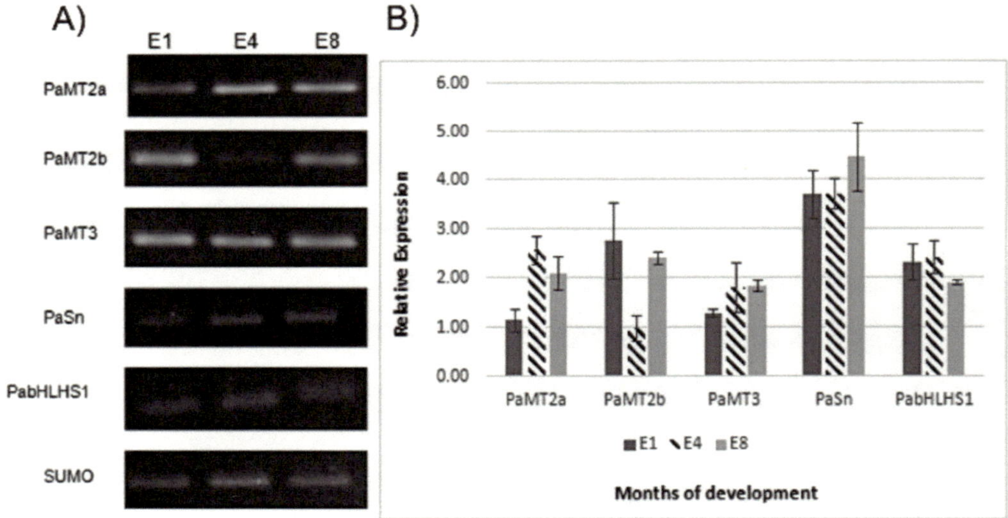


**Figure 10.** Transcript profiling of PaMT2a, PaMT2b, PaMT3, PaSn and PabHLHS1 during seed development. (A) Total RNA was extracted from avocado seeds of 1, 4 and 8 months of development. cDNAs were synthetized and used for sqPCR , the normalizing gene was avocado ubiquitin SUMO. (B) The amount of mRNA of different genes is expressed as the ratio of the densitometric measurement of the sample RT-PCR product to the SUMO corresponding product.

development used slightly above at the peak of expression compared to the endogenous gene but not for PaMT2a which has an initial level of expression low in first month of development, having its maximum expression peak in the stage of 4 months; PaMT2b has its maximum expression peak at the beginning of the fruit formation in the first month, decaying this by month 4 and recovering expression levels for ripening; it should be noted that metallothioneins have been directly involved within various roles within the functions. Some of them as carriers or facilitators of metal ions for processes of defense, synthesis or hydrolysis of reserve components to make them more bioavailable [46]; however, the authors do not reach an agreement to say that the different types or families of metallothioneins play a specific role. The snakin gene has a similar behavior throughout the development of the fruit emphasizing its role within the defense against pathogens or as the first barrier of protection or signaling of attack, making it therefore important to maintain the levels of expression throughout development and possibly after this for fruit protection (**Figure 10**). The expression patterns of the selected genes identified by sqPCR and the different expression patterns of avocado seed transcriptome suggested various roles of these genes in response to seed development and protection in avocado fruit.

## 4. Conclusions

In this work, we identified and characterized three novel metallothioneins and one transcription factor gene from avocado nativo mexicano seeds, which are expressed abundantly during seed development. This suggests that they can have a protagonic paper during seed development and probably form a network to protect the embryo for drought stress. More studies are necessary to elucidate the paper of these genes during avocado seed-fruit development.

## Author details

Luis María Suárez-Rodríguez, Fernando Sánchez-Albarrán, Huber León-Corona and Rodolfo López-Gómez*

*Address all correspondence to: lgomez@umich.mx

Institute of Chemical-Biological Studies, Michoacana University of Saint Nicolás of Hidalgo, Morelia Michoacán, México

## References

[1] Chanderbali AS, Albert VA, Ashworth VE, Clegg MT, Litz RE, Soltis DE, Soltis PS. *Persea americana* (avocado): Bringing ancient flowers to fruit in the genomics era. BioEssays: News and Reviews in Molecular, Cellular and Developmental Biology. 2008;**30**:386-396. DOI: 10.1002/bies.20721

[2] Rohwer JG. Lauraceae. In: Kubitzki K, Rohwer JG, Bittrich V, editors. The Families and Genera of Flowering Plants. Springer-Verlag; 1993. p. 366-391. DOI: 10.1007/978-3-662-02899-5

[3] Ibarra-Laclette E, Méndez-Bravo A, Pérez-Torres CA, Albert VA, Mockaitis K, Kilaru A, López-Gómez R, Cervantes-Luevano J, Herrera-Estrella L. Deep sequencing of the Mexican avocado transcriptome, an ancient angiosperm with a high content of fatty acids. BMC Genomics. 2015;**16**:599. DOI: 10.1186/s12864-015-1775-y

[4] Ben-Ya'acov A, Michelson E. Avocado rootstocks. Horticultural Review. 1995;**17**:381-429

[5] Barrientos PAF. El Aguacate. Biodiversitas. 2010;**88**:1-7

[6] Andrade HP, De León C, Espíndola BMC, Alvarado RD, López JA, García ER. Selección de portainjertos de aguacate para tolerancia-resistencia a *Phytoptora cinnamomi* Rands. usando temperaturas controladas. Spanish Journal of Rural Development. 2012;**4**:1-8 ISSN: 2171-1216

[7] Guzmán-Rodríguez JJ, Ibarra-Laclette E, Herrera-Estrella L, Ochoa-Zarzosa A, Suárez-Rodríguez LM, Rodríguez-Zapata LC, Salgado-Garciglia R, Jimenez-Moraila B, López-Meza JE, López-Gómez R. Analysis of expressed sequence tags (ESTs) from avocado seed (*Persea americana* var. drymifolia) reveals abundant expression of the gene encoding the antimicrobial peptide snakin. Plant Physiology and Biochemistry. 2013;**70**:318-324. DOI: 10.1016/j.plaphy.2013.05.045

[8] Rivas-Gil A, Lares VC, Amaro-Luis JM, Bahsas BA. 2-Alquil-4-Hidroxy-Tetrahidrofuranos de la pulpa del aguacate. *Avances en Química*. 2006;**1**(3):35-37

[9] Birkbeck J. Health benefits of avocado oil. Food New Zealand. 2002;40-42

[10] QY L, Arteaga JR, Qifeng Z, Huerta S, Go VLW, Heber D. Inhibition of prostate cancer cell growth by an avocado extract: Role of lipid-soluble bioactive substances. The Journal of Nutritional Biochemistry. 2005;**16**:23-30. DOI: 10.1016/j.jnutbio.2004.08.003

[11] Ding H, Chin YW, Kinghorn AD, D'Ambrosio SM. Chemopreventive characteristics of avocado fruit. Seminars in Cancer Biology. 2007;**17**:386-394. DOI: 10.1016/j.semcancer.2007.04.003

[12] Dabas D, Shegog RM, Ziegler GR, Lambert JD. Avocado (*Persea americana*) seeds as source of bioactive phytochemicals. Current Pharmaceutical Design. 2013;**19**:6133-6140. DOI: 10.2174/13816128113193400007

[13] Cummings K, Schroeder CA. Anatomy of avocado fruit. California Avocado Society Yearbook. 1942;**27**:56-64 Record Number: 19430300995

[14] Peng FY, Reid KE, Liao N, Schlooser J, Lijavetzky D, Holt R, Martínez Zapater JM, Jones S, Marra M, Bohlmann J. Generation of ESTs in *Vitis vinifera* wine grape (cabernet sauvignon) and table grape (Muscat Hamburg) and discovery of new candidate genes with potential roles in berry development. Gene. 2007;**402**:40-50. DOI: 10.1016/j.gene.2007.07.016

[15] Teh SL, Chan WS, Abdullah JO. Development of expressed sequence tag resources for Vanda Mimi palmer and data mining for EST-SSR. Molecular Biology Reports. 2011;**38**:3903-3909. DOI: 10.1007/s11033-010-0506-3

[16] Low ET, Alias H, Boon SH, Shariff E, Tan CY, Ooi L, Cheah SC, Raha AR, Wan KL, Singh R. Oil palm (*Elaeis guineensis* Jacq.) tissue culture ESTs: Identifying genes associated with callogenesis and embryogenesis. BMC Plant Biology. 2008;**8**:62-81. DOI: 10.1186/1471-2229-8-62

[17] Okubo K, Hori N, Matoba R, Niiyama T, Fukushima A, Kojima Y, Matsubara K. Large scale cDNA sequencing for analysis of quantitative and qualitative aspects of gene expression. Nature Genetics. 1992;**2**:173-179. DOI: 10.1038/ng1192-173

[18] Franco GR, Adams MD, Soares MB, Simpson AJ, Venter JC, Pena SD. Identification of new *Schistosoma mansoni* genes by the EST strategy using a directional cDNA library. Gene. 1995;**15**:141-147. DOI: 10.1016/0378-1119(94)00747-G

[19] Lee H, Lee HK, An G, Lee YK. Analysis of expressed sequence tags from the red alga *Griffithsia okiensis*. Journal of Microbiology. 2007;**45**(6):541-546

[20] López-Gómez R, Gómez-Lim MA. A method for extracting intact RNA from fruits rich in polysaccharides using ripe mango mesocarp. HortScience. 1992;**27**(5):440-442

[21] Ewing B, Green P. Base-calling of automated sequencer traces using *Phred*. II. Error probabilities. Genome Research. 1998;**8**:186-194. DOI: 10.1101/gr.8.3.186

[22] Li S, Chou HH. Lucy2: An interactive DNA sequence quality trimming and vector removal tool. Bioinformatics. 2004;**20**(16):2865-2866. DOI: 10.1093/bioinformatics/bth302

[23] Foissac S, Bardou P, Moisan A, Cros MJ, Schiex T. EuGene'Hom: A generic similarity-based gene finder using multiple homologous sequences. Nucleic Acids Research. 2003;**31**:3742-3745. DOI: 10.1093/nar/gkg586#

[24] Ashburner M, Ball CA, Blake JA, Botstein D, Butler H, Cherry JM, Davis AP, Dolinski K, Dwight SS, Eppig JT. Gene ontology: Tool for the unification of biology. The Gene Ontology Consortium Nat. Genet. 2000;**25**:25-29. DOI: 10.1038/75556

[25] Frishman D. Protein annotation at genomic scale: The current status. Chemical Reviews. 2007;**107**:3448-3466. DOI: 10.1021/cr068303k

[26] López-Gómez R, Suáres-Rodríguez LM, Ibarra-Laclette E, Guzmán-Rodríguez JJ, López-Meza JE, Ochoa-Zarzosa A, Salgado-Garciglia R, Rodríguez-Zapata LC, Jimenéz-Moraila B, Herrera-Estrella L. Transcriptome (ESTs) of native Mexican avocado fruit is dominated by stress and innate immunity genes. Acta Horticulturae. 2016;**1144**:43-48. DOI: 10.17660/ActaHortic.2016.1144.6

[27] Margoshes M, Vallee BL. A Cadmium protein from equine kidney cortex. Journal of the American Chemical Society. 1957;**79**(17):4813-4814. DOI: 10.1021/ja01574a064

[28] Kagi JHR, Schaffer A. Biochemistry of metallothionein. Biochemistry. 1988;**27**(23):8509-8515. DOI: 10.1021/bi00423a001

[29] Cobbett C, Goldsbrough P. Phytochelatins and metallothioneins: Roles in heavy metal detoxification and homeostasis. Annual Review of Plant Biology. 2002;**53**(1):159-182. DOI: 10.1146/annurev.arplant.53.100301.135154

[30] Zhou Y, Chu P, Chen H, Li Y, Liu J, Ding Y, Tsang EWT, Jiang L, Wu K, Huang S. Overexpression of *Nelumbo nucifera* metallothioneins 2a and 3 enhances seed germination vigor in Arabidopsis (electronic material). Planta. 2012;**235**(3):523-537. DOI: 10.1007/s00425-011-1527-4

[31] Mir G, Doménech J, Huguet G, Guo WJ, Goldsbrough P, Atrian S, Molinas M. A plant type 2 metallothionein (MT) from cork tissue responds to oxidative stress. Journal of Experimental Botany. 2004;**55**:2483-2493. DOI: 10.1093/jxb/erh254

[32] Soda N, Sharan A, Gupta BK, Singla-Pareek S, Pareek A. Evidence for nuclear interaction of a cytoskeleton protein (OsIFL) with metallothionein and its role in salinity stress tolerance. Scientific Reports. 2016;**6**:34762. DOI: 10.1038/srep34762

[33] Murre C, McCaw P, Vaessin H, Caudy M, Jan L, Jan Y, Cabrera C, Buskin J, Hauschka S, Lassar A, Weintraub H, Baltimore D. Interactions between heterologous helix-loop-helix proteins generates complexes that bind specifically to a common DNA sequence. Cell. 1989;**58**:537-544. DOI: 10.1016/0092-8674(89)90434-0

[34] Massari ME, Murre C. Helix-Loop-Helix proteins: Regulators of transcription in eukaryotic organisms. Molecular and Cellular Biology. 2000;**20**(2):429-440

[35] Jones S. An overview of the basic helix-loop-helix proteins. Genome Biology. 2004;**5**:226

[36] Robinson KA, Lopes JM. *Saccharomyces cerevisiae* basic helix-loop-helix proteins regulate diverse biological processes. Nucleic Acids Research. 2000;**28**(7):1499-1505

[37] Ramsay NA, Glover BJ. MYB–bHLH–WD40 protein complex and the evolution of cellular diversity. Trends in Plant Science. 2005;**10**:63-70

[38] Atchley WR, Fitch WM. A natural classification of the basic helix-loop-helix class of transcription factors. Proceedings of the National Academy of Sciences. 1997;**94**:5172-5176

[39] Ledent V, Vervoort M. The basic helix–loop–hélix protein family: Comparative genomics and phylogenetic analysis. Genome Research. 2001;**11**:754-770

[40] Fairman R, Chao HG, Mueller L, Lavoie TB, Shen L, Novotny J, Matsueda GR. Characterization of a new four-chain coiled-coil: Influence of chain length on stability. Protein Science. 1995;**4**:1457-1469

[41] Stevens JD, Roalson EH, Skinner MK. Phylogenetic and expression analysis of the basic helix-loop-helix transcription factor gene family: Genomic approach to cellular differentiation. Differentiation. 2008;**76**(9):1006-1022

[42] Skinner MK, Manikkam M, Guerrero-Bosagna C. Epigenetic transgenerational actions of environmental factors in disease etiology. Trends in Endocrinology and Metabolism. 2010;**21**:214-222

[43] Ichihashi Y, Horiguchi G, Gleissberg S, Tsukaya H. The bHLH transcription factor SPATULA controls final leaf size in *Arabidopsis thaliana*. Plant and Cell Physiology. 2010;**51**:252-261

[44] Nicolas P, Lecourieux D, Gomès E, Delrot S, Lecourieux F. The grape berry-specific basic helix–loop–helix transcription factor VvCEB1 affects cell size. Journal of Experimental Botany. 2013;**64**(4):991-1003. DOI: 10.1093/jxb/ers374

[45] Tani E, Tsaballa T, Stedel C, Kalloniati C, Papaefthimiou D, Polidoros A, Darzentas N, Ganopoullos I, Flemetakis E, Katinakis P, Tsaftaris A. The study of a *SPATULA*-like bHLH transcription factor expressed during peach (*Prunus persica*) fruit development. Plant Physiology and Biochemistry. 2011;**49**:654-663

[46] Hassinen VH, Tervahauta AI, Schat H, Kärenlampi SO. Plant metallothioneins – Metal chelators with ROS scavenging activity? Plant Biology. 2011;**13**:225-232. DOI: 10.1111/j.1438-8677.2010.00398.x

Chapter 4

# Identification and Mapping of Phosphorylated Isoforms of the Major Storage Protein of Potato Based on Two-Dimensional Electrophoresis

Javier Bernal, María López-Pedrouso, Daniel Franco,
Susana Bravo, Lucio García and Carlos Zapata

Additional information is available at the end of the chapter

http://dx.doi.org/10.5772/intechopen.70400

**Abstract**

Protein phosphorylation plays a key role in the synthesis and degradation of dry seed storage proteins. In contrast, no evidence for phosphorylation has been reported to date in vegetative storage proteins (VSPs). The patatin multigene family encodes the major VSP of the potato, *Solanum tuberosum* L. This study addresses for the first time the identification and mapping of phosphorylated patatin forms based on high-resolution two-dimensional electrophoresis (2-DE) profiles. Patatin isoforms from mature tubers of cultivar Kennebec were separated by 2-DE and subsequently identified by tandem mass spectrometry. In-gel identification and mapping of phosphorylated isoforms were performed using the multiplex phosphoprotein-specific staining Pro-Q DPS. We found that phosphorylation is a ubiquitous post-translational protein modification associated with isoforms of patatin. In addition, protein dephosphorylation with hydrogen fluoride-pyridine coupled to 2-DE was used for quantitative profiling of phosphorylated patatin. This experimental approach showed that patatin comprises multiple isoforms with very different phosphorylation level. Thus, phosphorylation rates over isoforms ranged from 4.6 to 52.3%. Overall, the identification and mapping of differentially phosphorylated patatin opens up new exploratory ways to unravel the molecular mechanisms underlying its mobilization along the tuber life cycle.

**Keywords:** patatin, *Solanum tuberosum*, seed storage proteins, storage protein mobilization, tuber phosphoproteome, vegetative storage proteins

## 1. Introduction

Potato storage proteins provide necessary nutrients for the development of tuber, mature-to-sprouting tuber transition and successful plant growth [1–4]. The patatin is the major VSP of

*S. tuberosum*, accounting for up to 45% of the total soluble protein [1, 4–7]. However, the molecular mechanism that triggers the cleavage of the patatin along the tuber life cycle has not to date been identified. The regulatory mechanisms involved in the synthesis and degradation of storage proteins are better known in dry seeds [8–13]. Phosphorylation has proven to be a key regulator mechanism in the maturation, dormancy and germination of seed storage proteins (SSPs). Thus, reverse phosphorylation of the phytohormone abscisic acid (ABA) seems to play a crucial regulatory role in the synthesis of SSPs at the transcriptional level [11, 12]. More specifically, phosphoproteome studies in rapeseed and rice reported that cruciferins and cupins achieve higher levels of phosphorylation at the late maturation stage [9, 13]. In addition, mobilization of the major SSP in the common bean, phaseolin, was found to occur in germinating seeds through degradation of highly phosphorylated isoforms [10]. It suggests that degradation of SSPs in dry-to-germinating seed transition occurs through a phosphorylation-dependent regulatory mechanism.

The past few years have witnessed a steady discovery of phosphorylated SSPs such as cruciferins, napins, cupins, legumins and vicilins [8–10, 14, 15], but no evidence of phosphorylated isoforms has been reported to date in patatin or other VSPs such as sporamins and ocatins. Therefore, elucidating the question of whether patatin can be phosphorylated is a mandatory initial step in follow-up research concerning the molecular processes underlying its mobilization. First of all, the term patatin applies to a group of glycoproteins encoded by a gene family constituted by ~10–18 genes per haploid genome, most of them organized as a single gene cluster at the end of the long arm of chromosome 8 [16–18]. Patatin gene family members exhibit a very high degree of nucleotide sequence identity [19, 20]. In addition, patatin is a family of immunologically indistinguishable isoforms with similar structural properties and thermal conformational stability [21, 22]. Overall, extensive heterogeneity in molecular mass (40–45 kDa) and isoelectric point (4.5–5.2) seems to be the most salient differential molecular features among isoforms [21–25].

The 2-DE has provided the most complete information about the heterogeneity in molecular mass ($M_r$) and isoelectric point (p*I*) of the patatin [23–25]. Specifically, a total of 17–23 spots with variations in $M_r$ and/or p*I* were detected in 2-DE patatin profiles obtained from different potato cultivars [25]. Variations in $M_r$ seem to be mainly due to differential N-glycosylation at three specific asparagine residues of the amino acid sequence by *N*-linked oligosaccharide side chains or glycans [21, 22, 25]. Charge differences among isoforms could be explained by variations in positively and negatively charged amino acids [22]. However, variable phosphorylation has potential to change the p*I* of proteins by substituting hydroxyl groups on amino acid residues with negatively charged phosphate groups [26]. Therefore, phosphorylation could be a plausible but unexplored factor contributes to explain charge heterogeneity among patatin isoforms on 2-DE gels.

In this study, we undertook a proteomic approach addressed to the identification and mapping of phosphorylated isoforms of the patatin multigene family based on high-resolution 2-DE. First, relatively abundant tuber proteins were successfully separated from low-abundance proteins by loading low amounts of total protein sample into 2-DE gels. Subsequently, high-abundance patatin proteins were identified and distinguished from other tuber abundant proteins on 2-DE gels using mass spectrometry (MS) techniques. Second, direct and rapid in-gel multiplex identification and mapping of phosphorylated isoforms of the patatin were achieved using the Pro-Q Diamond phosphoprotein stain (Pro-Q DPS), which specifically binds to the phosphate moieties of phosphoproteins [27]. Third, quantitative profiling of phosphorylated patatin isoforms was assessed by chemical dephosphorylation of phosphoproteins with hydrogen fluoride-pyridine (HF-P) [28–30]. For this purpose, the volume difference between phosphorylated and dephosphorylated 2-DE patatin spots was used to quantify protein phosphorylation levels. This experimental pipeline is a highly valuable top-down proteomic approach for the identification and mapping of phosphorylated isoforms of high-abundance storage proteins. It has been instrumental in unravelling the quantitative profiling of phosphorylated phaseolin isoforms in common bean seeds as well as their dynamic changes in dry-to-germinating seed transition [10]. Proteomic analyses were performed from total protein extracts of mature tubers of cultivar Kennebec. The obtained results will facilitate follow-up studies for better understanding of the regulatory mechanisms underlying patatin degradation and its biochemical status along the tuber life cycle.

## 2. Materials and methods

### 2.1. Plant material

Proteomic analyses were performed from mature potato tubers of cv. Kennebec (2n = 4x = 48). Larger pieces of lyophilized tuber were homogenized with a pre-cooled mortar and pestle. The samples were stored at −80°C until protein extraction. Four biological replicates were used for experiments.

### 2.2. Protein extraction and quantification

Total tuber proteins were extracted using the phenol extraction method. A 200 mg sample of lyophilized tuber was transferred to an extraction buffer (500 mM Tris-HCl, 500 mM EDTA, 700 mM sucrose, 100 mM KCl pH 8.0, 2% DTT and 1 mM PMSF). Tris-HCl (pH 6.6–7.9) saturated phenol was added and the phenol phase was collected using centrifuging (4500 rpm at 4°C). Protein precipitation solution of 0.1 M ammonium acetate in cold methanol was added. Protein pellet was washed with 0.1 M ammonium acetate and 10 mM DTT, and with 80% acetone and 10 mM DTT. The resuspended protein pellet was then diluted in lysis buffer (7 M urea; 2 M thiourea; 4% CHAPS; 10 mM DTT, and 2% Pharmalyte™ pH 3–10, GE Healthcare, Uppsala, Sweden). Protein concentration was evaluated using the commercial CB-X protein assay kit (G-Biosciences, St. Louis, MO, USA) according to the instructions of the manufacturer for interfering agent removal and use with a microplate reader. The bovine serum albumin (BSA) was used as standard protein to generate calibration curves.

### 2.3. Two-dimensional electrophoresis (2-DE)

High-resolution 2-DE profiles of patatin isoforms were obtained following the procedure described in López Pedrouso et al. [10]. Briefly, total protein samples (75 μg of protein) of each biological replicate were loaded into immobilized pH gradient (IPG) strips of 24-cm long and 4–7 pH linear gradient (Bio-Rad Laboratories, Hercules, CA, USA). First dimensional isoelectric focusing (IEF) was performed in a PROTEAN IEF Cell System (Bio-Rad Laboratories) after IPG strip rehydration for 12 h at 50 V. Rapid voltage ramping was subsequently applied to reach a total of 70 kVh. Equilibration of IEF strips was performed before running second dimension using equilibration buffers. The second dimension (SDS-PAGE) was performed on 10% polyacrylamide gels using an Ettan DALTsix large vertical electrophoresis system (GE Healthcare). Second-dimension gels were run using a constant electric current of 16 mA per gel for 15 h at 25°C.

### 2.4. Enzymatic deglycosylation of patatin

Patatin deglycosylation was performed with the enzyme protein-N-glycosidase F (PNGase F, New England Biolabs, Ipswich, MA, USA) according to the manufacturer specifications. A 75 μg sample of total protein extract from mature tuber was incubated with PNGase F (25 U/mL) and diluted in reaction buffer (New England Biolabs) until a final volume of 20 μL. The mixture was incubated for 12 h at 37°C. Patterns of deglycosylated patatin isoforms on 2-DE gels were obtained as described earlier.

### 2.5. Pro-Q staining for phosphoproteins

Pro-Q Diamond phosphoprotein stain (Pro-Q DPS, Molecular Probes, Leiden, The Netherlands) was used for the in-gel detection of phosphorylated patatin polypeptides, as described previously [31]. Briefly, gels were fixed with 50% methanol and 10% acetic acid for 60 min and washed twice for 15 min each with distilled water. The gels were subsequently incubated for 120 min with two-fold water-diluted Pro-Q DPS, destained four times (30 min per wash) with 50 mM sodium acetate and 20% ACN pH 4.0, and washed again twice with distilled water (5 min per wash). The PeppermintStick™ (Molecular Probes) phosphoprotein marker was added to tuber protein extracts before 2-DE. Phosphorylated (ovalbumin, 45.0 kDa; and β-casein, 23.6 kDa) and unphosphorylated (β-galactosidase, 116.25 kDa; bovine serum albumin, 66.2 kDa; avidin, 18.0 kDa; and lysozyme, 14.4) PeppermintStick proteins were used as positive and negative controls of phosphorylation, respectively.

### 2.6. Chemical dephosphorylation of patatin

The chemical dephosphorylation of patatin was performed with hydrogen fluoride-pyridine (HF-P) as previously described [28, 29], with some modification [10]. An amount of 1 mg of total protein extract from tuber of cv. Kennebec was dissolved in 250 μL of 70% HF-P and incubated on ice for 2 h. The mixture was neutralized by addition of 10 M sodium hydroxide solution. Proteins were desalinated using Amicon Ultra-4 centrifugal filter devices (Millipore, Billerica, MA, USA) and then eluted in 300 μL of lysis buffer. Prior to 2-DE, protein purification was performed using the Clean-up kit (GE Healthcare).

### 2.7. SYPRO Ruby staining for total protein

2-DE gels were stained with SYPRO Ruby fluorescent stain (Lonza, Rockland, ME, USA), for total protein density following the manufacturer's indications. Pro-Q DPS-stained gels were also post-stained with SYPRO Ruby to detect total protein.

### 2.8. Image analysis

The 2-DE images from gels stained with Pro-Q DPS or SYPRO Ruby fluorescent dyes were acquired using a Gel Doc XR+ system (Bio-Rad Laboratories). Digitalized gels were analyzed with PDQuest Advanced software v. 8.0.1 (Bio-Rad Laboratories). Gel matching, spot identification and quantification of spot volumes were performed after background subtraction and normalization based on total density in valid spots. Automatic matches were manually checked. Only the reproducibly detected patatin spots across replicates were selected for quantitative analyses. Experimental p*I*-values over spots were determined using the linear scale of IEF-strips as reference, whereas $M_r$-values were assessed using PeppermintStick (Molecular Probes) molecular weight markers and standard molecular mass markers ranging from 15 to 200 kDa (Fermentas, Ontario, Canada).

### 2.9. In-gel digests

Protein spots of interest were excised from polyacrylamide gels and subjected to in-gel digestion with trypsin as described previously [32]. Briefly, disulfide reduction and alkylation of the excised protein spots were performed with 10 mM DTT (Sigma-Aldrich, St. Louis, MO, USA) in 50 mM ammonium bicarbonate (Sigma-Aldrich) and 55 mM iodoacetamide (Sigma-Aldrich) in 50 mM ammonium bicarbonate, respectively. The gel pieces were washed with 50 mM ammonium bicarbonate in 50% methanol (HPLC grade, Scharlau, Barcelona, Spain), dehydrated with acetonitrile (ACN, HPLC grade) and subsequently dried in a SpeedVac (Thermo Fisher Scientific, Waltham, MA, USA). Dry gel pieces were incubated with modified porcine trypsin (Promega, Madison, WI, USA) at a concentration of 20 ng/μL in 20 mM ammonium bicarbonate, at 37°C for 16 h. After digestion, peptides were recovered by incubation (three times/20 min) in 40 μL of 60% ACN in 0.5% formic acid. The resulting tryptic peptides were concentrated in a SpeedVac and stored at −20°C.

### 2.10. Mass spectrometry (MS)

Protein identification was performed by MALDI-TOF and MALDI-TOF/TOF MS as reported by López-Pedrouso et al. [10]. Peptides were dissolved in 4 μL 0.5% formic acid and then were mixed with an equal volume (0.5 μL) of matrix solution, containing 3 mg of α-Cyano-4-hydroxycinnamic acid (CHCA) dissolved in 1 mL of 50% ACN in 0.1% trifluoroacetic acid (TFA). The mixture was deposited using the thin layer method, onto a 384 Opti-TOF MALDI plate (Applied Biosystems, Foster City, CA, USA). Peptide MS and MS/MS data were acquired with a 4800 MALDI-TOF/TOF mass spectrometer (Applied Biosystems). MS spectra were acquired in positive-ion reflector mode with an Nd:YAG laser (355 nm wavelength) and an average number of 1000 laser shots. Each spectrum was internally calibrated with at least three trypsin autolysis products. All MS/MS spectra were performed by selecting precursors ions with a relative resolution of 300 full width at half maximum (FWHM) and metastable suppression. The 4000 Series Explorer Software v. 3.5 (Applied Biosystems) was used for mass data analysis. Combined peptide mass fingerprinting (PMF) and MS/MS fragment-ion spectra were interpreted with GPS Explorer Software v. 3.6 using Mascot software v. 2.1 (Matrix Science, Boston, MA, USA) to search against the *S. tuberosum* UniProtKB/Swiss-Prot databases. Mascot database search parameters were: precursor mass tolerance of 50 ppm, MS/MS fragment tolerance of 0.6 Da, one missed cleavage allowed, carbamidomethyl cysteine (CAM) as fixed modification and oxidized methionine as variable modification. Identification of patatin phosphopeptides was also performed from spectrum data allowing phosphor-serine (PhosphoS), phosphor-tyrosine (PhosphoY) and phosphor-threonine (PhosphoT) residues as variable modification to search against the UniProtKB/Swiss-Prot databases. Analysis of phosphorylation sites was implemented using the Plant Protein Phosphorylation DataBase ($P^3DB$) [33]. All identifications and spectra were manually checked for validation. Proteins with at least two matched peptides and statistically significant ($p$-value $< 0.05$) MASCOT scores were selected as positively identified.

### 2.11. Data analysis

The phosphorylation rate at each spot was quantified using the measure *PR* [10]. It is defined as $PR = [(T - D)/T] \times 100$, where $T$ and $D$ are the volumes of a given spot on 2-DE gels untreated (total protein volume) and treated (dephosphorylated protein volume) with HF-P, respectively. Non-parametric bootstrap confidence intervals (CIs) were obtained for mean values of *PR* across four biological replicates by the bias-corrected percentile method [34]. For each observed mean of *PR*, 2000 bootstrap samples of size $N = 4$ were drawn with replacement by applying a Monte Carlo algorithm. The 95 and 99% CIs for the observed mean of *PR* were constructed from distribution of 2000 bootstrap mean replications. The bootstrap estimate of bias was obtained from the proportion of bootstrap mean replications lower than the original estimate of the mean, and bias-corrected CIs were then calculated using the theoretical normal distribution as described by Efron [34]. *PR* data were clustered by using the unweighted pair-group method with arithmetic averaging (UPGMA). The UPGMA dendrogram derived from the matrix of pairwise *PR*-values was generated using NTSYSpc v. 2.1 software (Applied Biostatistics, Setauket, NY, USA). Descriptive statistics and Spearman's correlation test were calculated with the IBM SPSS Statistics 20 (SPSS, Chicago, IL, USA) statistical software package.

## 3. Results and discussion

### 3.1. Map of patatin isoforms based on 2-DE

Patatin isoforms in mature potato tuber of cv. Kennebec were first recognized on our 2-DE gels according to the previously reported studies on 2-DE patatin profiles [23–25]. We found that patatin profiles were constituted by a complex constellation of different spots showing large variations in $M_r$ and/or p$I$ (**Figure 1**). Specifically, 2-DE resolved a total of 20 spots distributed

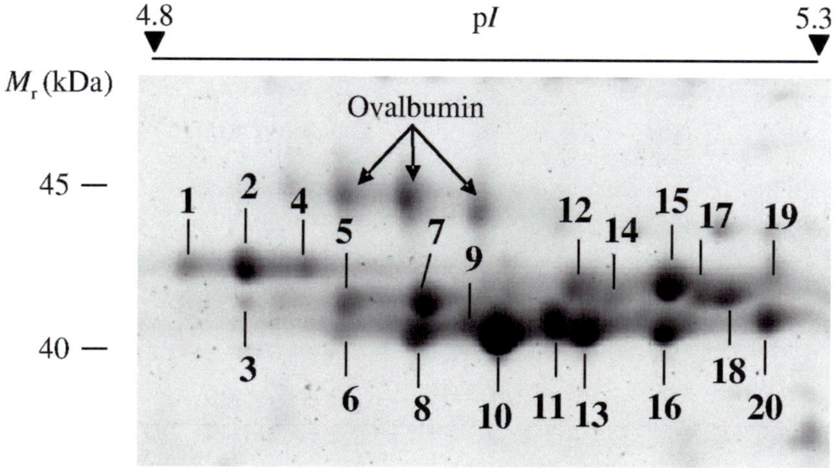


**Figure 1.** High-resolution 2-DE reference map of the patatin isoforms in mature tuber of cv. Kennebec. The enlarged gel image shows patatin spots consecutively numbered in the order of the lower to the higher p*I*. 2-DE was performed using a 24-cm long IPG strip of linear pH 4–7 gradient in the first dimension and SDS-PAGE (10% by mass) in the second. The protein loading was 75 μg and the gel was stained with SYPRO Ruby fluorescent stain. The arrows indicate ovalbumin (45.0 kDa) marker position on the gel. The $M_r$ of spots was assessed from ovalbumin and standard molecular mass markers ranging from 15 to 200 kDa and their p*I*s from strips of linear pH.

into three main levels with $M_r$ between 40.1 and 43.0 kDa and p*I* range varying from 4.8 to 5.3. A total of 20 spots were excised from gel and identified by MALDI-TOF and MALDI-TOF/TOF MS. The identification results are listed in **Table 1**. All but one small and weakly stained spot (spot 3) were confidently identified. MS analyses confirmed that, indeed, those spots contained only patatin polypeptides. However, most of the identifications were ambiguous and compatible with the occurrence of different types of patatin. This uncertainty is a consequence of the well-known high degree of sequence homology (at least 90%) among isoforms [19, 20]. Only protein spots with higher $M_r$ (spots 1, 2 and 4) were unambiguously identified as Patatin-3-Kuras 1 (PT3K1). It can be understood by considering that the pt3k1 gene exhibits the most differentiated sequence from other patatin genes according to the phylogenetic tree inferred from cDNA sequence analysis [24].

The two-dimensional map of the patatin was implemented with the location of glycosylated isoforms using the enzyme PNGase F. It is an effective enzymatic method for removing almost all *N*-linked oligosaccharides (glycans) from glycoproteins through the hydrolysis of the glycosamide linkage between the terminal GlcNAc and the Asn amide nitrogen [35]. We found that the three main spot levels in $M_r$ on 2-DE gels obtained from untreated samples merged to a single spot level after incubation with PNGase F, with an apparent decrease in $M_r$ (not shown). It indicates that variable degrees of glycosylation are a major contributor to the $M_r$ heterogeneity detected on 2-DE gels. High $M_r$ difference among patatin isoforms has been explained by the presence of up to three potential N-glycosylation sites at Asn residues [22, 24, 25, 36]. The mapping of glycosylated isoforms on 2-DE gels can be useful in future studies investigating the functional role of this post-translational protein modification (PTM) of the patatin.

| Spot no.[a] | Exp. p*I* [b] | Match./cov. (%)[c] | Mascot score | Protein name (type)[d] | No. phosphopeptides/ phosphosites |
|---|---|---|---|---|---|
| 1 | 4.84 | 7/24 | 191 | Patatin (PT3K1) | 1/1 |
| 2 | 4.88 | 13/58 | 571 | Patatin (PT3K1) | 7/17 |
| 3 | 4.90 | – | – | Unidentified | – |
| 4 | 4.93 | 5/16 | 158 | Patatin (PT3K1) | – |
| 5 | 4.96 | 3/8 | 106 | Patatin (various) | – |
| 6 | 4.96 | 5/21 | >60–187 | Patatin (various) | – |
| 7 | 5.02 | 11/40 | >60–200 | Patatin (various) | – |
| 8 | 5.02 | 6/25 | >60–297 | Patatin (various) | – |
| 9 | 5.05 | 7/28 | >60–324 | Patatin (various) | 1/1 |
| 10 | 5.08 | 5/11 | >60–199 | Patatin (various) | 1/1 |
| 11 | 5.12 | 9/32 | >60–337 | Patatin (various) | 3/7 |
| 12 | 5.13 | 8/30 | >60–298 | Patatin (various) | – |
| 13 | 5.14 | 9/31 | >60–475 | Patatin (various) | 1/1 |
| 14 | 5.16 | 9/31 | >60–267 | Patatin (various) | – |
| 15 | 5.20 | 2/11 | >60–103 | Patatin (various) | 4/9 |
| 16 | 5.20 | 10/40 | >60–440 | Patatin (various) | 4/10 |
| 17 | 5.23 | 5/25 | >60–123 | Patatin (various) | – |
| 18 | 5.25 | 9/31 | >60–259 | Patatin (various) | 5/9 |
| 19 | 5.29 | 3/9 | >60–129 | Patatin (various) | – |
| 20 | 5.27 | 10/41 | >60–550 | Patatin (various) | 4/8 |

[a]Gel position of assigned spots is shown in **Figure 1**.

[b]Experimental p*I* value.

[c]Matched peptides and percentage of the polypeptide sequence covered by matched peptides.

[d]PT3K1, abbreviation for Patatin-3-Kuras 1.

**Table 1.** Protein, phosphopeptides and phosphosites along 2-DE patatin spots of cv. Kennebec, identified from MALDI-TOF and MALDI-TOF/TOF MS data.

## 3.2. In-gel identification of phosphorylated patatin isoforms

Pro-Q DPS was used for in-gel multiplex identification of phosphorylated patatin isoforms. Representative 2-DE images of patatin in mature tuber of cv. Kennebec on the same gel stained with Pro-Q DPS and post-stained with SYPRO Ruby are shown in **Figure 2**. The PeppermintStick markers used as positive and negative controls of protein phosphorylation validated the specificity of the recognition of phosphoproteins by Pro-Q DPS under our experimental conditions. It was found that all 20 patatin spots of the reference pattern exhibited Pro-Q DPS fluorescent signal. Similar result was obtained for patatins from mature tubers of cvs. Agria, Amanda and Ivory Russet (not shown). We can conclude, therefore, that phosphorylation is a ubiquitous PTM associated with isoforms of the patatin.

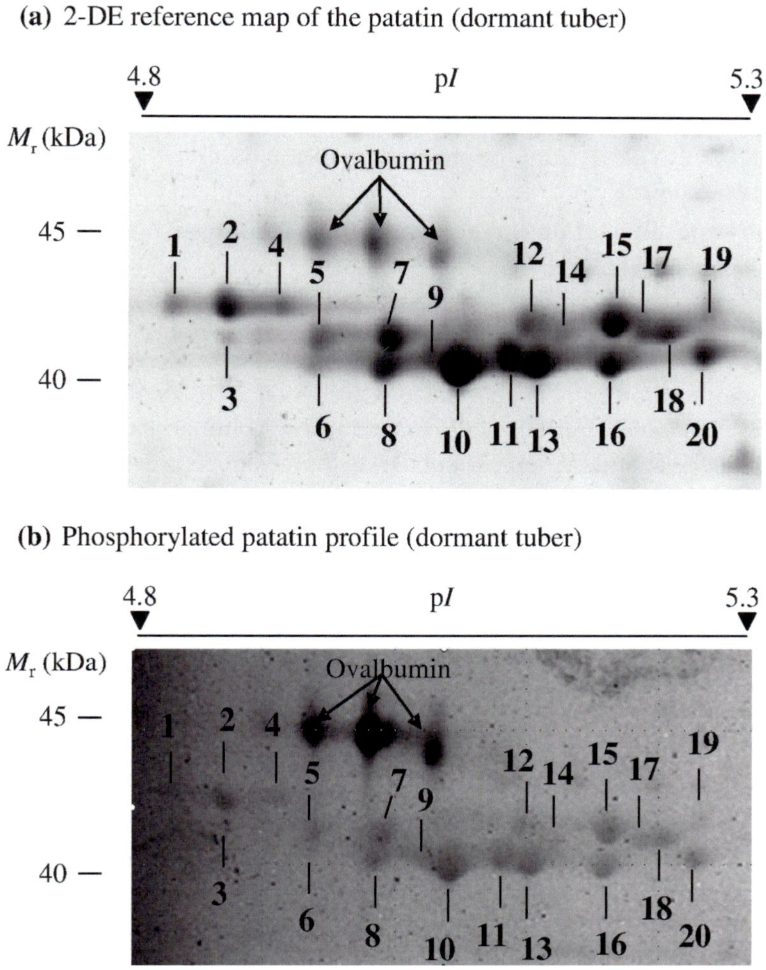


**Figure 2.** Mapping of phosphorylated patatin spots in mature tubers (cv. Kennebec) on 2-DE gel. (a) Reference profile of patatin spots on gel stained with the non-specific-protein SYPRO Ruby stain. (b) Profile of phosphorylated patatin spots from the same gel stained with the specific-phosphoprotein Pro-Q DPS fluorescent dye. The phosphoprotein ovalbumin was used as a marker of the reliability of Pro-Q DPS under our experimental conditions.

A prospective identification of phosphopeptides and phosphosites by MASCOT search using spectra data from MALDI-TOF and MALDI-TOF/TOF MS revealed 22 non-redundant patatin phosphopeptides containing 49 non-redundant phosphorylation sites (**Table 1**). Comparison with large-scale phosphoproteomic screens in other species using the Plant Protein Phosphorylation DataBase ($P^3DB$) [33] suggests that most phosphorylation sites identified are novel to this study. Thus, no phosphorylated ortholog sites were identified in other plant phosphoproteomics data for *Arabidopsis thaliana, Brassica napus, Glycine max, Medicago truncatula, Oryza sativa* and *Zea mays*. It is noteworthy that enrichment methods of underrepresented phosphorylated proteins or peptides can be conducted prior to high-resolution MS analysis to precisely identify and map phosphorylation sites, but the amount of protein collected in a spot is often insufficient for downstream enrichment methods [37].

Consequently, it would be difficult to assign phosphorylation sites to the specific isoforms found along 2-DE patatin patterns. At the present time, the 2-DE map of phosphorylated isoforms appears to be more informative than the exact identification of phosphosites in order to evaluate their biological meaning. Regardless of this, phosphorylation site prediction analysis is an additional evidence for phosphorylation of patatin.

### 3.3. Quantitative profiling of phosphorylated patatin isoforms

Changes in the phosphorylation level across patatin spots of cv. Kennebec were assessed by chemical dephosphorylation of total tuber protein extracts with HF-P coupled to 2-DE. This experimental approach provides more efficient information than Pro-Q DPS to the identification and quantification of phosphorylated proteins on 2-DE gels [10]. The reason is that the Pro-Q DPS fluorescent signal of spots containing low-abundance phosphopeptides is seriously suppressed by abundant non-phosphorylated phosphopeptides. The chemical dephosphorylation method has the advantage of using SYPRO Ruby stain, which combines good sensitivity with excellent linearity [38].

Representative 2-DE gel images of the patatin pattern before and after HP-F treatment are shown in **Figure 3**. First of all, note that spots of the protein phosphorylation marker, ovalbumin, underwent a basic shift on p*I* after HF-P treatment. This indicates that HF-P had sufficient time to achieve a complete dephosphorylation of polypeptides. With respect to the 2-DE profiles of dephosphorylated patatin, we can highlight two important observations. First, all spots observed in untreated samples were also present after dephosphorylation, but with an apparent decrease in volume. This suggests that patatin spots contained a mixture of phosphorylated and unphosphorylated isoforms. Accordingly, other factors together with protein phosphorylation must be contributing to charge heterogeneity along 2-DE patatin patterns such as difference in charged amino acids over isoforms [22]. Second, newly arisen spots (spots 21–27) found in dephosphorylated patatin patterns appeared on more basic positions of 2-DE gels. MS analysis confirmed that these new spots contained patatin (data not shown), and thereby they are isoforms that underwent a basic shift on p*I* after dephosphorylation with HF-P.

The phosphorylation level of each spot was evaluated with the measure *PR* using volumes obtained by PDQuest software from phosphorylated and dephosphorylated profiles. Mean (±SE, standard error) values of *PR* for each spot together with bias-corrected 95 and 99% bootstrap CIs are shown in **Table 2**. Interestingly, we found that spots were not uniformly phosphorylated: mean *PR*-values across spots were in the range of 4.6–52.3% and averaged (±SE) 34.4 ± 2.8%. The bootstrapped 95 and 99% CIs revealed statistically significant differences (p-value $< 0.05$) between many pairs of spots. Patatin spots were subsequently grouped into clusters from *PR*-values using a dendrogram UPGMA. The resulting dendrogram showed that spots cluster in three main groups with statistically significant mean differences in *PR* ($p < 0.01$) assessed by bias-corrected 99% bootstrap CIs (**Figure 4**). In particular, spots of the group 3 (spots 13 and 20) formed a well-separated cluster (mean *PR* = 8.3%) from the two remaining groups (the mean *PR* of groups 1 and 2 was 44.0 and 30.1%, respectively).

http://dx.doi.org/10.5772/intechopen.70400

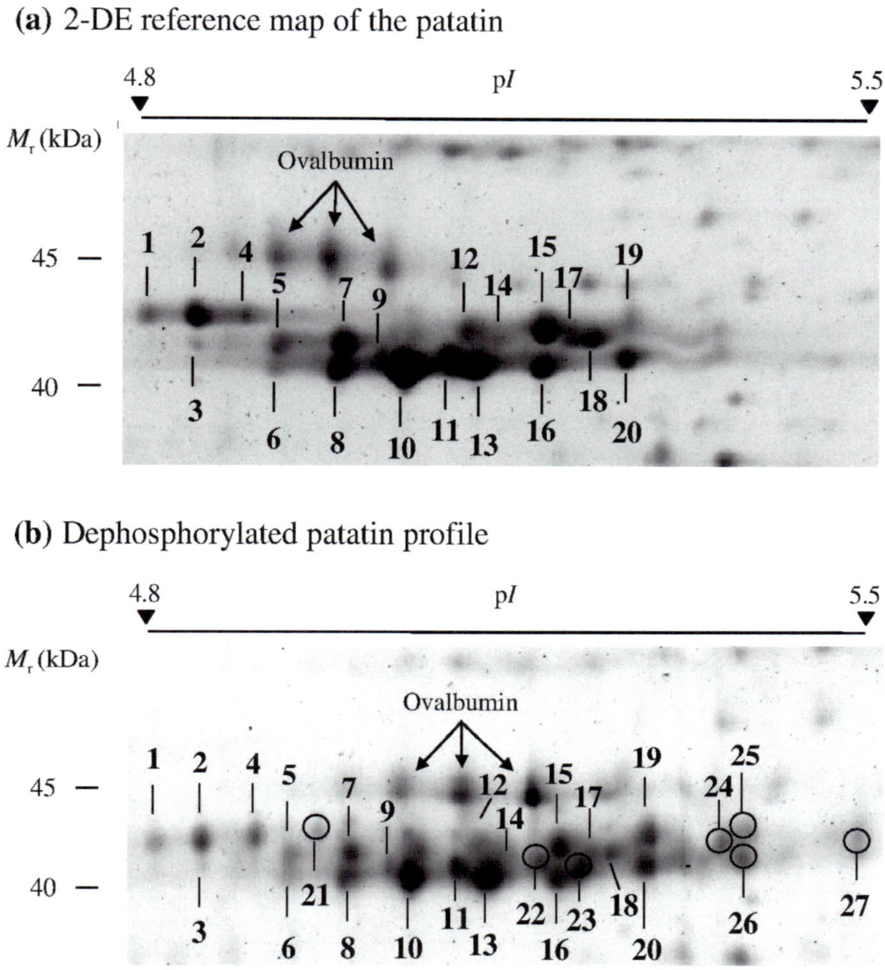


**Figure 3.** 2-DE profile of dephosphorylated patatin with HF-P in mature tuber (cv. Kennebec). (a) Reference profile of patatin without dephosphorylation treatment on gel stained with SYPRO Ruby stain. (b) Profile of patatin after chemical dephosphorylation on gel stained with SYPRO Ruby. Closed circles represent newly arisen spots on gel after dephosphorylation as compared to the reference profile. Identification of new spots as patatin was performed by MALDI-TOF and MALDI-TOF/TOF MS.

Elucidating whether changes in abundance of protein phosphorylation reflect either changes in phosphorylation status or changes in the abundance of the protein itself is a major challenge in the interpretation of quantitative phosphoproteomics studies [39, 40]. Thus, phosphopeptide enrichment methods prior to high-resolution MS permit the identification of low-abundance phosphoproteins but prevent joint quantitation of phosphorylation status and abundance of proteins [39]. However, our experimental approach allowed us to successfully tackle this problem. Thus, we have detected a statistically significant negative relationship between patatin spot volumes and their corresponding *PR*-values by Spearman's non-parametric correlation test ($r_s = -0.42$, $p < 0.001$, n = 70). In addition, *PR*-values were negatively correlated with spot p*I*s ($r_s = -0.31$, $p < 0.01$, n = 70). As expected under a differential phosphorylation

| Spot no.[a] | p$I$ | Mean (± SE) $PR$ | P ($\hat{\theta}_B \leq \hat{\theta}$)[b] | 95% bootstrap CI (CL, CU)[c] | 99% bootstrap CI (CL, CU)[c] |
|---|---|---|---|---|---|
| 1 | 4.84 | 39.75 ± 2.53 | 0.53 | 35.6, 44.1 | 34.6, 44.3 |
| 2 | 4.88 | 41.21 ± 6.07 | 0.57 | 31.9, 52.9 | 30.1, 54.5 |
| 3 | 4.90 | 43.05 ± 2.88 | 0.55 | 37.5, 47.9 | 37.4, 49.3 |
| 4 | 4.93 | 28.11 ± 4.35 | 0.76 | 23.8, 32.5 | 23.8, 32.5 |
| 5 | 4.96 | 42.83 ± 2.02 | 0.56 | 40.1, 46.9 | 39.4, 48.3 |
| 6 | 4.96 | 52.34 ± 4.10 | 0.57 | 46.4, 60.4 | 44.5, 61.8 |
| 7 | 5.02 | 39.01 ± 4.11 | 0.51 | 32.4, 45.6 | 30.4, 45.7 |
| 8 | 5.02 | 27.19 ± 5.61 | 0.52 | 16.9, 35.0 | 15.3, 35.0 |
| 9 | 5.05 | 32.48 ± 3.30 | 0.51 | 26.7, 37.3 | 22.9, 38.0 |
| 10 | 5.12 | 30.48 ± 4.72 | 0.57 | 23.3, 40.2 | 22.6, 42.4 |
| 11 | 5.12 | 25.13 ± 4.31 | 0.57 | 18.8, 33.1 | 16.6, 34.1 |
| 12 | 5.13 | 51.39 ± 5.52 | 0.55 | 41.2, 60.6 | 40.0, 62.1 |
| 13 | 5.14 | 4.60 ± 2.04 | 0.75 | 2.6, 6.6 | 2.6, 6.6 |
| 14 | 5.16 | 44.34 ± 9.37 | 0.52 | 25.7, 57.3 | 25.0, 59.0 |
| 15 | 5.20 | 41.87 ± 3.80 | 0.52 | 35.7, 49.3 | 34.7, 51.2 |
| 16 | 5.20 | 26.99 ± 8.37 | 0.58 | 18.1, 37.0 | 10.4, 37.0 |
| 17 | 5.23 | 35.96 ± 2.42 | 0.53 | 32.9, 40.9 | 32.2, 42.8 |
| 18 | 5.25 | 34.16 ± 4.09 | 0.53 | 26.2, 39.6 | 25.6, 40.0 |
| 19 | 5.29 | N/A[d] | N/A | N/A | N/A |
| 20 | 5.27 | 11.95 ± 4.63 | 0.55 | 6.8, 20.9 | 5.6, 20.9 |

Volume of spots for untreated and dephosphorylated protein samples with HF-P were assessed by PDQuest software.
[a]Gel position of assigned spots is shown in **Figure 1**.
[b]The bootstrap distribution is median biased if the probability (P) of ($\hat{\theta}_B \leq \hat{\theta}$) ≠ 0.50, which was calculated from 2000 bootstrap replicates; $\hat{\theta}_B$ and $\hat{\theta}$ are the bootstrap mean and the sample mean estimates, respectively.
[c]CI—Confidence interval; CL—lower bound; CU—upper bound.
[d]N/A = not available, weakly stained spot with a volume below the limit of detection.

**Table 2.** Mean (±SE) values of $PR$ for patatin spots estimated from four replicates from dormant tubers of cv. Kennebec.

hypothesis, isoforms located on acidic gel positions tended to be more highly phosphorylated than those of basic positions. It is also noteworthy that phosphorylation levels were estimated using the measure $PR$, which takes into account the amount of protein at each spot. Therefore, differential phosphorylation along patatin spots seems to be genuine and cannot be explained only by changes in protein abundance.

The control of tuber sprouting is a major target in potato breeding because premature tuber sprouting during their lengthy storage leads to important quality and economic loss [41–43]. However, the molecular mechanisms controlling dormancy release and tuber sprouting are not yet sufficiently known [42–44]. The identification and mapping of phosphorylated isoforms

http://dx.doi.org/10.5772/intechopen.70400

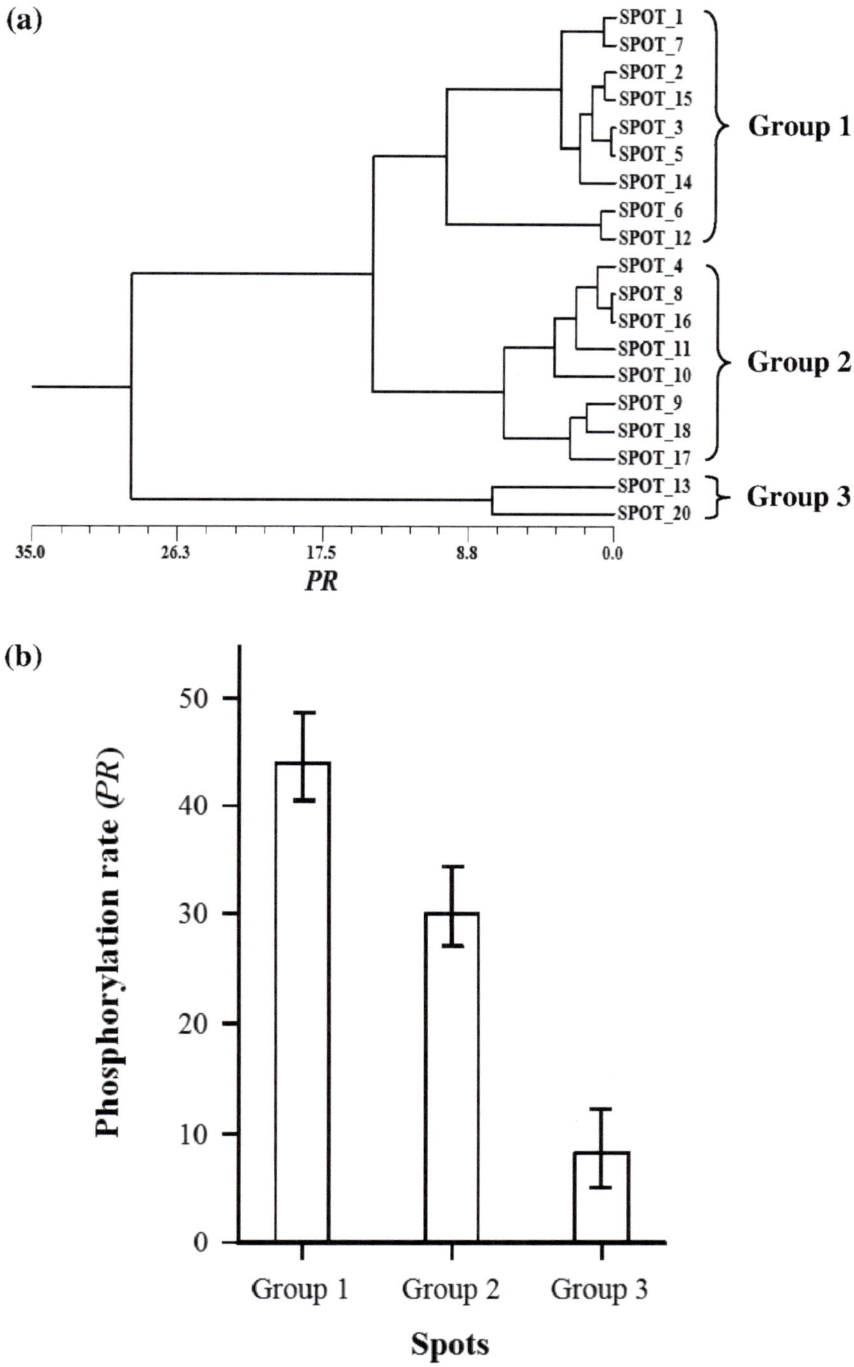


**Figure 4.** Evaluation of the differential *PR* along 2-DE patatin spots (mature tuber). (a) UPGMA dendrogram from the matrix of mean differences in *PR* between pairs of patatin spots. Spot numbers refer to numbers in **Figure 1**. (b) Mean *PR* values for each of the three main spot groups clustered by UPGMA. Bootstrapping (2000 replicates) was used to determine 99% CIs for mean *PR*-values at each group. *PR*-values over spots were calculated using the formula $PR = [(T - D)/T] \times 100$, where $T$ and $D$ represent the gel-spot volume in reference and dephosphorylated patatin profiles, respectively. Spot volumes over quadruplicate 2-DE gels were determined using the PDQuest software.

of the patatin opens up new exploratory ways to unravel the molecular mechanisms underlying mobilization of VSPs. The finding of differentially phosphorylated isoforms is particularly relevant because increase (or decrease) in phosphorylation status without a parallel change in the amount of protein has been considered to be a useful indicator for a specific functional change [39, 40, 45]. In this regard, systematic follow-up studies on VSPs will be needed to assess whether their degradation takes place through a phosphorylation-dependent regulatory mechanism, as it occurs in common bean during dry-to-germinating seed transition [10]. The establishment of a 2-DE-based reference map of patatin can be a very efficient tool to address this challenge in potato by monitoring changes in the phosphorylation status along the tuber life cycle.

## Acknowledgements

This research was supported by funds from the Consellería do Medio Rural (Xunta de Galicia, Spain).

## Author details

Javier Bernal[1], María López-Pedrouso[1], Daniel Franco[2], Susana Bravo[3], Lucio García[2] and Carlos Zapata[1*]

*Address all correspondence to: c.zapata@usc.es

1 Department of Zoology, Genetics and Physical Anthropology, University of Santiago de Compostela, Santiago de Compostela, Spain

2 Meat Technology Center of Galicia, Ourense, Spain

3 Proteomics Laboratory, CHUS, Santiago de Compostela, Spain

## References

[1] Liu Y-W, Han C-H, Lee M-H, Hsu F-L, Hou W-C. Patatin, the tuber protein of potato (*Solanum tuberosum* L.) exhibits antioxidant activity in vitro. Journal of Agricultural and Food Chemistry. 2003;**51**:4389-4393. DOI: 10.1021/jf030016j

[2] Racusen D. Lipid acyl hydrolase of patatin. Canadian Journal of Botany. 1984;**62**:1640-1644. DOI: 10.1139/b84-220

[3] Shewry PR. Tuber storage proteins. Annals of Botany. 2003;**91**:755-769. DOI: 10.1093/aob/mcg084

[4] de Souza Cândido E, Pinto MFS, Pelegrini PB, Lima TB, Silva ON, Pogue R. Plant storage proteins with antimicrobial activity: Novel insights into plant defense mechanisms. The FASEB Journal. 2011;**25**:3290-3305. DOI: 10.1096/fj.11-184291

[5] Müntz K. Deposition of storage proteins. Plant Molecular Biology. 1998;**38**:77-99. DOI: 10.1023/A:1006020208380

[6] Racusen D, Foote M. A major soluble glycoprotein of potato tubers. Journal of Food Biochemistry. 1980;**4**:43-52. DOI: 10.1111/j.1745-4514.1980.tb00876.x

[7] Paiva E, Lister RM, Park WD. Induction and accumulation of major tuber proteins of potato in stems and petioles. Plant Physiology. 1983;**71**:161-168. DOI: 10.1104/pp.71.1.161

[8] Agrawal GK, Thelen JJ. Large-scale identification and quantitative profiling of phosphoproteins expressed during seed filling in oilseed rape. Molecular & Cellular Proteomics. 2006;**5**:2044-2059. DOI: 10.1074/mcp.M600084-MCP200

[9] Meyer LJ, Gao J, Xu D, Thelen JJ. Phosphoproteomic analysis of seed maturation in *Arabidopsis*, rapeseed, and soybean. Plant Physiology. 2012;**159**:517-528. DOI: 10.1104/pp.111.191700

[10] López-Pedrouso M, Alonso J, Zapata C. Evidence for phosphorylation of the major seed storage protein of the common bean and its phosphorylation-dependent degradation during germination. Plant Molecular Biology. 2014;**84**:415-428. DOI: 10.1007/s11103-013-0141-1

[11] Hirayama T, Shinozaki K. Perception and transduction of abscisic acid signals: Keys to the function of the versatile plant hormone ABA. Trends in Plant Science. 2007;**12**:343-351. DOI: 10.1016/j.tplants.2007.06.013

[12] Ghelis T, Bolbach G, Clodic G, Habricot Y, Miginiac E, Sotta B, Jeannette E. Protein tyrosine kinases and protein tyrosine phosphatases are involved in abscisic acid-dependent processes in *Arabidopsis* seeds and suspension cells. Plant Physiology. 2008;**148**:1668-1680. DOI: 10.1104/pp.108.124594

[13] Han C, Wang K, Yang P. Gel-based comparative phosphoproteomic analysis on rice during germination. Plant & Cell Physiology. 2014;**55**:1376-1394. DOI: 10.1093/pcp/pcu060

[14] Irar S, Oliveira E, Pagès M, Goday A. Towards the identification of late-embryogenic-abundant phosphoproteome in *Arabidopsis* by 2-DE and MS. Proteomics. 2006;**6**:S175-S185. DOI: 10.1002/pmic.200500387

[15] Wan L, Ross ARS, Yang J, Hegedus DD, Kermode AR. Phosphorylation of the 12 S globulin cruciferin in wild-type and abi1-1 mutant *Arabidopsis thaliana* (thale cress) seeds. The Biochemical Journal. 2007;**404**:247-256. DOI: 10.1042/BJ20061569

[16] Ganal MW, Bonierbale MW, Roeder MS, Park WD, Tanksley SD. Genetic and physical mapping of the patatin genes in potato and tomato. Molecular & General Genetics. 1991;**225**:501-509. DOI: 10.1007/BF00261693

[17] Stupar RM, Beaubien A, Jin W, Song J, Lee M-K, Wu C, Zhang H-B, Han B, Jiang J. Structural diversity and differential transcription of the patatin multicopy gene family during potato tuber development. Genetics. 2006;**172**:1263-1275. DOI: 10.1534/genetics.105.051219

[18] The Potato Genome Sequencing Consortium. Genome sequence and analysis of the tuber crop potato. Nature. 2011;**475**:189-195. DOI: 10.1038/nature10158

[19] Mignery GA, Pikaard CS, Hannapel DJ, Park WD. Isolation and sequence analysis of cDNAs for the major tuber protein, patatin. Nucleic Acids Research. 1984;**12**:7987-8000. DOI: 10.1093/nar/12.21.7987

[20] Mignery GA, Pikaard CS, Park WD. Molecular characterization of the patatin multigene family of potato. Gene. 1988;**62**:27-44. DOI: 10.1016/0378-1119(88)90577-X

[21] Park WD, Blackwood C, Mignery GA, Hermodson MA, Lister RM. Analysis of the heterogeneity of the 40,000 molecular weight tuber glycoprotein of potatoes by immunological methods and by $NH_2$-terminal sequence analysis. Plant Physiology. 1983;**71**:156-160. DOI: 10.1104/pp.71.1.156

[22] Pots AM, Gruppen H, Hessing M, van Boekel MAJS, Voragen AGJ. Isolation and characterization of patatin isoforms. Journal of Agricultural and Food Chemistry. 1999;**47**:4587-4592. DOI: 10.1021/jf981180n

[23] Lehesranta SJ, Davies HV, Shepherd LVT, Nunan N, McNicol JW, Auriola S, Koistinen KM, Suomalainen S, Kokko HJ, Kärenlampi SO. Comparison of tuber proteomes of potato varieties landraces, and genetically modified lines. Plant Physiology. 2005;**138**:1690-1699. DOI: 10.1104/pp.105.060152

[24] Bauw G, Nielsen HV, Emmersen J, Nielsen KL, JØrgensen M, Welinder KG. Patatin, Kunitz protease inhibitors and other major proteins in tuber of potato cv. Kuras. The FEBS Journal. 2006;**273**:3569-3584. DOI: 10.1111/j.1742-4658.2006.05364.x

[25] Bárta J, Bártová V, Zdráhal Z, Šedo O. Cultivar variability of patatin biochemical characteristics: Table versus processing potatoes (*Solanum tuberosum* L). Journal of Agricultural and Food Chemistry. 2012;**60**:4369-4378. DOI: 10.1021/jf3003448

[26] Zhu K, Zhao J, Lubman DM. Protein pI shifts due to posttranslational modifications in the separation and characterization of proteins. Analytical Chemistry. 2005;**77**:2745-2755. DOI: 10.1021/ac048494w

[27] Romero-Rodríguez MC, Abril N, Sánchez-Lucas R, Jorrín-Novo V. Multiplex staining of 2-DE gels for an initial phosphoproteome analysis of germinating seeds and early grown seedlings form a non-orthodox specie: *Quercus ilex* L. subsp. *ballota* [Desf.] Samp. Frontiers in Plant Science. 2015;**6**:620. DOI: 10.3389/fpls.2015.00620

[28] Kuyama H, Toda C, Watanabe M, Tanaka K, Nishimura O. An efficient chemical method for dephosphorylation of phosphopeptides. Rapid Communications in Mass Spectrometry. 2003;**17**:1493-1496. DOI: 10.1002/rcm.1078

[29] Kita K, Okumura N, Takao T, Watanabe M, Matsubara T, Nishimura O, Nagai K. Evidence for phosphorylation of rat liver glucose-regulated protein 58, GRP58/ERp57/ER-60, induced by fasting and leptin. FEBS Letters. 2006;**580**:199-205. DOI: 10.1016/j.febslet.2005.11.074

[30] Woo EM, Fenyo D, Kwok BH, Funabiki H, Chait BT. Efficient identification of phosphorylation by mass spectrometric phosphopeptide fingerprinting. Analytical Chemistry. 2008;**80**:2419-2425. DOI: 10.1021/ac702059p

[31] Agrawal GK, Thelen JJ. Development of a simplified, economical polyacrylamide gel staining protocol for phosphoproteins. Proteomics. 2005;**5**:4684-4688. DOI: 10.1002/pmic.200500021

[32] Jensen ON, Wilm M, Shevchenko A, Mann, M. Sample preparation methods for mass spectrometric peptide mapping directly from 2-DE gels. In: Link AJ, editor. Methods in Molecular Biology: 2-D Proteome Analysis Protocols. Totowa, NJ, USA: Humana Press; 1999. pp. 513-530. DOI: 10.1385/1-59259-584-7:513

[33] Yao Q, Bollinger C, Gao J, Xu D, Thelen JJ. P$^3$DB: An integrated database for plant protein phosphorylation. Frontiers in Plant Science. 2012;**3**:206. DOI: 10.3389/fpls.2012.00206

[34] Efron B. The Jackknife, the Bootstrap and Other Resampling Plans. CBMS-NSF Regional Conference Series in Applied Mathematics No. 38. Philadelphia, Pennsylvania: Society for Industrial and Applied Mathematics; 1982. pp. 75-92

[35] Tarentino AL, Gómez CM, Plummer TH. Deglycosylation of asparagine-linked glycans by peptide-N-glycosidase F. Biochemistry. 1985;**24**:4665-4671. DOI: 10.1021/bi00338a028

[36] Sonnewald U, Sturm A, Chrispeels MJ, Willmitzer L. Targeting and glycosylation of patatin, the major potato tuber in leaves of transgenic tobacco. Planta. 1989;**179**:171-180. DOI: 10.1007/BF00393687

[37] Silva-Sánchez C, Li H, Chen S. Recent advances and challenges in plant phosphoproteomics. Proteomics. 2015;**156**:1127-1141. DOI: 10.1002/pmic.201400410

[38] Rabilloud T, Lelong C. Two-dimensional gel electrophoresis in proteomics: A tutorial. Journal of Proteomics. 2011;**74**:1829-1841. DOI: 10.1016/j.prot.2011.05.040

[39] Wu R, Dephoure N, Haas W, Huttlin EL, Zhai B, Sowa ME, Gygi SP. Correct interpretation of comprehensive phosphorylation dynamics requires normalization by protein expression changes. Molecular & Cellular Proteomics. 2011;**10**:M111.009654. DOI: 10.1074/mcp.M111.009654

[40] Kim M-S, Zhong J, Pandey A. Common errors in mass spectrometry-based analysis of post-translational modifications. Proteomics. 2016;**16**:700-714. DOI: 10.1002/pmic.201500355

[41] Suttle JC. Involvement of endogenous gibberellins in potato tuber dormancy and early sprout growth: A critical assessment. Journal of Plant Physiology. 2004;**161**:157-164. DOI: 10.1078/0176-1617-01222

[42] Aksenova NP, Sergeeva LI, Konstantinova TN, Golyanovskaya SA, Kolachevskaya OO, Romanov GA. Regulation of potato tuber dormancy and sprouting. Russian Journal of Plant Physiology. 2013;**60**:301-312. DOI: 10.1134/S1021443713030023

[43] Sonnewald S, Sonnewald U. Regulation of potato tuber sprouting. Planta. 2014;**239**:27-38. DOI: 10.1007/s00425-013-1968-z

[44] Mani F, Bettaieb T, Doudech N, Hannachi C. Physiological mechanisms for potato dormancy release and sprouting: A review. African Crop Science Journal. 2014;**2**:155-174

[45] Myernik JA, Hajduch M. Seed proteomics. Journal of Proteomics. 2011;**74**:389-400. DOI: 10.1016/j.jprot.2010.12.004

Section 2

# Seed Dormancy and Germination

Chapter 5

# Seed Dormancy

Mustafa Yildiz, Ramazan Beyaz, Mehtap Gursoy,
Murat Aycan, Yusuf Koc and Mustafa Kayan

Additional information is available at the end of the chapter

http://dx.doi.org/10.5772/intechopen.70571

**Abstract**

Dormancy is when there is a lack of germination in seeds or tubers even though the required conditions (temperature, humidity, oxygen, and light) are provided. Dormancy is based on hard seed coat impermeability or the lack of supply and activity of enzymes (internal dormancy) necessary for germination. Dormancy is an important factor limiting production in many field crops. Several physical and chemical pretreatments are applied to the organic material (seeds/tubers) to overcome dormancy. Physical and physiological dormancy can be found together in some plants, and this makes it difficult to provide high-frequency, healthy seedling growth, since the formation of healthy seedlings from the organic material (seeds/tubers) sown is a prerequisite for plant production. This chapter will focus on the description of four different methods we have not seen reported elsewhere for overcoming dormancy.

**Keywords:** dormancy, magnetic field strength, squirting cucumber fruit juice, sodium hypochlorite, gamma radiation

## 1. Introduction

Seeds germinate to grow and survive from seedlings at a favorable time and place. The prevention of germination in unfavorable circumstances is described as dormancy. Dormancy is where there is a lack of germination in a seed/tuber even though the required conditions (temperature, humidity, oxygen, and light) are provided [1]. Dormancy is a trait gained during evolution to survive in adverse conditions such as heat, cold, drought, and salinity. Dormancy enables plant species to adapt to different geographical regions, showing variations in precipitation and temperature [2]. Dormancy has a significant role in the development of new species and the successful dispersal of existing species [2].

There are two types of seed dormancy in general: seed coat (physical) dormancy and internal dormancy. In seed coat dormancy, the seed coat prevents oxygen and/or water permeating into the seed. Sometimes, dormancy is caused by inhibiting chemicals inside the seed. Seeds with seed coat dormancy can remain on/in the ground without germinating until the seed coat allows water and oxygen to enter the seed or eliminate the inhibiting chemicals. Seed coat dormancy is common in California lilac (*Ceanothus*), manzanita (*Arctostaphylos*), sumac (Rhus), and members of the legume family. Scarification, hot water, dry heat, fire, acid and other chemicals, mulch, and light are the methods used for breaking seed coat dormancy [3].

Physiological conditions causing internal dormancy arise from the presence of germination inhibitors inside the seed. The adverse effects of these inhibitors should be eliminated in order to start germination by using germination-promoting substances such as gibberellic acid (GA3) and potassium nitrate ($KNO_3$). The most common inhibitor is abscisic acid (ABA). Sugar maple [4], Norway maple (*Acer platanoides* L.) [5], planetree maple (*Acer pseudoplatanus*) [6], European hazel (*Corylus avellana* L.) [7], white ash (*Fraxinus americana* L.) [8], apple (*Malus pumila* Mill.) [9], northern red (*Quercus rubra* L.), and English oaks [10] are the species that have ABA as an internal inhibitor. Another type of internal dormancy is caused by lack of enzymes, which is required for complete physiological maturation.

Seed coat and internal dormancy can be found together in a species. Seeds with this combined dormancy should be treated by overcoming the problems raised by the impermeable seed coat first, and then overcoming the internal dormancy [3].

In this chapter, the effects of magnetic field strength squirting cucumber (*Ecballium elaterium* (L.) A. Rich.) fruit juice, sodium hypochlorite solutions, and gamma radiation on overcoming dormancy are discussed.

## 2. New methods for overcoming dormancy

### 2.1. Magnetic field strength

All living creatures are exposed to magnetic field throughout their lives. Exposing seeds to magnetic field is one of the physical treatments to increase seed germination and plant development [11–13]. It was reported that seed germination was improved by physiological changes in seeds, such as faster water assimilation and higher photosynthesis, under the effect of magnetic field [14]. Many researchers have reported that exposing seeds to a magnetic field increased seed vigor and germination [12, 13, 15–18]. Although there are many reports on the effects of magnetic field on seed germination, plant growth, protein biosynthesis, and root development, to our knowledge the effect of magnetic field on overcoming dormancy has not been reported before.

In the studies conducted with lentil (*Lens culinaris* Medik.), grass pea (*Lathyrus sativus* L.), and potato (*Solanum tuberosum* L.), magnetic field rapidly increased seed/tuber germination and seedling growth by overcoming dormancy. Seeds/tubers from lentil (cv. "Çiftçi"), grass pea (cv. "Gürbüz") and potato (cv. "Marabel") were sown in soil after keeping them in different magnetic field strengths (0-control, 75, 150, and 300 mT) for different period of times (0-control, 24, 48, and 72 h), and then lentil and grass pea were incubated for 14 days at 24 ± 1°C for a photoperiod of 16 h of light/8 h of darkness under white fluorescent light (27 μmol $m^{-2}$ $s^{-1}$), while potatoes were grown in a greenhouse at 24°C for 2 months. Tubers weighing 40–60 g were used in the study. Five replications were used for the lentil and grass pea, and 10 replications for the potatoes. Pots containing 10 seeds for lentil and grass pea and 1 tuber for potato were considered as an experimental unit. The study used two parallel treatments according to the "Completely Randomized Design" concept. Data were statistically analyzed with Duncan's multiple range test using IBM SPSS Statistics 22 software. Values presented in percentages were subjected to arcsine ($\sqrt{X}$) transformation before statistical analysis [19].

Observations were performed on the characteristics of germination percentage, seedling growth percentage, plant height, and total chlorophyll content in lentil; germination percentage, seedling growth percentage, seedling fresh and dry weights in grass pea; and day of emergence, plant height, and total chlorophyll content in potato. Seed germination percentage was determined at the end of the 5th day in lentil and the 4th day in grass pea, while seedling growth percentage was noted at the end of the 10th day in lentil and the 14th day in grass pea [20]. For all other characteristics, observations were performed at the end of the 14th day in lentil and grass pea, and the 2nd month from the study initiation for the potato.

The total chlorophyll content was determined in leaves of plants according to the protocol of Curtis and Shetty [21]. Fresh tissue from 50 mg of leaves was put in 3 ml of methanol and kept in total darkness at 23°C for 2 h. In this way, chlorophyll in the fresh tissue passed through into the methanol. After 2 h, absorbencies were determined at wavelengths of 665 and 650 nm using UV spectrophotometer. The total chlorophyll content was calculated as μg chlorophyll/g fresh tissue.

In lentil, the lowest values were noted in the control treatment where no magnetic field was used. Lower results in germination and seedling growth percentages were the indicators of dormancy. On the other hand, the highest results in all characteristics were recorded from seeds treated with a magnetic field with a strength of 300 mT for 24 h. At this strength, the highest germination, seedling growth percentages, plant height, and total chlorophyll content were noted as 96.50 %, 100.00%, 7.16 cm, and 586.32 μg chlorophyll/g fresh tissue, respectively (**Table 1**).

There is a close relationship between photosynthesis and chlorophyll content [22–25]. The chlorophyll content of a leaf is accepted as an indicator of the photosynthetic capacity of tissues [26–28]. The total chlorophyll content was determined to be 586.32 μg of chlorophyll/g fresh tissue when treated with a magnetic field strength of 300 mT for 24 h, while it was 125.56 μg of chlorophyll/g fresh tissue in the control sample (**Table 1**).

In grass pea, the lowest results were again recorded for the control treatment, while the highest values were obtained from seeds treated with a magnetic field strength of 300 mT for 72 h. At the end of the study, the germination and seedling growth percentages, plant height, plant fresh, and dry weights were recorded as 0.00%, 75.00%, 21.66 cm, 0.556 g, and 0.131 g, respectively, for the control treatment, whereas they were 100.00%, 100.00%, 28.50 cm, 0.798 g, and 0.160 g, respectively, from seeds treated with a magnetic field strength of 300 mT for 72 h. The

| Magnetic field strength (mT) | Treatment period (h) | Germination (%) | Seedling growth (%) | Plant height (cm) | Total chlorophyll content (µg chlorophyll/g fresh tissue) |
|---|---|---|---|---|---|
| 0—control | 0 | 5.00 g | 15.00 f | 2.00 e | 125.56 g |
| 75 | 24 | 25.00 f | 37.50 ef | 2.54 e | 205.48 f |
| | 48 | 32.25 e | 42.25 e | 2.88 de | 238.56 ef |
| | 72 | 44.50 d | 52.00 d | 3.22 d | 289.75 e |
| 150 | 24 | 55.75 cd | 63.75 c | 3.76 cd | 378.56 d |
| | 48 | 58.00 c | 67.25 c | 4.45 c | 399.74 d |
| | 72 | 62.50 c | 72.50 b | 4.85 bc | 445.21 c |
| 300 | 24 | 96.50 a | 100.00 a | 7.16 a | 586.32 a |
| | 48 | 92.25 ab | 98.25 a | 6.46 ab | 541.00 ab |
| | 72 | 88.75 b | 95.75 ab | 5.85 b | 510.32 b |

Each value is the mean of five replications. All experiments were repeated two times.
Values within a column followed by different letters are significantly different at the 0.01 level.

**Table 1.** The effect of different magnetic field strengths applied to lentil seeds for different periods of time on seed germination and seedling growth in cv. "Çiftçi".

significant effect of magnetic field strength on overcoming dormancy was easily seen when the results of the control sample using a magnetic field strength of 300 mT for 72 h were compared. Germination percentage that was 0.00% in the control sample increased to 100.00% when treated with a magnetic field strength of 300 mT for 72 h (**Table 2**).

In both genotypes (lentil and grass pea), high biomass formation above ground was observed after magnetic field treatments, where the highest values were recorded, compared with the control where no magnetic field was applied (**Figure 1**). Leaf numbers were higher for the magnetic field treatment compared with the control, and it is the main reason for higher photosynthetic activity that achieves a higher yield. Higher biomass accumulation above ground gives a higher food supply for livestock. In our case, the plant fresh weight was 0.798 g when treated with a magnetic field strength of 300 mT for 72 h, while it was only 0.556 g in the control sample at the end of the study (**Table 2**). This means more than a 50% increase in fresh weight and also more than a 50% increase in food supply for livestock.

In potatoes, sprouting emerged above ground 17 days after planting with tubers treated with a magnetic field strength of 300 mT for 72 h. In the control treatment, sprouts emerged on day 39.50. There was 22.50 days difference observed between the treatment with a magnetic field and the control. In the control sample where no magnetic field was used, the lowest results in plant height and total chlorophyll content were found to be 25.56 cm and 1127.46 µg chlorophyll/g fresh tissue, respectively, at the end of the study. On the other hand, the highest values in plant height and total chlorophyll content were recorded as 90.78 cm and

http://dx.doi.org/10.5772/intechopen.70571

| Magnetic field strength (mT) | Treatment period (h) | Germination (%) | Seedling growth (%) | Plant height (cm) | Plant fresh weight (g) | Plant dry weight (g) |
|---|---|---|---|---|---|---|
| 0—control | 0 | 0.00 g | 75.00 d | 21.66 c | 0.556 c | 0.131 c |
| 75 | 24 | 27.50 f | 80.00 c | 23.66 c | 0.603 bc | 0.130 bc |
| | 48 | 35.00 f | 85.00 bc | 24.16 bc | 0.613 b | 0.133 bc |
| | 72 | 70.00 c | 90.00 b | 24.66 bc | 0.626 b | 0.134 bc |
| 150 | 24 | 30.00 f | 90.00 b | 25.76 b | 0.631 b | 0.136 bc |
| | 48 | 45.00 e | 95.00 ab | 25.96 b | 0.636 b | 0.138 b |
| | 72 | 80.00 b | 97.50 a | 26.25 b | 0.638 b | 0.142 b |
| 300 | 24 | 40.00 e | 97.50 a | 27.33 ab | 0.720 ab | 0.145 b |
| | 48 | 55.00 d | 100.00 a | 27.66 ab | 0.725 ab | 0.150 ab |
| | 72 | 100.00 a | 100.00 a | 28.50 a | 0.798 a | 0.160 a |

Each value is the mean of five replications. All experiments were repeated two times.
Values within a column followed by different letters are significantly different at the 0.01 level.

**Table 2.** The effect of different magnetic field strengths applied to grass pea seeds for different periods of time on seed germination and seedling growth in cv. "Gürbüz".

**Figure 1.** Plant development of grass pea in control (a) and in a magnetic field strength of 300 mT for 72 h (b) at the end of the 14th day.

2105.74 μg of chlorophyll/g fresh tissue, respectively, for the treatment with a magnetic field strength of 150 mT for 72 h (**Table 3** and **Figure 2**).

**Figure 2.** The effect of a magnetic field strength of 150 mT applied to potato tubers for different periods (a. 0 h, b. 24 h, c. 48 h and d. 72 h) on plant development in cv. "Marabel".

| Magnetic field strength (mT) | Treatment period (h) | Day of emergence of sprouts | Plant height (cm) | Total chlorophyll content (μg chlorophyll/g fresh tissue) |
|---|---|---|---|---|
| 0—control | 0 | 39.50 e | 25.56 e | 1127.46 g |
| 75 | 24 | 31.25 d | 39.00 d | 1214.56 f |
| | 48 | 30.75 d | 40.25 d | 1327.69 e |
| | 72 | 29.25 d | 42.45 d | 1321.12 e |
| 150 | 24 | 26.00 c | 80.35 b | 1927.36 b |
| | 48 | 21.25 b | 88.27 a | 2000.27 ab |
| | 72 | 17.00 a | 90.78 a | 2105.74 a |
| 300 | 24 | 31.00 d | 75.76 c | 1500.48 d |
| | 48 | 26.25 c | 77.27 b | 1558.34 d |
| | 72 | 20.50 b | 78.00 b | 1789.46 c |

Each value is the mean of five replications. All experiments were repeated two times.

Values within a column followed by different letters are significantly different at the 0.01 level.

**Table 3.** The effect of different magnetic field strengths applied to potato tubers for different periods of time on the day of emergence of sprouts, plant height, and total chlorophyll content in cv. "Marabel".

## 2.2. Squirting cucumber (*Ecballium elaterium* (L.) A. Rich.) fruit juice

Squirting cucumber (*Ecballium elaterium* (L.) A. Rich.), which is commonly found in Turkey, is an important medicinal plant [29]. It contains different compounds such as α-elaterin (cucurbitacin E), β-elaterin (cucurbitacin B), elatericine A [30], and elatericine B (cucurbitacin I) in different plant organs [31]. It also contains sterols, phenolic compounds, vitamins, flavonoids, alkaloids, resin, starch, amino acids, and fatty acids [31]. *In vitro* regeneration was affected significantly by squirting cucumber fruit juice [32]. It was also reported that squirting cucumber fruit juice stimulated rapid germination and seedling growth in rapeseed [33].

Mature squirting cucumber fruits were collected from their natural growing areas around Ankara. Fruit juice was extracted and then filtered in order to remove the bigger parts. The fruit juice was subject to sterile filtration and kept in the refrigerator at −20°C.

In a study conducted for overcoming dormancy in cv. "Marabel" potato tubers using squirting cucumber mature fruit juice, tubers at 40–60 g in weight were rinsed for 6 h at 180 rpm in bottles containing water with different concentrations of mature squirting cucumber fruit juice (0-control, 200, 400, 800, and 1600 μl/L). Then, the tubers were planted in pots. Ten pots were used for each concentration, and one tuber was put in each pot. A pot was considered an experimental unit. The study used two parallel treatments according to the Completely Randomized Design concept. Data were statistically analyzed by Duncan's multiple range test using IBM SPSS Statistics 22 software. Values presented in percentages were subjected to arcsine ($\sqrt{X}$) transformation before statistical analysis [19].

Results clearly showed that squirting cucumber fruit juice had a significant role in overcoming dormancy in potato tubers. The best result was recorded at a concentration of 800 μl/L of fruit juice as 16.25 days. That meant that the sprouts of tubers treated with 800 μl/L of fruit juice emerged above the ground 16.25 days after planting. In contrast, in the control treatment where no fruit juice was used, sprouts emerged above ground later than the other treatments. From the results of the study, it could be concluded that squirting cucumber fruit juice stimulated sprout development by overcoming dormancy in potato tubers (**Table 4**).

In **Figure 3**, plant development was recorded for tubers treated with different concentrations of squirting cucumber fruit juice at the end of the 45th day. For a concentration of 800 μl/L of fruit juice, the plants grew better than any of the others. Squirting cucumber fruit juice at a concentration of 800 μl/L encouraged plants to develop faster and with more energy by overcoming dormancy in tubers. At this concentration, plants had more branches and leaves.

## 2.3. Sodium hypochlorite solutions

Sodium hypochlorite (NaOCl) has been most widely used for surface sterilization. NaOCl is highly effective against all kinds of bacteria, fungi, and viruses [34, 35]. Moreover, NaOCl has strong oxidizing properties that make it highly reactive with amino acids [36, 37], nucleic acids [38], amines, and amides [39, 40].

The most important treatment prior to culture initiation is perhaps surface-sterilization of the explant. Since *in vitro* conditions provide bacteria and fungi with an optimal growth

| Concentration of squirting cucumber fruit juice (µl/L) | Day of emergence of sprouts |
|---|---|
| 0—control | 31.50 c |
| 200 | 23.00 bc |
| 400 | 22.00 bc |
| 800 | 16.25 a |
| 1600 | 20.00 b |

Each value is the mean of five replications. All experiments were repeated two times.
Values within a column followed by different letters are significantly different at the 0.01 level.

**Table 4.** The effect of different concentrations of squirting cucumber fruit juice on the day of emergence of sprouts in potato cv. "Marabel".

**Figure 3.** Plant development at the 45th day for potato tubers of the cv. "Marabel" treated with different concentrations of squirting cucumber fruit juice.

environment, unsuccessful sterilization hinders the progress of tissue culture studies. On the one hand, sterilization of the tissue aims to eliminate all microorganisms that can easily grow *in vitro* conditions; on the other hand, it should guarantee the explant's viability and regeneration capacity, which are known to be affected by the concentration, treatment period [41], and temperature of the disinfectant [42].

NaOCl can also be used for overcoming dormancy [43–45] by decomposing germination inhibitors [46], scarifying the seed coat [42, 47], and increasing $\alpha$-amylase activity [48].

In the study conducted by Telci et al. [49], sodium hypochlorite (NaOCl) solution was successfully used to overcome dormancy in the seeds of *Lathyrus chrysanthus* Boiss., which is used as an ornamental plant with its big, colorful flowers. In the study, *L. chrysanthus* Boiss. seeds of an ecotype (Diyarbakir) from southeast Turkey were treated with a 3.75% NaOCl solution at three different temperatures (25, 35, and 45°C) for 15 min with continuous stirring. This was followed by rinsing three times with sterile water. Seeds were then germinated on a basal medium containing Murashige and Skoog's (MS) mineral salts and vitamins [50], 3%

sucrose, and 0.7% agar in Magenta vessels (15 × 15 cm). The pH of the medium was adjusted to 5.8 prior to autoclaving. For seed germination, cultures were incubated at 15±1°C in the dark for 5 days. Then, all cultures were transferred to a growth chamber for incubation at 25±1°C under cool white fluorescent light (27 μmol $m^{-2}$ $s^{-1}$) with a 16-h light/8-h dark photoperiod. Seed germination and seedling growth percentages were recorded after 5 and 14 days following culture initiation, whereas seedling height and root length, seedling fresh and dry weights, chlorophyll a, chlorophyll b, and total chlorophyll contents were noted 28 days after culture initiation. The chlorophyll contents were determined in the leaves of seedlings according to the protocol of Curtis and Shetty [21]. All statistical analyses were performed using SPSS for Windows software. Three replicates were tested. Petri dishes (100 × 10 mm) were considered the units of replication, and there were 30 seeds per replication. All experiments were repeated twice. One-way Analysis of Variance (ANOVA) was used to test the effect of the 3.75% NaOCl solution at different temperatures. Duncan's multiple range test was used for comparing the means. Data given in percentages were subjected to arcsine (√X) transformation before statistical analysis [19].

The lowest values were recorded in a 3.75% NaOCl solution at a 45°C temperature. Low results at 45°C could be attributed to the fact that the activity of NaOCl increases [51], and it penetrates more easily through the seed coat [52]. The highest results in all characteristics examined were obtained from a 3.75% NaOCl solution at 35°C. Seed germination and seedling growth percentages decreased to 67.76 and 53.53% at 45°C, while they were 88.74 and 77.74% in a 3.75% NaOCl solution at 35°C. Seedling height and root length were 6.77 and 9.26 cm in a 3.75% NaOCl solution at 35°C temperature (**Table 5** and **Figure 4**). These findings were parallel to those of Hsiao and Hans [53], Hsiao and Quick [54], and Yildiz and Er [42] who reported that disinfectants at high concentrations and high temperatures affected seed germination and seedling growth negatively.

The highest seedling fresh and dry weights and tissue water content were recorded when seeds were treated with a 3.75% NaOCl solution at 35°C for 15 min (**Table 5**). The fresh weight

| Dis. temp (°C) | Germination (%) | Seedling growth[1] (%) | Seedling height (cm) | Root length (cm) | Seedling | | Chlorophyll contents (μg chlorophyll/g fresh tissue) | | |
|---|---|---|---|---|---|---|---|---|---|
| | | | | | Fresh weight (g) | Dry weight (g) | Chl. a | Chl. b | Total Chl. |
| 25 | 69.17 b | 53.05 b | 3.90 b | 6.20 b | 0.218 c | 0.037 b | 1297.89 b | 561.28 ab | 1020.81 ab |
| 35 | 88.74 a | 77.74 a | 6.77 a | 9.26 a | 0.370 a | 0.043 a | 1546.14 a | 760.61 a | 1296.42 a |
| 45 | 67.76 b | 53.53 b | 4.87 b | 7.11 b | 0.323 b | 0.039 b | 1125.98 b | 504.52 b | 900.81 b |

Values in a column followed by different letters are significantly different at the 0.01 level in germination percentage, root length, seedling fresh, and dry weights and chlorophyll a content, while significantly different at the 0.05 level in seedling growth, seedling height, chlorophyll b, and total chlorophyll contents.

[1]Seedling growth means seedlings developed out of the total germinated seeds.

**Table 5.** The effects of a 3.75% NaOCl solution at different temperatures on *in vitro* seed germination, seedling growth, seedling height, root length, seedling fresh and dry weights, chlorophyll a, chlorophyll b, and total chlorophyll contents in the leaves of *L. chrysanthus* Boiss. seedlings.

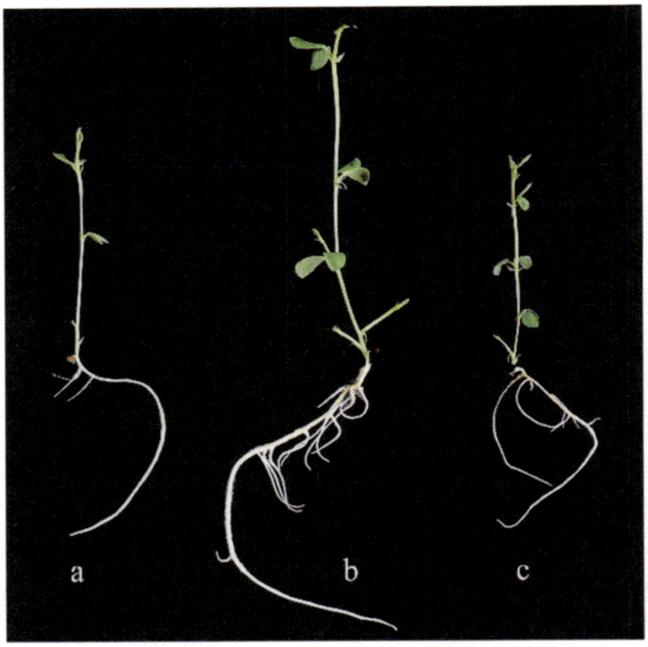


**Figure 4.** *In vitro* seedling growth from *L. chrysanthus* Boiss. seeds treated with a 3.75% NaOCl solution at temperatures of (a) 25°C, (b) 35°C and (c) 45°C for 15 min.

increase could be attributed to cell enlargement [55]. The increase in dry weight was due to cell division and new material synthesis [56]. Higher results in seedlings grown were from seeds treated with 3.75% NaOCl solution at 35°C for 15 min and could be caused by higher tissue water content as reported that *in vitro* explant growth and plantlet establishment have been affected significantly by tissue water content [57].

In the study, the highest chlorophyll a, chlorophyll b, and total chlorophyll contents were seen with a 3.75% NaOCl solution at 35°C temperature (**Table 5**).

### 2.4. Gamma radiation

Gamma rays have an ionizing radiation effect on plant growth and development by inducing cytological, biochemical, physiological, and morphological changes in cells and tissues by producing free radicals in cells [58–60]. Higher doses of gamma radiation have been reported to be inhibitory [61, 62], whereas lower doses are stimulatory. Low doses of gamma rays have been reported to increase seed germination and plant growth, cell proliferation, germination, cell growth, enzyme activity, stress resistance, and crop yields [63–69]. Stimulation of plant growth at low gamma radiation doses is known as hormesis [70]. The hormesis phenomenon is described as a stimulating effect on any factor in the growth of an organism [71].

In the study conducted by Beyaz et al. [72], the effects of gamma radiation on overcoming dormancy in seeds of *L. chrysanthus* Boiss. under *in vitro* conditions were examined. In the study, *L. chrysanthus* Boiss., seeds of an ecotype "Diyarbakir" were first irradiated with different doses (0-control, 50, 100, 150, 200, and 250 Gy) of $^{60}Co$ $\gamma$ rays at 0.8 kGy $h^{-1}$ at the Sarayköy Nuclear Research and Training Center of the Turkish Atomic Energy Authority at Sarayköy, Ankara. Seeds were surface-sterilized with a 3.75% NaOCl solution at 35°C temperature for 15 min. as reported by Telci et al. [49]. The seeds were then placed between filter papers in Petri dishes each containing 6 ml of distilled water. The Petri dishes were incubated for 7 days at 15±1°C in the dark for seed germination. The pre-germinated seeds were then transferred to Magenta vessels (12 × 12 cm) containing a basal medium of Murashige and Skoog's (MS) mineral salts and vitamins [50], 3% sucrose, and 0.7% agar 14 days after the study initiation. The pH of the medium was adjusted to 5.8 prior to autoclaving. Then, all cultures were transferred to a growth chamber for incubation at 25±1°C under cool white fluorescent light (27 μmol $m^{-2}$ $s^{-1}$) with a 16-h light/8-h dark photoperiod. The seed germination percentage was determined at the end of the 7th day, while seedling growth percentage, seedling height, and root length were recorded 14 days after culture initiation [20]. All statistical analyses were performed using SPSS for Windows software. Three replicates were tested, and there were 30 seeds per replication. All experiments were repeated twice. One-way Analysis of Variance (ANOVA) was used to test the effect of different doses of gamma radiation on seed germination and seedling growth. Duncan's multiple range test was used for comparison of the means. Data given in percentages were subjected to arcsine ($\sqrt{X}$) transformation before statistical analysis [19].

The stimulatory effect of low gamma doses was observed in the study at a radiation dose of 150 Gy. The best results in seed germination percentage at the end of the 7th day and in seedling growth percentage, seedling height, and root length at the end of the 14th day were observed at a dose of 150 Gy of gamma radiation (**Table 6** and **Figure 5**). In doses over 150 Gy, the inhibitory effect of gamma radiation was seen. Seed germination percentage was 62.4% at a gamma radiation dose of 150 Gy, while it was 14.3% for a gamma radiation dose of 250 Gy (**Table 6**). The highest seedling growth percentage, seedling height, and root length were again recorded for a 150 Gy gamma radiation dose as 75.7%, 1.2 cm, and 2.9 cm, respectively.

| Gamma doses (Gy) | Day 7 | Day 14 | | |
|---|---|---|---|---|
| | Seed germination (%) | Seedling growth (%)[1] | Seedling height (cm) | Root length (cm) |
| 0 | 35.0 c | 40.0 b | 0.5 d | 1.8 cd |
| 50 | 26.7 c | 45.0 b | 0.7 c | 2.1 bc |
| 100 | 48.6 b | 74.6 a | 0.9 b | 2.4 b |
| 150 | 62.4 a | 75.7 a | 1.2 a | 2.9 a |
| 200 | 35.4 c | 68.8 a | 0.6 cd | 1.6 d |
| 250 | 14.3 d | 37.5 b | 0.5 cd | 1.3 e |

Values in a column followed by the different letters are significantly different at the 0.01 level.
[1]Seedling growth percentage means seedlings developed out of the total seed number.

**Table 6.** Effects of different gamma doses on *in vitro* seed germination, seedling growth, seedling height, and root length in *L. chrysanthus* Boiss.

**Figure 5.** *In vitro* seed germination and seedling growth in *L. chrysanthus* Boiss. seeds irradiated with (a) 0, (b) 50, (c) 100, (d) 150, (e) 200, and (f) 250 Gy gamma doses (white vertical bar = 1 cm).

The root length obtained from seeds irradiated with 150 Gy of gamma radiation was significantly increased by 63.2% from 1.8 cm in the control treatment (0 Gy) to 2.9 cm, which has been confirmed by Melki and Marouani [73].

## 3. Conclusion

Dormancy is a state of lack of germination in seeds/tubers even though the required conditions (temperature, humidity, oxygen, and light) have been provided and is based on hard seed coat impermeability or a lack of supply and activity of the enzymes necessary for germination. Dormancy is an important factor limiting production in many field crops. Several physical and chemical pretreatments are applied to organic material (seeds/tubers) for overcoming dormancy. Physical and internal dormancy can be found together in some plant species and this makes it difficult to provide high-frequency healthy seedling growth, whereas the sprouting of seeds/tubers sown and the formation of healthy seedlings is a prerequisite for plant production. This chapter focuses on four different methods that have not been reported elsewhere for overcoming dormancy. We think that these newly described methods will help growers and researchers to overcome dormancy problem in plant production.

## Author details

Mustafa Yildiz[1]*, Ramazan Beyaz[2], Mehtap Gursoy[3], Murat Aycan[4], Yusuf Koc[4] and Mustafa Kayan[4]

*Address all correspondence to: myildiz@ankara.edu.tr

1 Department of Field Crops, Faculty of Agriculture, Ankara University, Ankara, Turkey

2 Department of Soil Science and Plant Nutrition, Faculty of Agriculture, Ahi Evran University, Kırsehir, Turkey

3 Guzelyurt Vocational School, Aksaray University, Aksaray, Turkey

4 Department of Field Crops, Graduate School of Natural and Applied Sciences, Ankara University, Ankara, Turkey

## References

[1] Bonner FT. Glossary of Seed Germination Terms for Tree Seed Workers. Gen. Tech. Rep. SO-49. New Orleans: USDA Forest Service, Southern Forest Experiment Station; 1984. p. 4

[2] Baskin CC, Baskin JM. Germination ecology of seeds with nondeep physiological dormancy. In: Baskin CC, Baskin JM, editors. Seeds: Ecology, Biography, and Evolution of Dormancy and Germination. San Diego, California: Academic Press; 1998. pp. 57-64

[3] Emery DE. Seed Propagation of Native California Plants. Santa Barbara, CA: Santa Barbara Botanic Garden; 1987

[4] Enu-Kwesi L, Dumbroff EB. Changes in abscisic acid in the embryo and covering structures of *Acer saccharum* during stratification. Zeitschrift für Pflanzenphysiologie. 1978;**86**:371-377

[5] Tillberg E, Pinfield NJ. Changes in abscisic acid levels during, after-ripening and germination of *Acer platanoides* L. seeds. New Phytologist. 1982;**92**:167-172

[6] Webb DP, Wareing PF. Seed dormancy in Acer: Endogenous germination inhibitors and dormancy in *Acer pseudoplatanus* L. Planta. 1972;**104**:115-125

[7] Williams PM, Ross JD, Bradbeer JW. Studies in seed dormancy: 4. The abscisic acid content of the seeds and fruits of *Corylus avellana* L. Planta. 1973;**110**:303-310

[8] Sondheimer E, Tzou DS, Galson EC. Abscisic acid levels and seed dormancy. Plant Physiology. 1968;**43**:1443-1447

[9] Singh Z, Browning G. The role of ABA in the control of apple seed dormancy re-appraised by combined gas chromatography-mass spectrometry. Journal of Experimental Botany. 1991;**42**:269-275

[10] Szczotka Z. Changes in the activity of indole acetic acid and abscisic acid in the embryo axes and cotyledons of *Quercus borealis* Michx. and *Quercus robur* L. acorns stored under controlled conditions. Arboretum Kornickie. 1977;**22**:257-273

[11] Florez M, Carbonell MV, Martinez E. Exposure of maize seeds to stationary magnetic fields: Effects on germination and early growth. Environmental and Experimental Botany. 2007;**59**:68-75

[12] Soltani F, Kashi A, Arghavani M. Effect of magnetic field on *Asparagus officinalis* L. seed germination and seedling growth. Seed Science and Technology. 2006;**34**:349-353

[13] Carbonell MV, Martinez E, Florez M, Maqueda R, Lopez-Pintor A, Amaya JM. Magnetic field treatments improve germination and seedling growth in *Festuca arundinacea* Schreb. and *Lolium perenne* L. Seed Science and Technology. 2008;**36**:31-37

[14] Podlesny J, Misiak L. Podlesna a concentration of free radicals in pea seeds afterpresowing treatment with magnetic field. International Agrophysics. 2004;**18**:261-267

[15] Martinez E, Carbonell MV, Amaya JM. A static magnetic field of 125 mT stimulates the initial growth stages of barley (*Hordenumvulgare L.*). Electro and Magnetobiology. 2000;**19**(3):271-277

[16] Florez M, Carbonell MV, Martinez E. Early sprouting and first stages of growth of rice seeds exposed to a magnetic field. Electro and Magnetobioloy. 2004;**23**(2):167-176

[17] Mano J, Nakahara T, Torii Y, Hirose H, Miyakoshi J, Takimoto K. Seed deterioration due to high humidity at high temperature is suppressed by extremely low frequency magnetic fields. Seed Science and Technology. 2006;**34**:189-192

[18] De Souza A, Garcia D, Sueiro L, Gilart F, Porras E, Licea L. Pre-sowing magnetic treatments of tomato seeds increase the growth and yield of plants. Bioelectromagnetics. 2006;**27**:247-257

[19] Snedecor GW, Cochran WG. Statistical Methods. Iowa, USA: The Iowa State University Press; 1967

[20] ISTA. International Rules for Seed Testing. Basserdorf: International Seed Testing Association; 2003

[21] Curtis OF, Shetty K. Growth medium effects on vitrification, total phenolics, chlorophyll, and water content of *in vitro* propagated oregano clones. Acta Horticulturae. 1996;**426**:498-503

[22] Emerson R. Chlorophyll content and the rate of photosynthesis. Proceedings of the National Academy of Sciences of the United States of America. 1929;**15**(3):281-284

[23] Rensburg LV, Kruger GHJ. Evaluation of components of oxidative stres metabolism for use in selection of drought tolerant cultivars of *Nicotiana tabacum* L. Journal of Plant Physiology. 1994;**143**:730-737

[24] Kyparissis A, Petropoulou Y, Manetas Y. Summer survival of leaves in a soft-leaved shrub (*Phlomis fruticosa* L., Labiatae) under Mediterranean field conditions: Avoidance of photoinhibitory damage through decreased chlorophyll contents. Journal of Experimental Botany. 1995;**46**:1825-1831

[25] Jagtap V, Bhargava S, Sterb P, Feierabend J. Comparative effect of water, heat and light stresses on photosynthetic reactions in *Sorghum bicolor* (L.). Moench. Journal of Experimental Botany. 1988;**49**:1715-1721

[26] Pal RN, Laloraya MM. Effect of calcium levels on chlorophyll synthesis in peanut and linseed plants. Biochemie and Physiologie der Pflanezen. 1972;**163**:443-449

[27] Wright GC, Nageswara RRC, Farquhar GD. Water use efficiency and carbon isotope discrimination in peanut under water deficit conditions. Crop Science. 1994;**34**:92-97

[28] Nageswara RRC, Talwar HS, Wright GC. Rapid assessment of specific leaf area and leaf nitrogen in peanut (*Arachis hypogaea* L.) using chlorophyll meter. Journal of Agronomy and Crop Science. 2001;**189**:175-182

[29] Sezik E. Research on the Turkish medicinal plant *Ecballium elaterium*. Chemistry of Natural Compounds. 1997;**33**:541-542

[30] Özcan SF, Yıldız M, Ahmed HAA, Aasim M. Effects of squirting cucumber (*Ecballium elaterium*) fruit juice on *Agrobacterium tumefaciens*-mediated transformation of plants. Turkish Journal of Biology. 2015;**39**:790-799

[31] Memişoğlu M, Toker G. Biological activities and traditional usage of *Ecballium elaterium* (L.) A. Rich. FABAD Journal of Pharmaceutical Science. 2002;**27**:157-164

[32] Yildiz M, Kayan M, Aycan M. The effect of squirting cucumber (*Ecballium elaterium* (L.) A. Rich.) fruit juice on *in vitro* shoot regeneration in durum wheat (*Triticum durum* Desf.). 18. National Biotechnology Congress, 19-18 December, Konya, Turkey; 2015

[33] Darcin S, Aycan M, Kayan M, Köm D, Taher M, Yildiz M. The effect of squirting cucumber (*Ecballium elaterium* (L.) A. Rich.) fruit juice on seed germination and seedling growth in rapeseed (*Brassica napus* L.). National Botanic Plant Science Congress, 25-28 October, Antalya, Turkey; 2014

[34] Smith CR. Mycobactericidal agents. In: Lawrence CA, Block SS, editors. Disinfection, Sterilization, and Preservation. Philadelphia: Lea and Febiger; 1968. pp. 504-514

[35] Spaulding EH. Chemical disinfection of medical and surgical materials. In: Lawrence CA, Block SS, editors. Disinfection, Sterilization, and Preservation. Philadelphia: Lea and Febiger; 1968. pp. 517-531

[36] Bietz JA, Sandford PA. Reaction of sodium hypochlorite with amines and amides: Automation of the method. Analytical Biochemistry. 1971;**44**:122-133

[37] Kantouch A, Ardel-Fattah SH. Action of sodium hypochlorite on a-amino acids. Chemicke Zvesti. 1971;**25**:222-230

[38] Hayatsu H, Pan S, Ukita T. Reaction of sodium hypochlorite with nucleic acids and their constituents. Chemical & Pharmaceutical Bulletin. 1971;**19**:2189-2192

[39] Sandford PA, Nafziger AJ, Jeanes A. Reaction of sodium hypochlorite with amines and amides: A new method for quantitating amino sugars in manomeric form. Analytical Biochemistry. 1971;**42**:422-436

[40] Sandford PA, Nafziger AJ, Jeanes A. Reactions of sodium hypochlorite with amines and amides: A new method for quantitating polysaccharides containing hexosamines. Analytical Biochemistry. 1971;**44**:111-121

[41] Allan A. Plant cell culture. In: Stafford A, Warren G, editors. Plant Cell and Tissue Culture. Milton Keynes: Open University Press; 1991. pp. 1-24

[42] Yildiz M, Er C. The effect of sodium hypochlorite solutions on *in vitro* seedling growth and shoot regeneration of flax (*Linum usitatissimum*). Naturwissenschaften. 2002;**89**:259-261

[43] Igbinnosa I, Okonkowo SNC. Stimulation of germination of seeds of cowpea witchweed (*Striga gesneriodes*) by sodium hypochlorite and some growth regulators. Weed Science. 1992;**40**:25-28

[44] Bewley JD, Seeds BM. Physiology of Development and Germination. New York: Plenum Press; 1994. p. 445

[45] Kayan N, Yildiz M, Ozgen M. Breaking dormancy and seedling growth in lentil (*Lens Culinaris* Medik.) under in vitro conditions. XIV. National Biotechnology Congress, Eskişehir, Turkey; 2005. pp. 348-351

[46] Ogawa K, Iwabuchi M. A mechanism for promoting the germination of *Zinnia elegans* by hydrogen peroxide. Plant & Cell Physiology. 2001;**42**:286-291

[47] Böhm J. Terrestrial orchid seed handling: Growing orchids of temperate climates from seed. In: Seed Pretreatment Protocol. Homnurg-Saar, Germany: Institute of Physiology, Medical Faculty, University of Saarland; 2003

[48] Kanecko Y, Morohashi Y. The effect of sodium hypochlorite treatment on the development of $\alpha$-amylase activity in mung bean cotyledons. Plant Science. 2003;**164**:287-292

[49] Telci C, Yıldız M, Pelit S, Önol B, Erkılıç EG, Kendir H. The effect of surface-disinfection process on dormancy-breaking, seed germination, and seedling growth of *Lathyrus chrysanthus* Boiss. under *in vitro* conditions. Propagation of Ornamental Plants. 2011;**11**:10-16

[50] Murashige T, Skoog F. A revised medium for rapid growth and bioassays with tobacco tissue cultures. Physiologia Plantarum. 1962;**15**:431-497

[51] Racoppi F. Domestic bleaches containing sodium hypochlorite. Procter and Gamble. Rome: Italia SPA Product Development Department; 1990

[52] Schull W. Temperature and rate of moisture intake in seeds. Botanical Gazette. 1920;**69**:361

[53] Hsiao AI, Hans JA. Application of sodium hypochlorite seed viability test to wild oat populations with different dormancy characteristics. Canadian Journal of Plant Science. 1981;**61**:115-122

[54] Hsiao AI, Quick AW. Action of sodium hypochlorite and hydrogene peroxide on seed dormancy and germination of wild oats, *Avena fatua* L. Weed Research. 1984;**24**:411-419

[55] Dale JE. The control of leaf expansion. Annual Review of Plant Physiology and Plant Molecular Biology. 1988;**39**:267-295

[56] Sunderland N. Cell division and expansion in the growth of the leaf. Journal of Experimental Botany. 1960;**11**:68-80

[57] Yildiz M, Özgen M. The effect of a submersion pretreatment on in vitro explant growth and shoot regeneration from hypocotyls of flax (*Linum usitatissimum*). Plant Cell, Tissue and Organ Culture. 2004;**77**:111-115

[58] Gunckel JE, Sparrow AH. In: Ruhland W, editor. Encycl. Plant Physiology, in Ionizing Radiation: Biochemical, Physiological and Morphological Aspects of their Effects on Plants. Berlin: Springer-Verlag; 1961. p. 555-611

[59] Kim JH, Chung BY, Kim JS, Wi SG. Effects of in Planta gamma-irradiation on growth, photosynthesis, and antioxidative capacity of red pepper (*Capsicum annuum* L.) plants. Journal of Plant Biology. 2005;**48**:47-56

[60] Wi SG, Chung BY, Kim JH, Baek MH, Yang DH, Lee JW, Kim JS. Ultrastructural changes of cell organelles in Arabidopsis stem after gamma irradiation. Journal of Plant Biology. 2005;**48**:195-200

[61] Radhadevi DS, Nayar NK. Gamma rays induced fruit character variations in Nendran, a varieties of banana (*Musa paradasiaca* L.). Geobios. 1996;**23**:88-93

[62] Kumari R, Singh Y. Effect of gamma rays and EMS on seed germination and plant survival of *Pisum sativum* L. and *Lens culinaris* medic. Neo Botanica. 1996;**4**:25-29

[63] Chaomei Z, Yanlin M. Irradiation induced changes in enzymes of wheat during seed germination and seedling growth. Acta Agriculturae Nucleatae Sinica. 1993;**7**:93-97

[64] Maherchandani N. Effects of gamma radiation the dormant seeds of *Avena fatua*. Radiation Botany. 1975;**15**:439-443

[65] Kuzin AM, Vagabova ME, Vilenchik MM, Gogvadze VG. Stimulation of plant growth by exposure to low-level gamma radiation and magnetic field, and their possible mechanism of action. Environmental and Experimental Botany. 1986;**26**:163-167

[66] Charbaji T, Nabulsi I. Effect of low doses of gamma irradiation on in vitro growth of grapevine. Plant Cell, Tissue and Organ Culture. 1999;**57**:129-132

[67] Baek MH, Kim JH, Chung BY, Kim JS, Lee IS. Alleviation of salt stress by low dose g-irradiation in rice. Biologia Plantarum. 2005;**49**(2):273-276

[68] Chakravarty B, Sen S. Enhancement of regeneration potential and variability by g-irradiation in cultured cells of *Scilla indica*. Biologia Plantarum. 2001;**44**:189-193

[69] Kim JS, Lee YK, Park HS, Back MH, Kim DH. Influence of low dose gamma radiation on the growth of maize (*Zea mays* L.) varieties. Korean Journal of Environmental Agriculture. 2000;**19**:328-331

[70] Sheppard SC, Evenden WG. Factors controlling the response of field crops to very low doses of gamma irradiation of the seed. Canadian Journal of Plant Science. 1986;**66**:431-441

[71] Szarek S. Use of concept of hormesis phenomenon to explain the law of disminishing returns part II. Electronic Journal of Polish Agricultural Universities. 2005;**8**(4):61

[72] Beyaz R, Telci Kahramanogullari C, Yildiz C, Darcin ES, Yildiz M. The effect of gamma radiation on seed germination and seedling growth of *Lathyrus chrysanthus* Boiss. under *in vitro* conditions. Journal of Environmental Radioactivity. 2016;**162-163**:1-5

[73] Melki M, Marouani A. Effects of gamma rays irradiation on seed germination and growth of hard wheat. Environmental Chemistry Letters. 2010;**8**:307-310

Chapter 6

# Characterization of Genotype by Planting Date Effects on Runner-Type Peanut Seed Germination and Vigor Response to Temperature

Timothy L. Grey, Charles Y. Chen, Russell Nuti,
Walter Scott Monfort and George Cutts III

Additional information is available at the end of the chapter

http://dx.doi.org/10.5772/intechopen.70584

**Abstract**

Experiments evaluated the genotype by environment effects on seed germination and vigor of the peanut runner-type cultivars 'Georgia Green', 'AT3085R0', 'AT271516', 'Georgia 03L', and 'FR458' grown under similar production practices, for three planting dates: April, May, and June in Georgia and Alabama. Objectives were to determine if time of planting and harvest dates would subsequently affect germination and vigor when tested using a thermal gradient devise (temperature range14 to 35 °C). Runner-type peanut seed grown in Dawson Georgia in 2008 had the strongest seed vigor with $Germ_{80}$ of 22 to 40 growing degree days (GDD), and maximum incidence of germination rate 84.8-95.7% when planted April, May, and June 2008 across 15 seed lots. In contrast, seed harvested from plantings of May 2009 at Dawson Georgia exhibited $Germ_{80}$ of 24 to 40 GDD with maximum incidence of germination rate 79.8-93.6%, but seed from April 2009 plantings had poor vigor of 56.8-72.8% and no amount of GDD could achieve $Germ_{80}$, with similar results for June 2009 plantings for this location. For Headland April, May, and June 2009 plantings of the same cultivars, all seed had poor vigor, ≤75.6% maximum incidence for germination rate, and none obtained a measurable $Germ_{80}$.

**Keywords:** *Arachis hypogaea*, germination, genotype, phenological development, thermal time

## 1. Introduction

Runner-type peanut, *Arachis hypogaea* (L.) producers often grow different cultivars in order to take advantage of genetic diversity of this Fabaceae crop. Production often centers on peanut runner-type cultivars (**Figure 1**), which have spreading indeterminate plant

**Figure 1.** Runner-type peanut -foreground, cotton (*Gossypium hirsutum* L.)middle, and maize (*Zea mays* L.) background.

morphology, can grow to 65 cm in height, and spread to over 1 m in width [1]. While genetic diversity is essential for the production of many crops for pest management, the cultivar 'Florunner' [2] dominated runner-type peanut production for more than 20 years in the Southeastern United States region with planting occurring in April and May [3, 4, 5, 6]. But what usually happens when there is over reliance on a single cultivar for production, a Tospovirus described as Tomato Spotted Wilt Virus (TSWV) increased rapidly across this runner-type peanut production region in the 1990s [7], eventually leading to the replacement of Florunner with tolerant cultivars [7, 8]. Since then, the utilization of newly released cultivars has been a constant factor in runner-type peanut production as many new genotypes have improved disease resistance, yield, quality, and economic value [9–14]. One recommendation for planting to avoid TSWV was to plant peanut after 15 May of each year. Peanut cultivars with tolerance to TSWV exhibited a linear decline in the disease incidence from greater than 50% for April plantings to less than 10% for June plantings [15]. This recommendation was in place for over 20 years and was practiced commonly from 2000 to 2010 (**Figure 2A**). Growers would delay planting until the 2nd–4th week of May with most planting completed by early June. However this created issues as this delayed planting pushed harvest windows into Oct and Nov, resulting in reduced yield and quality [8, 15, 16].

## 2. Importance

Seed quality issues occurred for some TSWV-resistant runner-type peanut cultivars in the 2000s [17]. Specifically, Georgia-01R [18] and York [19] cultivars had germination and stand establishment failures when planted for production. When tested and evaluated in field settings, advanced breeding lines of these cultivars did not have stand establishment and germination issues, but when planted in producer fields, some cultivars did not perform as expected with respect to stand establishment. This led to some TSWV resistant cultivars not to be accepted by growers. While peanut seeds were certified via individual state's standard

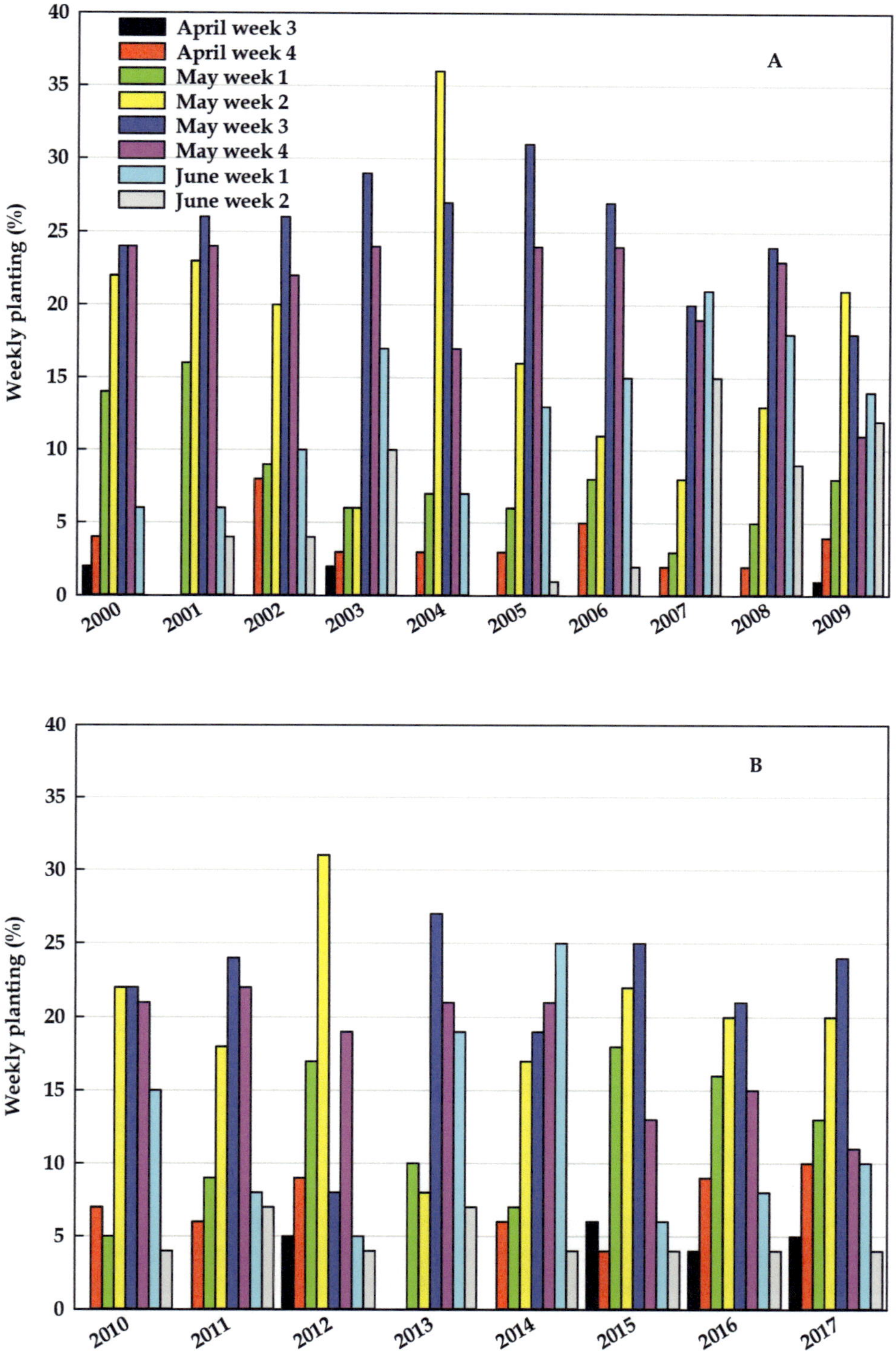


**Figure 2.** Weekly planting of runner-type peanut from 2000 to 2009 (A) and 2010 to 2017. (B) in Georgia, United States (National Agricultural Statistics Service, USDA. Data available at https://www.Nass.Usda.Gov).

seed germination testing, this did not always guarantee adequate stand establishment in the field. Seeds can be a substantial cost of growing peanuts due to its large size, often requiring greater than 170 kg $ha^{-1}$. When there is poor stand establishment, replanting can be expensive due to additional seed expenditures and trips through the field, and can reduce yield due to planting outside optimum time for peanut production. This can result in difficulty in determining optimum maturity, can promote weed escapes due to soil disturbance, incur greater disease opportunities, and can potentially incur additional pesticide costs. Therefore, planting cultivars with high germination and vigor to maximize net returns on input costs is essential.

## 3. Background information on peanut seed

It has been noted that runner-type peanut has an indeterminate growth habit which hampers the identification of an optimum harvest date [20, 21]. Due to this indeterminate growth, just prior to harvest there will be a range of pods with different maturities present on the same plant. When harvested all pods that have reached a given size and weight, regardless of maturity, are collected. This can result in high levels of immature pods in the harvested product. After shelling, this mix of seed maturities has critical implications for overall seed quality. Seed germination is often the trait that is commonly utilized to establish seed quality; but this presents significant limitations because there is often a difference between germination and overall seed vigor [20]. The maturity pattern of runner-type peanuts will vary by cultivar and one from one year to another. For virginia-type peanut, it has been demonstrated that seed maturity impacted not only germination capacity, but overall seed vigor [21]. However information about the effect on maturity for runner-type peanut germination and vigor has not reported.

### 3.1. Runner-type cultivars

One distinct aspect of runner-type peanut is that almost all cultivars are releases from public institutions from the Southeast region including the University of Georgia, University of Florida, Auburn University, and United States Department of Agriculture Research Service (USDA-ARS) [14]. These institutions have maintained constant releases over the past 25 years, and as previously noted, some cultivars have not been successful due to poor stand establishment [17], leading to their rapid demise from production even though they had desirable traits for disease resistance and improved quality [22]. In contrast, some cultivars have been rapidly adapted and garnered greater than 80% of field planting in some years. For example, 'Georgia Green' [23] was released in 1996 and from then to the mid-2000s was widely adapted and in some growing seasons, planted to greater than 90% of production hectares in the Southeast [24]. 'Georgia-06G' [25] replaced Georgia Green with plantings of greater than 75% (estimated) of the Southeast hectares. One advantage to using the aforementioned cultivars is their demonstrated resistance to TSWV. This has prompted the recommendation that growers begin planting peanut again in late April and early May [26].

This change has occurred as over the past few years as noted by a shift to late April-early May peanut planting (**Figure 2B**). For example, comparing the May week 1 in 2007–2012 and 2015 there is a clear indication that producers made that shift to planting earlier after Georgia-06G and other more TSWV cultivars were released. In contrast, for the May week 4 planting in the early 2000's, over 20% of the crop was planted the last week of May [27]. As prior noted, planting in late May and early June often reduces yield and quality of runner-type peanut due to maturity issues [8, 15, 16]. By 2015 and 2016, 15% or less was planted in late May and less than 10% in June. These changes and rapid acceptance of new cultivars has led to a need to ensure high quality seed and prevent any future stand establishment issues.

### 3.2. Seed quality

In reviews of factors that contribute to seed deterioration over time during storage, it has been noted that seed moisture content, mechanical and insect damage, pathogen attack, seed maturity, relative humidity, and temperature can have negative impacts [17]. These impacts have been quantified by previous research [14, 28]. Vigor testing can be utilized to evaluate seed for successful field establishment under different environmental conditions [29]. Strong primary seedling development in standard germination testing is regarded as an indicator of strong vigor [30], but this does not always translate into adequate field performance.

### 3.3. Seed testing

One method of testing seed quality, germination, and vigor is the use of a thermal gradient device [31–34]. This method has been used for weed and other crop seed evaluations [35, 36] to determine germination speed and vigor. A thermogradient allows investigators to examine a single seed lot, or multiple seed lots, at different temperatures simultaneously without the use of growth media, such as soil or growth chambers. Previous research using this process has demonstrated grower stock-dependent differences in seed vigor [22]. Quantifying genotypic by phenotypic differences in vigor have also been evaluated where site-specific differences in environmental factors during seed development kept all management variables equal (irrigation, pesticides, fertility) in order to minimize environmental variation [37]. Data indicated that eight runner-type cultivars exhibited phenotypic vigor variation by year, over the course of six years during the experiment, with genotype consistency across years.

## 4. Research

In order to quantify if phenotypic differences in peanut cultivar seed production could occur over different planting dates in the same year, multiple cultivars were planted over a 40-day planting window. After harvest and processing, evaluations of runner-type peanut seed germination and vigor were conducted from one location in 2008 (Dawson Georgia) and two locations in 2009 (Dawson Georgia and Headland Alabama). Initial data from this research was used to quantify TSWV effects on pod yield and quality [15], but seeds were also saved. Thus, the objectives of this research were to evaluate the same seed for different peanut cultivars grown using the same management practices each year to determine if there were differences in seed vigor when planted at different times in the same field. Multiple cultivars were evaluated for seed viability and vigor for two consecutive growing seasons using a thermal gradient device (**Figure 3**).

### 4.1. Field trials

Field experiments were conducted near Dawson Georgia in 2008 and 2009 and repeated near Headland Alabama in 2009. Soils were a Tifton, Fine-loamy, kaolinitic, thermic Plinthic Kandiudults at Dawson and a Dothan, Fine-loamy, kaolinitic, thermic plinthic kandiudults at Headland. Five peanut cultivars were planted and included Georgia Green, as the agronomic standard because of its tolerance to TSW. Other cultivars included Georgia-03 L [38], AT 3085RO and AT271516 [39] were considered to have greater field tolerance to TSW compared to Georgia Green. Cultivar Flavor Runner 458 was also included as a susceptible check [40]. For runner-type peanut seed production for eventual evaluation, seeds were planted at three different dates each year starting in April. The earliest planting date in each year and location was determined when the 10-cm soil temperature reached 18.3°C for 3 consecutive days after 10 April. Once the initial planting date was determined, planting was repeated in 20-day intervals two times for three in total (**Table 1**).

All plantings were on single rows on 0.9 m centers, with planting date at each location arranged as a randomized complete block design in a separate block [41]. All treatments were replicated four times in each planting date block. Agronomic management inputs and irrigation were

**Figure 3.** Thermal gradient table (left) and peanut seed after evaluation for germination (right).

| Dawson Georgia | | | | | | | | Headland Alabama | | | |
|---|---|---|---|---|---|---|---|---|---|---|---|
| 2008 | | | | 2009 | | | | 2009 | | | |
| Month | R[b] | SR | GDD | Month | R | SR | GDD | Month | R | SR | GDD |
| | mm | MJ $m^{-2}$ | | | mm | MJ $m^{-2}$ | | | mm | MJ $m^{-2}$ | |
| April | 20 | 227 | 44 | April | 0 | 239 | 62 | April | 0 | 205 | 8 |
| May | 30 | 686 | 250 | May | 188 | 526 | 235 | May | 233 | 455 | 281 |
| June | 114 | 685 | 360 | June | 39 | 688 | 377 | June | 53 | 620 | 416 |
| July | 228 | 662 | 373 | July | 100 | 639 | 352 | July | 223 | 540 | 391 |
| Aug | 409 | 533 | 346 | Aug | 352 | 568 | 261 | Aug | 190 | 523 | 382 |
| Sept | 59.9 | 541 | 279 | Sept | 44 | 490 | 292 | Sept | 89 | 397 | 334 |
| Oct | 0.0 | 51 | 15 | Oct | 41 | 98 | 58 | — | | | |
| Total by planting date | | | | | | | | | | | |
| 21 Apr | 802 | 2824 | 1398 | 21 Apr | 681 | 2724 | 1322 | 20 Apr | 736 | 2394 | 1518 |
| 12 May | 841 | 2907 | 1533 | 11 May | 708 | 2811 | 1464 | 11 May | 702 | 2437 | 1619 |
| 2 June | 799 | 2452 | 1361 | 1 June | 577 | 2484 | 1340 | 1 June | 555 | 2384 | 1523 |

[a]Rainfall, solar radiation, and GDD are reported from initial planting to digging. For Dawson, in 2008 digging dates were 2 Sept, 10 Sept, and 3 Oct, respectively; 2009 digging dates were 4 Sept, 23 Sept, and 8 Oct, respectively. For Headland, in 2009 digging dates were 4 Sept, 22 Sept, and 1 Oct.

[b]R, rainfall; SR, solar radiation; GDD, growing degree days.

**Table 1.** Monthly and total rainfall, solar radiation, and total growing degree days for various peanut planting dates up to time of digging from University of Georgia and Auburn University weather stations 1 km from experiments[a].

applied according to University of Georgia [42] and Auburn University recommendations. All cultivars were considered to have similar maturity requirements, thus digging date was determined separately by location and planting date according to the hull scrape method [43] of Georgia Green in each respective planting date block. Border rows were planted to Georgia Green to use for maturity determination. Plots were 6.1 m by two rows in 2008 and 6.1 m by four rows in 2009. Peanut vines were threshed with a stationary harvester (Kingaroy Engineering Works, Kingaroy, Australia) after sufficient field drying, since all pods and seeds can be cleaned-out between plots to prevent mixtures and maintain cultivar purity with this machine. A uniform sample was obtained with divider, then this was graded and these seed were saved for germination and vigor testing.

### 4.2. Seed screening

After threshing, peanut pods were dried with forced 30–40°C warm air to 7% moisture. All samples of pods were then hand-cleaned over a screen table. For each cultivar and planting date sample, grades were determined according to Federal State Inspection Service procedures [44] for runner-type peanut. Pods were mechanically shelled on a small scale unit and seeds were sized according to diameter via round holed screens with selection based on what passed through a size 8.3 mm (Screen No. 21) but retained over 7.1 mm (Screen 18). Further screening was then conducted and sound mature seeds retained over a 7.1 by 19.1 mm slotted screen from each plot (four plots for each cultivar and planting date combination each year) were then evaluated for seed size based on Georgia Federal-State Inspection Service regulations [44]. Seeds were then stored at 16–18°C at approximately 30% humidity for up to 7 months prior to testing. Seed response to temperature and time for all seed to germinate were then evaluated on a thermal gradient table [31, 37].

### 4.3. Thermal gradient testing

The thermal gradient table was constructed from solid aluminum block measuring 2.4 m long by 0.9 m wide by 7.6 cm thick with a mass of 470 kg (**Figure 3**). On each end of the aluminum block, a 1.0 cm hole was drilled across the side section to allow fluid to be pumped into the table. On each end of the table, ethylene glycol plus water (1:10 mixture) at 14 or 35°C were pumped at 3.8 L per min to generate the thermogradient. Approximately 1.0°C increments occurred every 10 cm along the length of the thermogradient with a constant temperature across the width. This produced 24 increments across the length to obtain different temperatures, with nine increments across the width at each temperature. Thermocouples made from duplex insulated wire (PR-T-24 wire, Omega Engineering, Inc. Stamford, CT) were mounted to the underside of the table from the hot to cold ends. These were inserted vertically into a hole on the bottom of the table. Holes measuring 8 mm wide by 7 cm deep were drilled to allow the thermocouple to be placed within 5 mm of the upper table surface, at 10 cm intervals along the length of the table. This created a continuous temperature gradient ranging from 14 to 35°C along the length of the table. Temperatures were monitored continuously for each thermocouple and recorded at 30 minute intervals with a Graphtec midi data logger (MicroDAQ,com Ltd., Contoocook, NH).Temperature data for each thermocouple was recorded daily.

### 4.4. Seed testing

Peanut seeds for the appropriate plot of each cultivar by planting date were evenly distributed on germination paper (SDB 86 mm, Anchor Paper Co., St. Paul, MN), which was placed in a 100 by 15 mm sterile plastic Petri dish (Fisher Scientific Education, Hanover Park, IL). Twenty seeds were placed in each Petri dish followed by 10 ml of distilled water. A single Petri dish was then placed at each 1.0°C increment every 10 cm along the length of the table for a total of 24 dishes per plot (**Figure 3**). Beginning within 68–72 hours after seeding, peanut seed germination was counted when the radicle extended more than 5 mm beyond the seed, and then the seed was removed from the dish. Peanut seed with radicles longer than 2 mm from the seed coat are considered germinated [45] but 5 mm was chosen as it has been used in previous research [46]. Distilled water in 5 ml increments was added as needed to maintain adequate moisture in each Petri dish, and varied by temperature increment. Tests were run for 7 days with counts taken daily. All counts were taken in less than one hour each day at approximately the same time, depending upon when an experiment was started on day zero. Counts were conducted from the cold end working toward the warm end. Seeds availability were limited each year, so individual field plots were considered replications with 24 Petri dishes for each replication (n = 480 seed per field replication, n = 1920 seed per cultivar by planting date each year). Germination data was converted to a percentage by day, and cumulative germination was determined for each Petri dish over the duration of that assay. Temperature data was recorded by the data loggers for each experiment. Data included temperature maximum and minimum (±0.5°C for each thermocouple) by individual Petri dish. Maximum and minimum temperatures were the highest and lowest measures, respectively taken during one germination experiment for a specific Petri dish.

### 4.5. Data analysis

Maximum and minimum temperatures were then used to determine the thermal time [30, 31] or growing degree day (GDD) accumulation for the following equation.

$$t_n = \sum_{i=1}^{n}\left[\frac{Ti_{max} + Ti_{min}}{2} - T_b\right] \quad (1)$$

where $t_n$ is the sum of GDD for $n$ days, $Ti_{max}$ and $Ti_{min}$ are the daily maximum and minimum temperature (°C) of Day $i$ [47], and $T_b$ is the base temperature for peanut, in this model $T_b$ was set at 15°C [48].

For all measurements, analysis of variance (ANOVA) was applied to the data combined across cultivar, planting date, experiment replication in time, and year to test for the differences among group means of variables and interactions. Years were regarded as random factors while cultivars (seed lot cultivar within a year) and seed germination thermal times were considered fixed effects. Interactions between cultivar and these factors were used as error terms.

Nonlinear regression using the logistics growth curve with three parameters was used to model data [49]. The equation

$$Y = \frac{a}{1 + [((a - b1)/b1)^{*} e^{(-b2x)}]} \quad (2)$$

with the parameters $a$ being the height of the horizontal asymptote at a very large $X$, $b1$ the expected value of $Y$ at time $X = 0$, $b2$ is the measure of growth rate, and $Y$ is the predicted seed germination. One indicator of seed vigor is the number of GDD required to reach the 80% germination rate ($Germ_{80}$). $Germ_{80}$ was then determined by solving the logistic growth curve equation using the parameter estimates for each seed lot cultivar setting $Y$ = 80%. Data for cultivar by planting date equations were subjected to ANOVA using the general linear models procedures with mean separation using 95% asymptotic confidence intervals. The 95% confidence limits of three parameters in the equations were used to compare the significant differences for Eq. (2). Nonlinear regressions were graphed using SigmaPlot 13.0 (SigmaPlot 13.0. SPSS Inc. 233 S. Wacker Dr., Chicago, Illinois).

## 5. Research

There were differences for environmental measures taken during the course of each experiment. All experiments were conducted at times when runner-type peanut seed production could normally occur and are thus representative of producer practices. Cumulative rainfall ranged from 555 to 841 mm between the times of the first and last planting (**Table 1**), which are representative for the region. Irrigation was applied as recommended when required (data not shown). Maximum solar radiation (MJ $m^{-2}$) and total GDD occurred each year with the May, as opposed to April or June, planting dates for all three site-year locations. Significant cultivar-by-year interactions prevented the data from being combined by cultivar across tests. Therefore, data for the Dawson 2008 and 2009, and Headland 2009 seed experiments were analyzed separately and presented by seed location and planting date for each cultivar (**Table 1**).

There are three primary requirements for seed germination: heat, water, and oxygen. Temperature was the only variable evaluated for the runner-type peanut seed in this research. It is an important factor influencing germination in the field [48, 50]. Germination patterns by day against temperature under thermal times were consistent from year to year (Data not shown). Patterns were nonlinear in progression with germination beginning slowly at low temperatures, followed by a rapid growth phase from 20 to 32°C and then remained constant. The optimum temperatures for experiments were >25°C for all intervals greater than 48 hours. Variation in radicle development occurred with respect to temperature, and therefore variation in vigor detected (**Figure 4**).

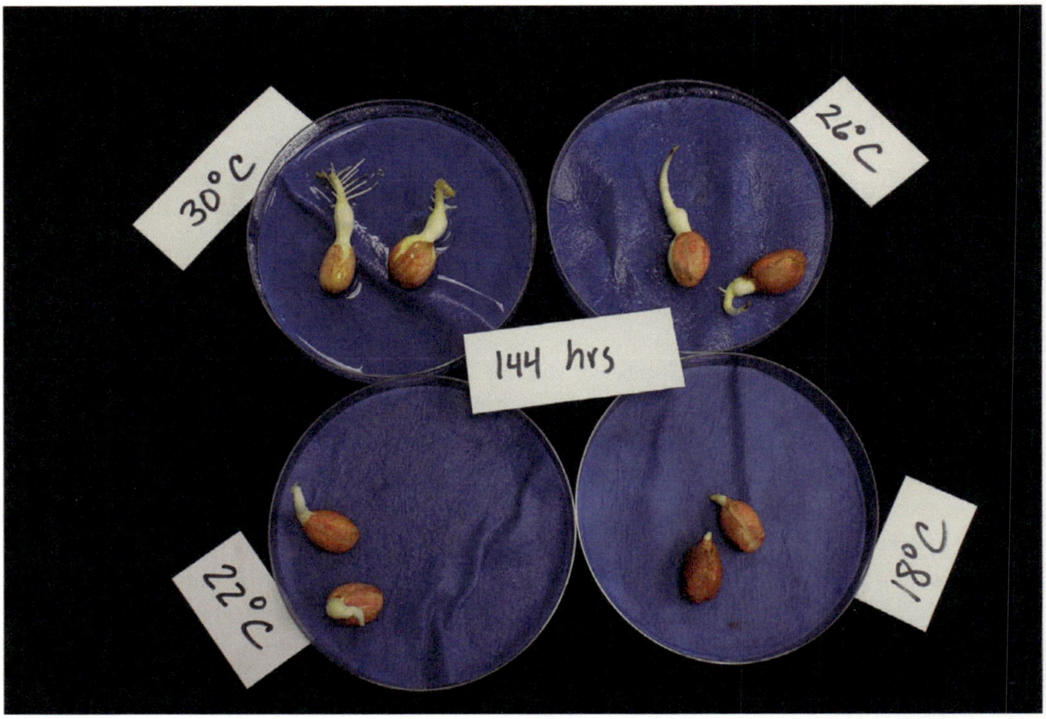


**Figure 4.** Runner-type peanut seed radicle length 144 hours after initiation for temperatures of 18, 22, 26 and 30°C. Photograph curiosity of Sidney Cromer.

Germination varied by year and planting date for each experiment. For Dawson 2008 runner-type peanut seed germination was 79–93% across all planting dates and cultivars when averaged over all temperatures (**Table 2**). While these differences are noted, they do not relate to seed vigor. For non-linear regression, the seed produced from all cultivars had a greater germination rate (parameter *a*) of 89–93% for the 12 May 2008 planting date, as opposed to the 21 April or 1 June plantings at 79–86%. Using 95% confidence intervals, the three parameters in the logistics growth curves were compared within cultivars over planting dates [51]. Maximum germination rate (parameter *a*) were different dependent on the cultivar and time of planting (**Table 2**). Overall runner-type peanut seed produced from 21 April, 12 May, and 2 June 2008 plantings maintained a high level of vigor when exposed to gradient temperatures ranging from 14 to 35°C. Overlap existed in parameters *b1* and *b2* in most cultivars, indicating that the initial germination rate and growth speed were similar (**Figure 5**), although some significant differences did occur (**Table 2**). Runner-type

| | | | Parameter *a*[c] | | | Parameter *b1*[c] | | | Parameter *b2*[c] | | | |
|---|---|---|---|---|---|---|---|---|---|---|---|---|
| **Cultivar** | **Planting date**[a] | **Germination**[a] | **Maximum rate** | **95% CL** | | **Estimate** | **95% CL** | | **Estimate** | **95% CL** | | $\mathbf{Germ_{80}}$ |
| | | % | | | | | | | | | | **GDD** |
| GA Green | 21 April | 85 | 90.5 | ±1.2 | b[d] | 0.74 | ±0.21 | a[d] | 0.20 | ±0.01 | a[d] | 34 |
| | 12 May | 93 | 95.7 | ±1.2 | a | 0.93 | ±0.31 | a | 0.28 | ±0.02 | b | 22 |
| | 2 June | 86 | 88.0 | ±0.9 | b | 0.20 | ±0.07 | b | 0.32 | ±0.02 | b | 26 |
| AT3085R0 | 21 April | 80 | 88.0 | ±1.2 | b | 0.80 | ±0.23 | a | 0.20 | ±0.01 | a | 35 |
| | 12 May | 91 | 94.0 | ±1.2 | a | 1.2 | ±0.35 | a | 0.25 | ±0.02 | b | 24 |
| | 2 June | 86 | 89.5 | ±1.0 | b | 0.40 | ±0.11 | b | 0.30 | ±0.02 | c | 25 |
| AT271516 | 21 April | 83 | 86.1 | ±1.0 | b | 0.50 | ±0.14 | b | 0.22 | ±0.01 | a | 35 |
| | 12 May | 92 | 93.7 | ±1.2 | a | 1.3 | ±0.40 | a | 0.24 | ±0.02 | a | 25 |
| | 2 June | 90 | 92.3 | ±0.8 | a | 0.17 | ±0.06 | c | 0.34 | ±0.02 | b | 24 |
| GA-03 L | 21 April | 84 | 90.1 | ±1.2 | a | 0.74 | ±0.22 | b | 0.20 | ±0.01 | a | 34 |
| | 12 May | 90 | 93.2 | ±1.4 | a | 2.0 | ±0.53 | a | 0.21 | ±0.02 | a | 27 |
| | 2 June | 84 | 86.3 | ±1.0 | b | 0.34 | ±0.11 | c | 0.29 | ±0.02 | b | 28 |
| FR458 | 21 April | 79 | 85.5 | ±1.3 | b | 0.90 | ±0.25 | a | 0.18 | ±0.01 | a | 40 |
| | 12 May | 89 | 94.0 | ±1.1 | a | 1.1 | ±0.30 | a | 0.22 | ±0.01 | b | 28 |
| | 2 June | 81 | 84.8 | ±1.3 | b | 0.38 | ±0.15 | b | 0.30 | ±0.02 | c | 27 |

[a]Year seed were grown, tested the following year after processing; n = 1920 seed.

[b]CL, confidence limit; $Germ_{80}$, cumulative growing degree day value at 80% germination; NA, not applicable as the seed lot of that cultivar did not achieve 80% germination over the duration of the assay; GDD, growing degree day.

[c]Parameter estimates calculated by nonlinear regression equation (2) for seed germination with respect to time based on GDD accumulation: *a* is the height of the horizontal asymptote at a very large X, *b1* is expected value of Y (cumulative germination) at time X = 0, and *b2* is a measure of growth rate.

[d]Values for each parameter within a column for each cultivar followed by the same letter are not significantly different at the 5% probability level. General linear models procedures were used with mean separation using 95% asymptotic confidence intervals. To obtain the equation for the respective regression line in **Figure 1**, the parameters from this table are used.

**Table 2.** Standard germination[a], logistic growth parameter estimates, 95% confidence limits (CL[b]), and vigor indices ($Germ_{80}$[b]) for germination of seed lots of runner-type peanut planted over 40-day period in 2008[a] at Dawson Georgia using a thermogradient germination assay.

peanut seed produced for all planting dates and cultivars in 2008 Dawson $Germ_{80}$ were 22–40 GDD. This was similar to other research comparing runner-type breeder seed over a 6-year period with $Germ_{80}$s of 24–42 GDD [37].

Data from 2009 varied by location, planting date, and cultivar. Germination was 59–75% for runner-type seed produced at Dawson (**Table 3**). Runner-type peanut seed produced from the Dawson 11 May 2009 plantings had the most consistent maximum germination rates

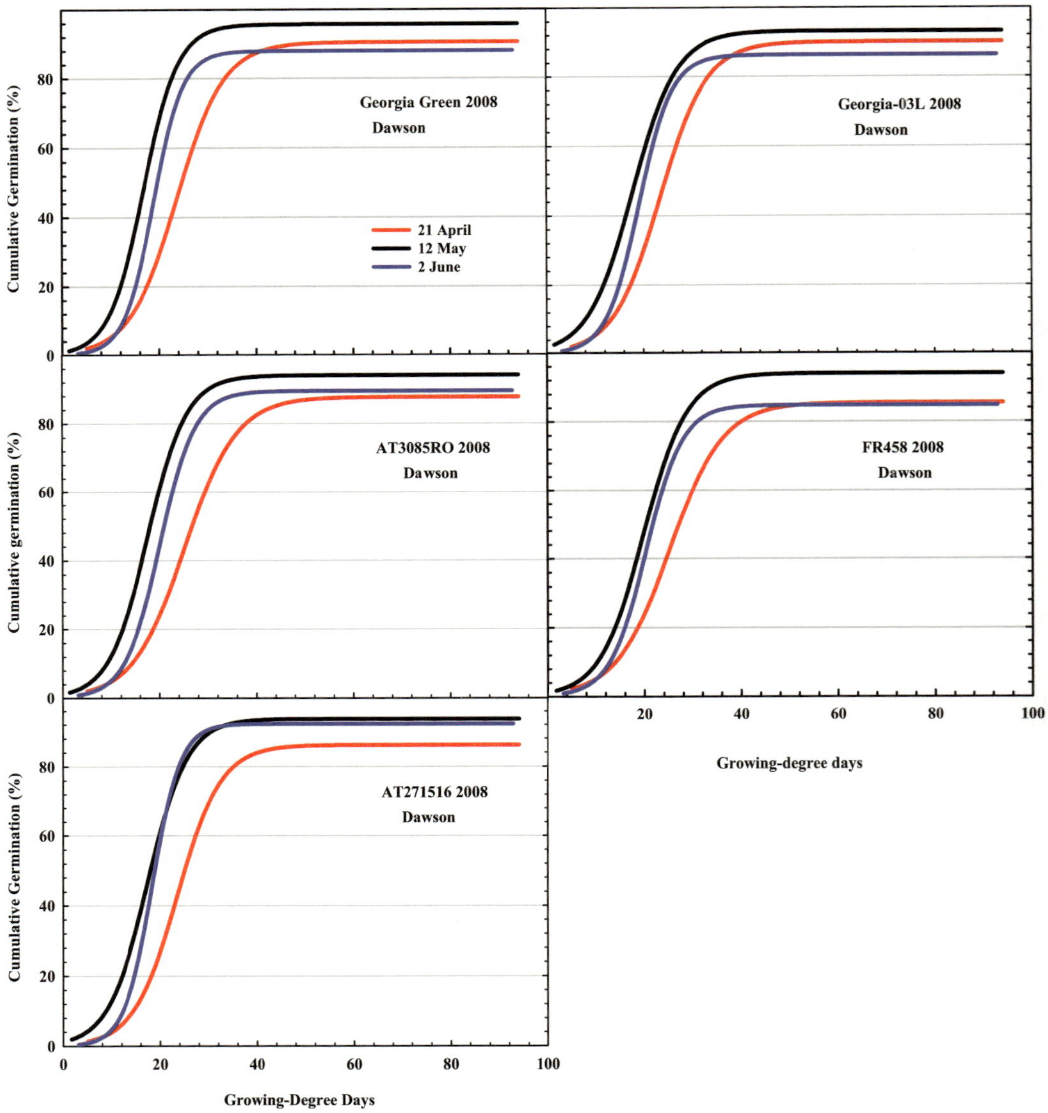


**Figure 5.** Cumulative germination patterns for runner-type peanut seed produced in 2008 at Dawson Georgia, based on nonlinear regression using growing-degree day (GDD) accumulation with a base temperature 15°C. To calculate the regression equation for the respective seed lot, the parameter estimates shown in **Table 2** for the Eq. (2) were used. Germination was measured on a thermal gradient.

| | | | Parameter $a$[c] | | | Parameter $b1$[c] | | | Parameter $b2$[c] | | | |
|---|---|---|---|---|---|---|---|---|---|---|---|---|
| Cultivar | Planting date[a] | Germination[a] | Maximum rate | 95% CL | | Estimate | 95% CL | | Estimate | 95% CL | | $Germ_{80}$ |
| | | % | | | | | | | | | | GDD |
| GA Green | 21 April | 70 | 67.3 | ±2.7 | b[d] | 6.9 | ±3.1 | a[d] | 0.21 | ±0.05 | a[d] | NA |
| | 11 May | 75 | 83.0 | ±2.3 | a | 1.9 | ±1.0 | b | 0.28 | ±0.04 | a | 25 |
| | 1 June | 73 | 69.9 | ±2.7 | b | 7.5 | ±3.2 | a | 0.22 | ±0.05 | a | NA |
| AT3085R0 | 21 April | 59 | 54.3 | ±3.0 | b | 10.3 | ±4.1 | a | 0.16 | ±0.05 | b | NA |
| | 11 May | 68 | 93.6 | ±1.1 | a | 1.2 | ±0.3 | b | 0.26 | ±0.01 | a | 24 |
| | 1 June | 60 | 58.2 | ±3.0 | b | 7.9 | ±3.5 | a | 0.17 | ±0.06 | b | NA |
| AT271516 | 21 April | 62 | 59.1 | ±3.1 | b | 10.6 | ±4.1 | a | 0.2 | ±0.04 | a | NA |
| | 11 May | 66 | 81.6 | ±2.5 | a | 4.0 | ±1.5 | b | 0.2 | ±0.02 | a | 33 |
| | 1 June | 69 | 77.7 | ±2.4 | a | 3.0 | ±1.3 | b | 0.21 | ±0.03 | a | NA |
| GA-03 L | 21 April | 62 | 72.8 | ±2.6 | c | 3.2 | ±1.3 | a | 0.18 | ±0.03 | a | NA |
| | 11 May | 72 | 79.8 | ±2.4 | b | 3.6 | ±1.4 | a | 0.22 | ±0.03 | a | 40 |
| | 1 June | 67 | 97.1 | ±1.5 | a | 3.4 | ±0.47 | a | 0.14 | ±0.01 | b | 35 |
| FR458 | 21 April | 61 | 56.8 | ±3.1 | b | 9.5 | ±3.9 | a | 0.16 | ±0.04 | a | NA |
| | 11 May | 68 | 93.1 | ±1.3 | a | 1.1 | ±0.27 | b | 0.25 | ±0.06 | a | 25 |
| | 1 June | 62 | 57.3 | ±2.6 | b | 7.6 | ±3.4 | a | 0.2 | ±0.05 | a | NA |

[a]Year seed were grown, tested the following year after processing; n = 1920 seed.

[b]CL, confidence limit; $Germ_{80}$, cumulative growing degree day value at 80% germination; NA, not applicable as the seed lot of that cultivar did not achieve 80% germination over the duration of the assay; GDD, growing degree day.

[c]Parameter estimates calculated by nonlinear regression equation (2) for seed germination with respect to time based on GDD accumulation: $a$ is the height of the horizontal asymptote at a very large X, $b1$ is expected value of Y (cumulative germination) at time X = 0, and $b2$ is a measure of growth rate.

[d]Values for each parameter within a column for each cultivar followed by the same letter are not significantly different at the 5% probability level. General linear models procedures were used with mean separation using 95% asymptotic confidence intervals. To obtain the equation for the respective regression line in **Figure 1**, the parameters from this table are used.

**Table 3.** Standard germination[a], logistic growth parameter estimates, 95% confidence limits (CL[b]), and vigor indices ($Germ_{80}$[b]) for germination of seed lots of runner-type peanut planted over 40-day period in 2009[a] at Dawson Georgia using a thermogradient germination assay.

(parameter *a*) of 79.8–93.6%, as compared to 21 April at 54.3–72.8%, and 1 June at 57.3–97.1% (**Table 3**). Vigor differences were noted as none of the seed produced from 21 April plantings achieved 80% maximum rate of germination. Similarly, all seed produced from 1 June planting at Dawson in 2009, except GA-03 L at 97.1%, had low vigor as determined by maximum rate of germination (**Figure 6**). The $Germ_{80}$ for Dawson 2009 seed were 25–40 GDD for the 11 May

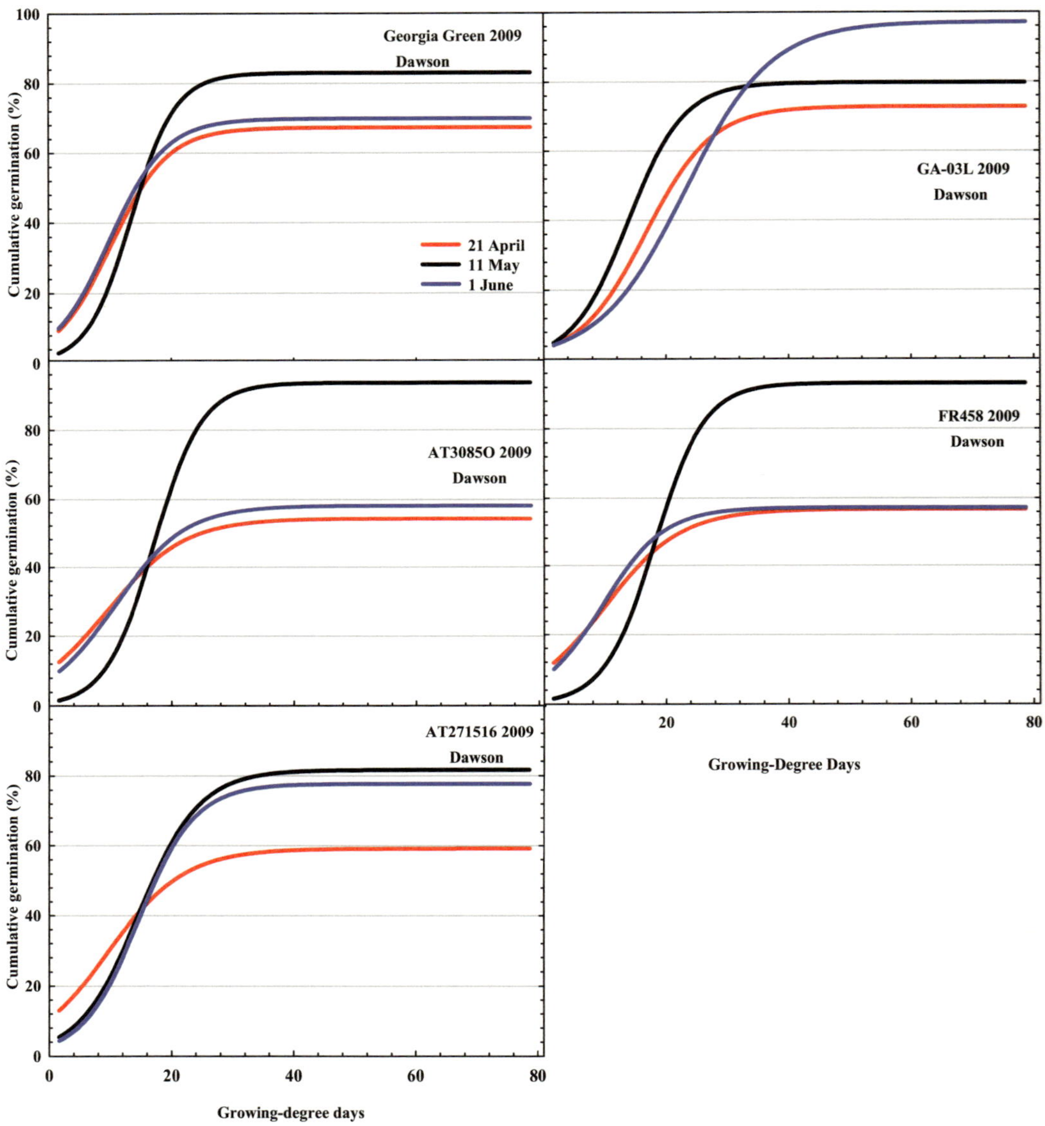


**Figure 6.** Cumulative germination patterns for runner-type peanut seed produced in 2009 at Dawson Georgia, based on nonlinear regression using growing-degree day (GDD) accumulation with a base temperature 15°C. To calculate the regression equation for the respective seed lot, the parameter estimates shown in **Table 3** for the Eq. (2) were used. Germination was measured on a thermal gradient.

| | | | Parameter $a$[c] | | | Parameter $b1$[c] | | | Parameter $b2$[c] | | | |
|---|---|---|---|---|---|---|---|---|---|---|---|---|
| Cultivar | Planting date[a] | Germination[a] | Maximum rate | 95% CL | | Estimate | 95% CL | | Estimate | 95% CL | | $Germ_{80}$ |
| | | % | | | | | | | | | | GDD |
| GA Green | 20 April | 68 | 66.9 | ±2.6 | a[d] | 3.7 | ±2.2 | a[d] | 0.26 | ±0.06 | b[d] | NA |
| | 11 May | 68 | 53.0 | ±2.8 | b | 18.0 | ±8.0 | b | 0.19 | ±0.09 | a | NA |
| | 1 June | 67 | 55.9 | ±2.7 | b | 18.0 | ±8.8 | b | 0.22 | ±0.10 | b | NA |
| AT3085R0 | 20 April | 65 | 59.1 | ±2.7 | a | 5.4 | ±2.7 | a | 0.21 | ±0.05 | a | NA |
| | 11 May | 57 | 43.7 | ±2.4 | b | 16.8 | ±11.6 | b | 0.28 | ±0.20 | a | NA |
| | 1 June | 65 | 61.5 | ±3.0 | a | 10.0 | ±3.7 | a | 0.16 | ±0.04 | a | NA |
| AT271516 | 20 April | 65 | 63.0 | ±2.7 | a | 5.4 | ±2.9 | a | 0.22 | ±0.05 | a | NA |
| | 11 May | 64 | 49.0 | ±2.8 | b | 20.7 | ±9.2 | b | 0.18 | ±0.11 | a | NA |
| | 1 June | 60 | 44.6 | ±2.0 | b | 3.2 | ±12.5 | b | 1.4 | ±1.8 | b | NA |
| GA-03 L | 20 April | 70 | 66.2 | ±2.7 | a | 6.2 | ±2.9 | a | 0.22 | ±0.05 | a | NA |
| | 11 May | 62 | 46.3 | ±2.6 | b | 18.9 | ±10.6 | b | 0.23 | ±0.15 | a | NA |
| | 1 June | 64 | 47.8 | ±2.0 | b | 7.0 | ±28.7 | b | 1.3 | ±2.2 | b | NA |
| FR458 | 20 April | 65 | 75.6 | ±2.8 | a | 2.8 | ±1.1 | a | 0.17 | ±0.02 | a | NA |
| | 11 May | 60 | 46.7 | ±2.6 | b | 18.5 | ±10.1 | b | 0.22 | ±0.14 | a | NA |
| | 1 June | 58 | 43.3 | ±1.9 | b | 2.2 | ±10.0 | b | 1.6 | ±2.1 | b | NA |

[a]Year seed were grown, tested the following year after processing; n = 1920 seed.

[b]CL, confidence limit; $Germ_{80}$, cumulative growing degree day value at 80% germination; NA, not applicable as the seed lot of that cultivar did not achieve 80% germination over the duration of the assay; GDD, growing degree day.

[c]Parameter estimates calculated by nonlinear regression equation (2) for seed germination with respect to time based on GDD accumulation: $a$ is the height of the horizontal asymptote at a very large X, $b1$ is expected value of Y (cumulative germination) at time X = 0, and $b2$ is a measure of growth rate.

[d]Values for each parameter within a column for each cultivar followed by the same letter are not significantly different at the 5% probability level. General linear models procedures were used with mean separation using 95% asymptotic confidence intervals. To obtain the equation for the respective regression line in **Figure 1**, the parameters from this table are used.

**Table 4.** Standard germination[a], logistic growth parameter estimates, 95% confidence limits (CL[b]), and vigor indices ($Germ_{80}$[b]) for germination of seed lots of runner-type peanut planted over 40-day period in 2009[a] at headland Alabama using a thermogradient germination assay.

plantings, however, seed from the 21 April plantings had poor vigor and never achieved $Germ_{80}$, while only GA-03 L had a $Germ_{80}$ of 35 GDD for seed from the 1 June planting date.

Runner-type seed produced by any planting date for Headland 2009 had very poor germination and vigor. Germination was less than 70% (**Table 4**). The maximum rate of germination (parameter *a*) for vigor was less than 75.6% for all cultivars and planting dates. $Germ_{80}$ was not achieved indicating low vigor (**Figure 7**). Previous research has indicated that there can

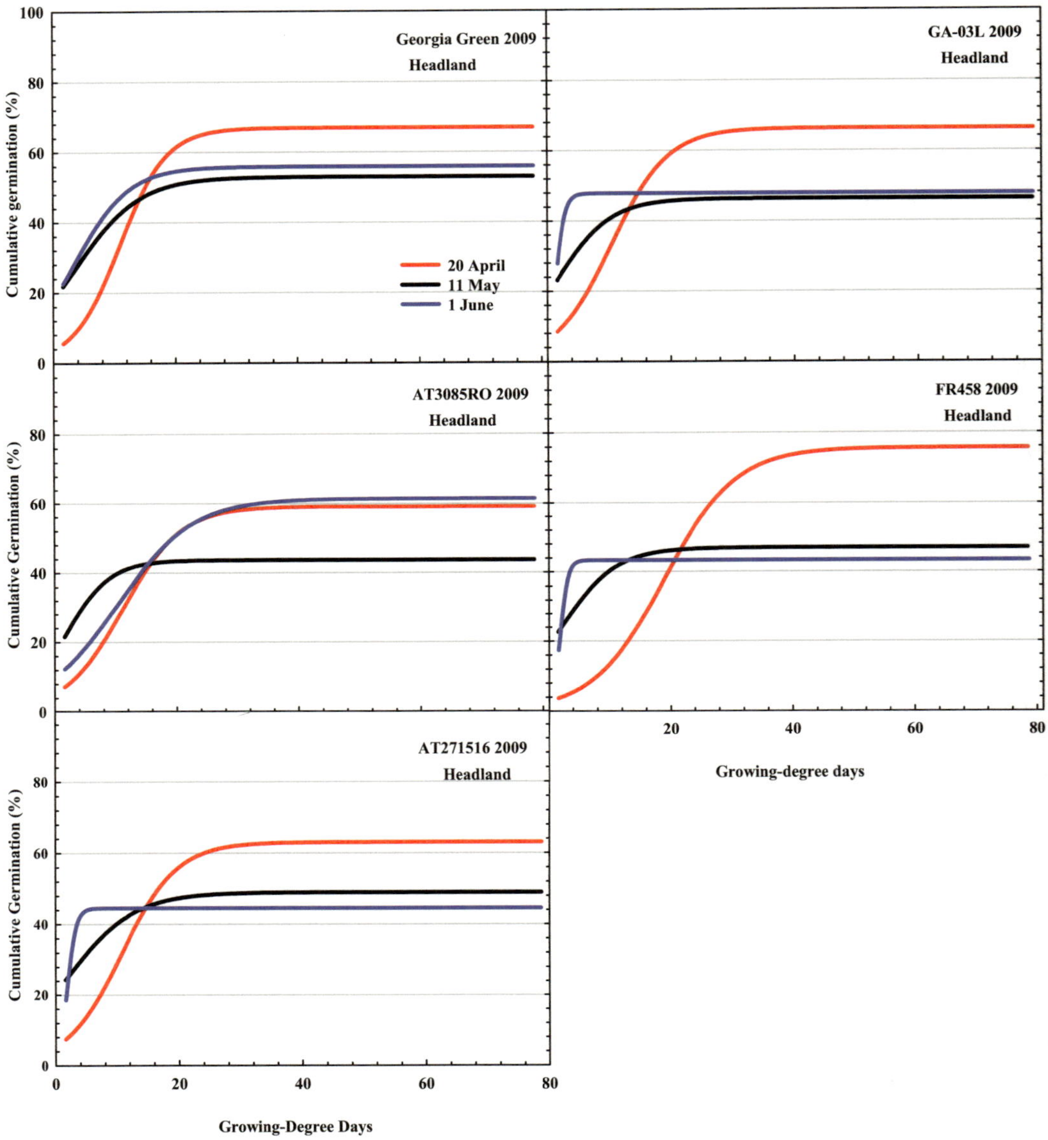


**Figure 7.** Cumulative germination patterns for runner-type peanut seed produced in 2009 at Headland Alablama, based on nonlinear regression using growing-degree day (GDD) accumulation with a base temperature 15°C. To calculate the regression equation for the respective seed lot, the parameter estimates shown in **Table 4** for the Eq. (2) were used. Germination was measured on a thermal gradient.

be variability of vigor for runner-type peanut seed of unknown origin, especially when the environmental condition under which that seed is produced is unknown [22].

## 6. Conclusion/recommendations

All cultivars exhibited phenotypic vigor variation by planting and harvest date across years. Comparing data generated from the thermal gradient using these growth curve models provided maximum germination rates with optimal temperatures (**Tables 2–4**). Cold germination testing can be used as a measure to stress peanut to evaluate vigor [52], using the thermal gradient apparatus used to evaluate peanut cultivars in this study established variation in seed vigor across a wide range of temperatures simultaneously. This method of seed evaluation provided an indication of vigor which assist peanut seed producers in determining the success of the cultivar over a range of temperatures, unlike the standard peanut germination test [30]. Seed produced from mid-May plantings each year were consistent with respect to germination, $Germ_{80}$, and GDD to reach maximum germination (*a*) among the five cultivars evaluated in Dawson for 2008 and 2009. Phenotypic differences were noted when these same cultivars were grown in Headland in 2009. These data assisted in determining phenotypic variation between planting dates when grown under known environmental conditions. This information will assist growers with making planting decisions based on these vigor testing methods.

## Author details

Timothy L. Grey[1]*, Charles Y. Chen[2], Russell Nuti[3], Walter Scott Monfort[1] and George Cutts III[2]

*Address all correspondence to: tgrey@uga.edu

1 Crop and Soil Sciences Department, University of Georgia, Tifton, Georgia, United States

2 Crop, Soil & Environmental Sciences, Auburn University, Auburn, Alabama, United States

3 PhytoGen Cottonseed, Dow Chemical Inc. Shellman, Georgia, United States

## References

[1] Moss JP, Rao VR. The peanut – reproductive development to plant maturity. In: Pattee HE, Stalker HT, editors. Advances in Peanut Science. Tifton GA 31793: American Peanut Research and Education Society, Inc.; 1995

[2] Norden AJ, Lipscomb RW, Carver WA. Registration of 'Florunner' peanuts. Crop Science. 1969;**9**:850

[3] Hammons RO. Peanuts: Genetic vulnerability and breeding strategy. Crop Science. 1976;**16**:527-530

[4] Kvien CK, Culbreath AK, Wilcut JW, Brown SL, Bell DK. Peanut production in systems restricting use of pesticides based on carcinogenicity or leachability. Peanut Science. 1993;**20**: 118-124

[5] McLean DE, Sullivan GA. Influence of cultural and harvest practices on peanut seed quality. Peanut Science. 1981;**8**:145-148

[6] Norden AJ, Smith OD, Gorbet DW. Breeding of the cultivated peanut. In: Pattee HE, Young CT, editors. Peanut Science and Technology. Tifton GA 31793: American Peanut Research and Education Society, Inc; 1982

[7] Culbreath AK, Todd JW, Gorbet DW, Brown SL, Baldwin J, Pappu HR, Shokes FM. Reaction of peanut cultivars to spotted wilt. Peanut Science. 2000;**27**:35-39

[8] Tillman BL, Gorbet DW, Andersen PC. Influence of planting date on yield and spotted wilt of runner market type peanut. Peanut Science. 2007;**34**:79-84

[9] Branch WD, Baldwin JA, Culbreath AK. Genotype by seeding rate interaction among TSWV-resistant, runner-type peanut cultivars. Peanut Science. 2003;**30**:108-111

[10] Branch WD, Fletcher SM. Evaluation of advanced Georgia peanut breeding lines with reduced-input and without irrigation. Crop Protection. 2004;**11**:1085-1088

[11] Culbreath AK, Todd JW, Isleib TG, Tillman B, Gorbet DW, Martinez-Ochoa N, Holbrook CC. High levels of field resistance to tomato spotted wilt virus in peanut breeding lines derived from hypogaea and hirsuta botanical varieties. Peanut Science 2005;**32**: 20-24

[12] Holbrook CC, Timper P, Culbreath AK. Resistance to tomato spotted wilt virus and root-knot nematode in peanut interspecific breeding lines. Crop Science. 2003;**43**:1109-1113

[13] Holbrook CC, Timper P, Dong W, Kvien CK, Culbreath AK. Development of near-isogenic peanut lines with-and-without resistance to the peanut root-knot nematode. Crop Science. 2008;**48**:194-198

[14] Isleib TG, Day JL, Coy AE, Beasley JP, Branch WD. Genotype-by-irrigation interaction in the Georgia peanut official variety test. Peanut Science. 2014;**41**:1-7

[15] Nuti RC, Chen CY, Dang PM, Harvey JE. Peanut cultivar response to tomato spotted wilt over five planting dates. Peanut Science. 2014;**41**:32-41

[16] Sarver J., Tubbs R.S., Beasley, Jr. J.P., Culbreath A.K., Grey T.L., Rowland D.L., Smith N.B. Evaluating plant population and replant method effects on peanut planted in twin rows. Peanut Science **44**:21-27

[17] Morton BR, Tillman BL, Gorbet DW, Boote KJ. Impact of seed storage environment on field emergence of peanut cultivars. Peanut Science. 2008;**35**:108-155

[18] Branch WD. Registration of 'Georgia-01R' peanut. Crop Science. 2002;**42**:1750-1751

[19] Gorbet DW, Tillman BL. Registration of 'Florida-07' peanut. Journal Plant Registrations. 2009;**3**:14-18

[20] Carter E, Rowland D, Tillman B, Erickson J, Grey TL, Gillett-Kaufman J, Clark M. Pod maturity in the shelling process. Peanut Science. 2017;**44**:28-36

[21] Spears JF, Sullivan GA. Relationship of hull mesocarp color to seed germination and vigor in large-seeded virginia-type peanuts. Peanut Science. 1995;**22**:22-26

[22] Grey TL, Beasley JM, Webster TM, Chen CY. Peanut seed vigor evaluation using a thermal gradient. International Journal of Agronomy. 2011; On line at http://hindawi.com/journals/ija/aip/202341/

[23] Branch WD. Registration of 'Georgia Green' peanut. Crop Science. 1996;**36**:806

[24] RS,Tubbs, Culbreath AK, Beasley JP, Smith NB, and Smith AR. Reducing Seeding Rates for Large-Seeded TSWV-Resistant Runner Peanut Cultivars. 2009; Online http://www.caes.uga.edu/Applications/ImpactStatements/index.cfm

[25] Branch WD. Registration of 'Georgia-06G' peanut. Journal Plant Registrations. 2007;**1**:120

[26] R Kemerait, Brenneman T, and Culbreath A. peanut disease update. In University of Georgia Extension 2015 Peanut Update. P 29-49. 2015; Online http://www.caes.uga.edu/commodities/fieldcrops/peanuts/documents/2015peanutupdate.pdf

[27] National Agricultural Statistics Service (NASS). National Agricultural Statistics Service U.S. Dept. of Agri. Published Estimates Database. Washington DC: NASS-USDA; 2016 http://www.nass.usda.gov/Statistics_by_Subject/Environmental/index.asp

[28] Tai PY, Hammons RO. Genotype-environment interaction effects in peanut variety evaluation. Peanut Science. 1978;**5**:72-74

[29] Spears JF, Jordan DL, Bailey JE. Peanut Seed Production: A Guide for Producers of Virginia-Type Peanut Production. North Carolina state Univ. Ext. Publication AG-622; 2002

[30] Association of Official Seed Analysts (AOSA). Rules for Seed Testing. Las Cruces, NM: Association of Official Seed Analysts; 2002

[31] Cardina J, Hook JE. Factors influencing germination and emergence of Florida beggarweed (Desmodium totruosum). Weed Technology. 1989;**3**:402-407

[32] Chatterton N, Kadish A. A temperature gradient germinator. Agronomy Journal. 1969;**61**:643-644

[33] Cole DF. Use of the thermogradient plate as an aid in determining the relative vigor of sweet corn (Zea mays). Agronomy Journal. 1972;**64**:749-751

[34] Mohapatra SC, Johnson WH. A thermogradient incubator for seed germination studies. Agronomy Journal. 1986;**78**:351-356

[35] Scott SJ, Jones RA. Quantifying seed germination response to low temperatures: Variation among Lycopersicon spp. Environmental and Experimental Botany. 1984;**25**:129-137

[36] Thompson PA. Characterization of the germination response to temperature of vegetable seeds. I. Tomatoes. Scientia Horticulturae. 1974;**2**:35-54

[37] Grey TL, Branch WD, Tubbs RS, Snider JL, Webster TM, Arnold J, Li X. The impact of genotype by environment effects on runner-type peanut seed vigor response to temperature. Agronomy Journal. 2016;**108**:1424-1433

[38] Branch WD. Registration of 'Georgia-03L' peanut. Crop Science. 2004;**44**:1485-1486

[39] Chen CY, Nuti RC, Rowland DL, Faircloth WH, Lamb MC, Harvey E. Heritability and genetic relationships for drought-related traits in peanut. Crop Science. 2013;**53**:1392-1402

[40] Brown SL, Culbreath AK, Todd JK, Gorbet DW, Baldwin JA, Beasley JP. Development of a method of risk assessment to facilitate integrated management of spotted wilt of peanut. Plant Disease. 2005;**89**:348-356

[41] Gomez KA, Gomez AA. Statistical Procedures for Agricultural Research. New York: John Wiley & Sons; 1984

[42] Beasley JP Jr. Seeding rate, row patterns, and planting date. In: Prostko EP, editor. Peanut update. CSS-03-0105. Georgia Cooperative Extension Service. Athens: University of Georgia; 2007

[43] Williams EJ, Drexler JS. A non-destructive method for determining peanut pod maturity. Peanut Science. 1981;**8**:134-141

[44] Anonymous. 1997. United States Standards for Grades of Shelled Runner Type Peanuts. Online at: http://www.ams.usda.gov/AMSv1.0/getfile?dDocName=STELPRDC5050496

[45] Ketring DL. Physiology of oil seeds. X. Seed quality of peanut genotypes as affected by ambient storage temperature. Peanut Science. 1992;**19**:72-77

[46] Ketring DL, Morgan PW. Ethylene as a component of the emanations from germinating peanut seeds and its effect on dormant virginia-type seeds. Plant Physiology. 1969;**44**:326-330

[47] Ketring DL, Wheless TG. Thermal time requirements for phonological development of peanut. Agronomy Journal. 1989;**81**:910-917

[48] Ketring DL, Brown RH, Sullivan GA, Johnson BB. Peanut physiology. In: Pattee HE, Young CT, editors. Peanut Science and Technology. Stillwater, OK: American Peanut Research and Education Society, Inc.; 1982

[49] Freund RJ, Littell RC. SAS System for Regression. Second ed. Cary NC: SAS Inst., Inc.; 1991

[50] Awal MA, Ikeda T. Effects of changes in soil temperature on seedling emergence and phenological development in field-grown stands of peanut. Environmental and Experimental Botany. 2002;**47**:101-113

[51] Chen CY, Nelson RL. Variation in early plant height in wild soybean. Crop Science. 2006;**46**:865-869

[52] Guerke WR. Evaluating peanut seed vigor. Seed Technology. 2005;**27**:121-126

Chapter 7

# Impact of Conditions of Water Supply on the Germination of Tomato and Pepper Seeds

Yekaterina Shapira, Edward Bormashenko,
Gene Whyman, Bat-Chen Lubin and Elyashiv Drori

Additional information is available at the end of the chapter

http://dx.doi.org/10.5772/intechopen.70386

**Abstract**

The influence of the cold radiofrequency air plasma treatment on the imbibition and germination of tomato (*Solanum lycopersicum*) and pepper (*Capsicum annuum*) seeds, exerted to conditions appearing in natural seedbeds, was investigated. Various conditions of water supply (from nonrestricted water supply to water drought stress) to the germinating seeds were studied. Plasma treatment markedly increased the water imbibition in the case of tomato seeds under limited water availability conditions. Cold radiofrequency plasma treatment had no noticeable impact on the germination of tomato and pepper seeds under the conditions of nonrestricted water supply. In the case of drought stress for both studied cultivars, the cold plasma treatment essentially influenced the germination rate and the kinetics of germination (the median of the Richards' curve was changed essentially under conditions of water drought).

**Keywords:** plasma treatment, tomato and pepper seeds, drought stress, water imbibition, germination

## 1. Introduction

Interaction of cold plasmas with plant seeds has drawn the evident attention of scientists in the last decade [1–8]. The investigators mainly concentrated on the possibility to control the germination by exposure of seeds to various types of plasma, including atmospheric and low-pressure plasma discharges. It should be stressed that experimental data related to the influence of plasma treatment (carried out with capacitive and inductive discharges) on the germination time and percentage are contradictory. The first profound investigation of the impact of plasma discharges on seed germination was reported by Volin et al. [1]. These authors exposed seeds of radishes (*Raphanus sativus*) and two pea cultivars (*Pisum sativum* cv. "Little Marvel" and *P. sativum* cv. "Alaska") to $CF_4$ and octadecafluorodecalin plasma and observed an essential delay in germination [1]. Since then, investigators focused on the following main fields: (1) decontamination of seeds by plasma, (2) breaking of dormancy with plasma, (3) the impact of plasma treatment (PT) on the rate and kinetics of germination, and (4) impact of PT on the root generation (sprouting).

Decontamination and inactivation of pathogenic microorganisms of seeds by PT have been communicated recently by various groups [2, 3, 5, 6]. Nishioka et al. reported the effectiveness of low-pressure plasma treatment in the inactivation of the seed-borne plant pathogenic bacteria [6]. Researchers reported the impact of PT on germination, sprouting, and dormancy breaking of seeds. Sera et al. investigated the influence of PT on wheat and oat germination. The authors reported that PT did not affect germination of oat seeds, but they did note accelerated root generation in plants grown from plasma-treated seeds [7]. The same group also communicated that PT did change seed germination in Lamb's Quarters seeds [8]. Similar results were reported for wheat seeds (*Triticum aestivum*) by Dobrin et al. [9].

In contrast, Ji et al. communicated the significant improvement of the germination rate of *Coriander sativum* under nonthermal atmospheric pressure treatment [10]. Dhayal et al. showed about 50% increase in the germination rate, and the activity was increased twice after plasma treatment of safflower [11]. The essential augmentation of the germination rate by low-temperature plasma treatment was registered by Stolárik et al. for pea (*Pisum sativum* L. var. Prophet). Stolárik et al. related the observed effect to the chemical modification of the pea surface by plasma [12]. These results support the observations of the modification of the physical structure of seed coat by the low-pressure argon gas discharge reported by Dhayal et al. [11].

A stimulating effect of cold plasma on both the germination and sprouting of tomato seeds (*Lycopersicon esculentum L. Mill.* cv. "Zhongshu No. 6") has been reported [13]. Similar results were reported for *Pauwlonia tomentosa* seeds [14]. Kitazaki et al. studied growth enhancement of radish sprouts (*Raphanus sativus* L.) induced by low-pressure $O_2$ radio-frequency plasma irradiation [15]. Dobrin et al. reported that the roots and sprouts of plasma-treated wheat seeds (*T. aestivum*) were longer and heavier than those of the non-treated seeds [9]. The improvement of the germination rate of the seeds of legumes and grain crops (*Lupinus angustifolius* (blue lupine), *Galega virginiana* (catgut), and *Melilotus albus* (honey clover and soy)) by low-pressure (5.28 MHz) plasma was reported by Filatova et al. [16].

The experimental results revealed that oxygen-related radicals strongly enhance growth, whereas ions and photons do not [15]. The positive effect of cold helium plasma treatment on seed germination, growth, and yield was reported recently for wheat [17]. Treatment of spinach seeds by magnetized arc plasma increased the germination rate by 137% [18]. It has been demonstrated that cold atmospheric plasma treatment had little effect on the final germination percentage of radish seeds, but it influenced their early growth [19]. The contradictory data related to the impact of plasmas on the seed germination were summarized in the recent review by Randeniya and de Groot [20]. Ji et al. suggested that plasma can enhance seed germination by triggering biochemical processes in seeds [4].

Basically, the investigators admit that cold plasma treatment is an efficient and "green," non-waste method to improve seed germination and crop yield. It plays essential roles in a broad spectrum of biological processes in plants, including reducing the bacterial bearing rate of seeds, modification of the seed coat chemical composition, hydrophilization of seed coats, homogenization of the kinetics of germination, and influence on the seedling growth [16, 21].

Our group recently reported the impact of cold air plasma on the surface properties of lentils (*Lens culinaris*) and beans (*Phaseolus vulgaris*) [22, 23]. We established that cold plasma treatment leads to essential hydrophilization of the cotyledon and tissues constituting the testa when they were separately exposed to the plasma discharge. Contrastingly, when the entire bean is exposed to plasma treatment, only the external surface of the bean has been hydrophilized by the cold plasma [23]. The pronounced hydrophilization of seeds by plasmas was reported also by other groups [9, 21]. Actually, the effect of hydrophilization of natural organic surfaces by plasma treatment has been already extensively studied [24, 25].

The cold plasma treatment, which inspired change in wettability, is followed by a consequent change in the water imbibition of the seeds [9, 22, 23]. At the same time, the relation between the change in the wettability, induced by the plasma treatment of seeds, and the parameters of germination (time and rate) remains obscure. In our paper we address the problem of the impact of the cold radiofrequency air plasma treatment, performed with the inductive plasma discharge [24], on the germination of seeds exerted to various conditions of water availability. The main goal of the research is the establishment of the influence of cold plasma treatment on the germination rate and kinetics under the conditions of limited water supply conditions.

## 2. Methods and procedures

### 2.1. Plasma device

Plasma treatment (PT) was carried out with the plasma unit EQ-PDC-326 manufactured by MTI Co., USA. The scheme of the experimental unit used for plasma treatment of the seeds is depicted in **Figure 1**. The unit generates the inductive plasma discharge [26].

### 2.2. Seed materials and plasma gases

Seeds of tomato (Efrat-70, *Solanum lycopersicum*) and pepper (Roni-272, *Capsicum annuum*) were supplied by Hazera Co. (Israel).

Dry air (99.999%) was supplied by Mifalei Hamzan Co. (Israel).

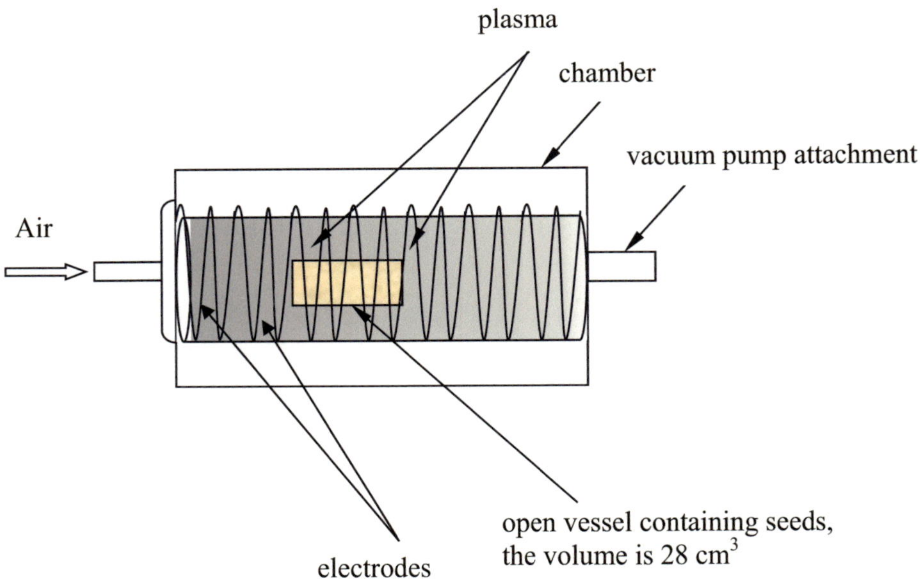


**Figure 1.** Plasma unit used for treatment of seeds.

### 2.3. Treatment conditions

Healthy uniform seeds without visible damages were selected and exposed to the inductive plasma discharge under the following parameters: the plasma frequency was 13.56 MHz (the RF matching is automatic), the pressure was 0.5 Torr, the supplied power of plasma discharge was 18 W, and the volume of the discharge chamber was 45 cm$^3$.

The parameters of plasma for the used pressure of 0.5 Torr were temperature of electrons (11.76 ± 0.51 eV) and concentration of ions $4.97 \pm 1.16 \times 10^{15} m^{-3}$ (as established with the double Langmuir probe (AlP150, Impedans Ltd. Plasma Measurement, Ireland). The time span of irradiation was varied from 30 to 60 s. During the plasma treatment, the temperature in the discharge chamber was ambient. The seedling measurements and content analyses mentioned in detail below were carried out immediately after the PT of seeds.

### 2.4. Modeling of water availability conditions

Water supply to seeds was controlled as follows: the seeds were germinated in 90-mm-diameter Petri dishes. Seedbed was composed in various series of experiments of three filter paper layers (well-watered (WW)), which means nonrestricted water supply); four filter paper layers (medium drought stress (MDS), below in the text); or five filter paper layers (harsh drought stress (HDS), below in the text) of the filter paper (500-H, the thickness was 112 μm, and the maximal pore size was 48 μm).

The stacks were moistened with 4 ml of distilled water ($dH_2O$) dripped on the stacks by a syringe (see **Figure 2**). The specific conductivity of distilled water was 18MΩ/cm.

http://dx.doi.org/10.5772/intechopen.70386

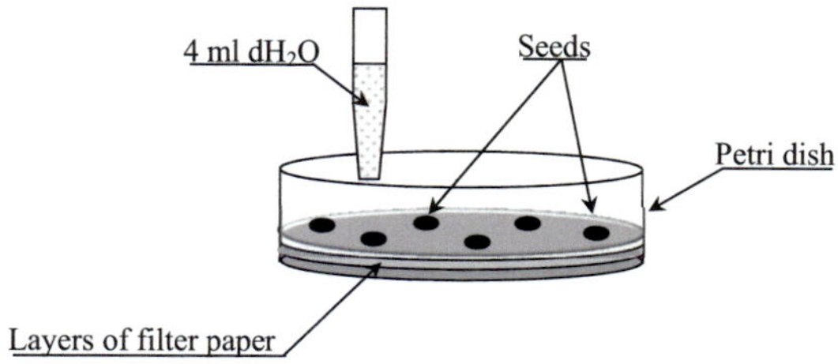


WW – Well watered – 3 layers of filter paper + 4 ml $dH_2O$;

MDS – Medium drought stress – 4 layers of filter paper + 4 ml $dH_2O$;

HDS – Harsh drought stress – 5 layers of filter paper + 4 ml $dH_2O$.

**Figure 2.** Experimental setup for treatment of seeds.

### 2.5. Study of the influence of water availability on imbibition

For the study of the time dependence of seed water absorption (imbibition) by irradiated and nonirradiated tomato and pepper seeds, 100 seeds were placed on humid filter paper stacks in different water conditions (for details see Section 2.4). Seeds were weighed at 1, 3, 6, 9, and 24 hours after plasma treatment with a MRC ASB-220-C2 analytical balance. The relative water imbibition (absorption) was defined as $\frac{\Delta m(t)}{m_0} 100\ \% = \frac{m(t) - m_0}{m_0} 100\%$, where $m_0$ is the total initial mass of seeds and $m(t)$ is the running total mass of seeds [23]. The comparative study of imbibition was carried out for non-treated pepper/tomato seeds and seeds treated by plasma during 30 and 60s. The experiment was planned with three replications for each treatment.

### 2.6. Study of influence of water drought and temperature on germination

For the study of influence of water availability and germination chamber temperature on germination by irradiated and nonirradiated tomato and pepper seeds, 15 pepper/tomato seeds were placed on the upper layer of the filter paper stack and germinated in different water conditions (for details see Section 2.4). The Petri dishes were covered with lids and sealed using a strip of Parafilm in order to prevent water evaporation. Seeds were placed in an incubator at 21 or at 27°C. The incubation cycle included 12 hours of darkness and 12 hours of light per day. The seeds were considered to be germinated when the radicals were half the seed length. The constant conditions of germination were provided by growth chamber model PGI-500H (Illumination 40 W × 5 tubes) (MRC, Israel). The germination percentage was recorded every 24 hours for 12 days in a case of pepper seeds and every 24 hours for 10 days in a case of tomato seeds. The experiment was planned as a completely randomized design with five replications.

The comparative study of germination was carried out for non-treated pepper/tomato seeds and seeds treated by plasma during 30 and 60s.

Germination rate (%) was defined as the number of seeds germinated in 10 or 12 days related to the total number of seeds.

## 3. Results

### 3.1. Seed water imbibition

The previous findings suggest acceleration in the water uptake of PT tomato seeds, under MDS and HDS water conditions compared to non-treated seeds [22, 23]. It is distinctly seen that the fastest water imbibition was observed for 30s plasma-treated seeds (see **Figure 3**). In the case of pepper seeds, the rate of water imbibition is almost the same in plasma-treated and non-treated samples (see **Figure 4**). Under free water supply conditions (WW), the imbibition of PT seeds and non-treated seeds is almost the same for both cultivars.

In addition, it is distinctly seen that the fastest water imbibition was observed under free water supply conditions (WW), and the slowest water absorption was registered under HDS conditions for both cultivars in non-treated and plasma-treated seeds (see **Figures 3** and **4**). The water availability is an important factor influencing the rate of water imbibition.

### 3.2. Seed germination

Positive influence of the cold plasma treatment on the germination rate and on the kinetics of germination was recorded for the pepper and tomato seeds under the conditions of medium and harsh drought stress (see **Figures 5** and **6**; **Table 1**) at temperature of 21°C. The change in the germination rate (denoted by *Vi* (viability) in **Table 1**) was much more pronounced for tomato seeds when compared to the pepper ones. Under MDS and HDS water conditions, the germination rate of non-treated seeds was significantly reduced by 8 and 11%, respectively, for tomato seeds and by 1 and 13% for pepper seeds, compared to the well-watered

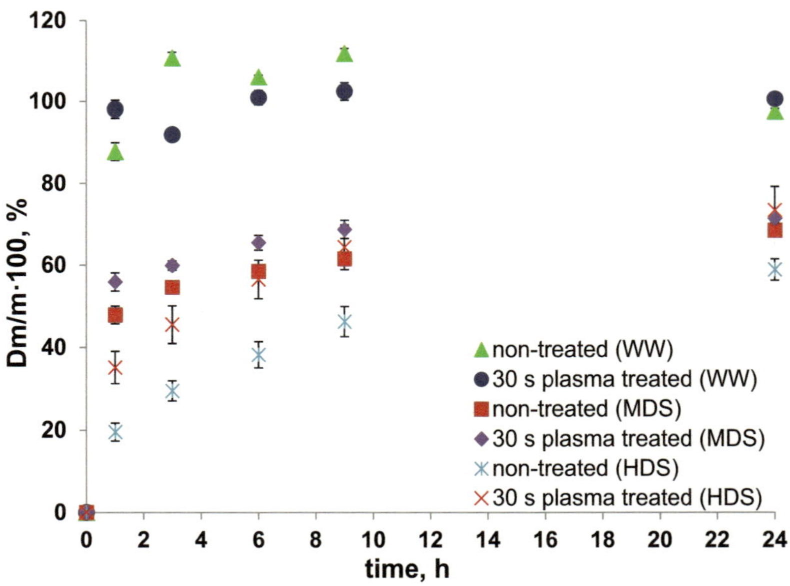


**Figure 3.** Comparative study of water imbibition by non-treated and 30s plasma-treated tomato seeds under WW, MDS, and HDS conditions.

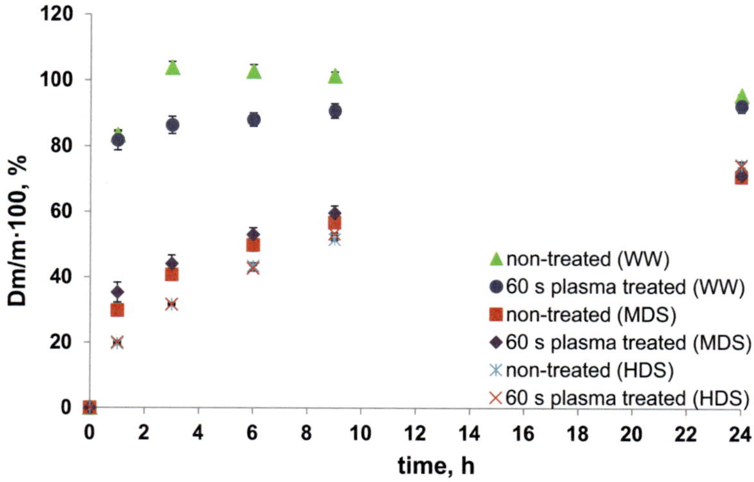


**Figure 4.** Comparative study of water imbibition by non-treated and 60s plasma-treated pepper seeds under WW, MDS, and HDS conditions.

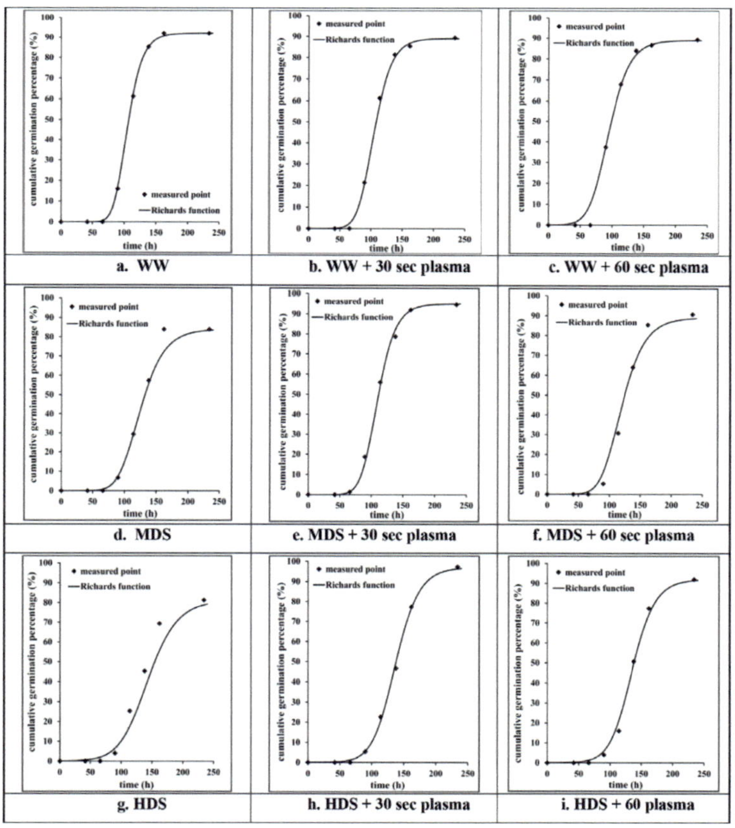


**Figure 5.** Germination curves calculated using Richards' fitting function [25] for tomato seeds (at 21°C). (a) WW, (b) WW + 30 s plasma, (c) WW + 60 s plasma, (d) MDS, (e) MDS + 30 s plasma, (f) MDS + 60 s plasma, (g) HDS, (h) HDS + 30 s plasma, and (i) HDS + 60 plasma.

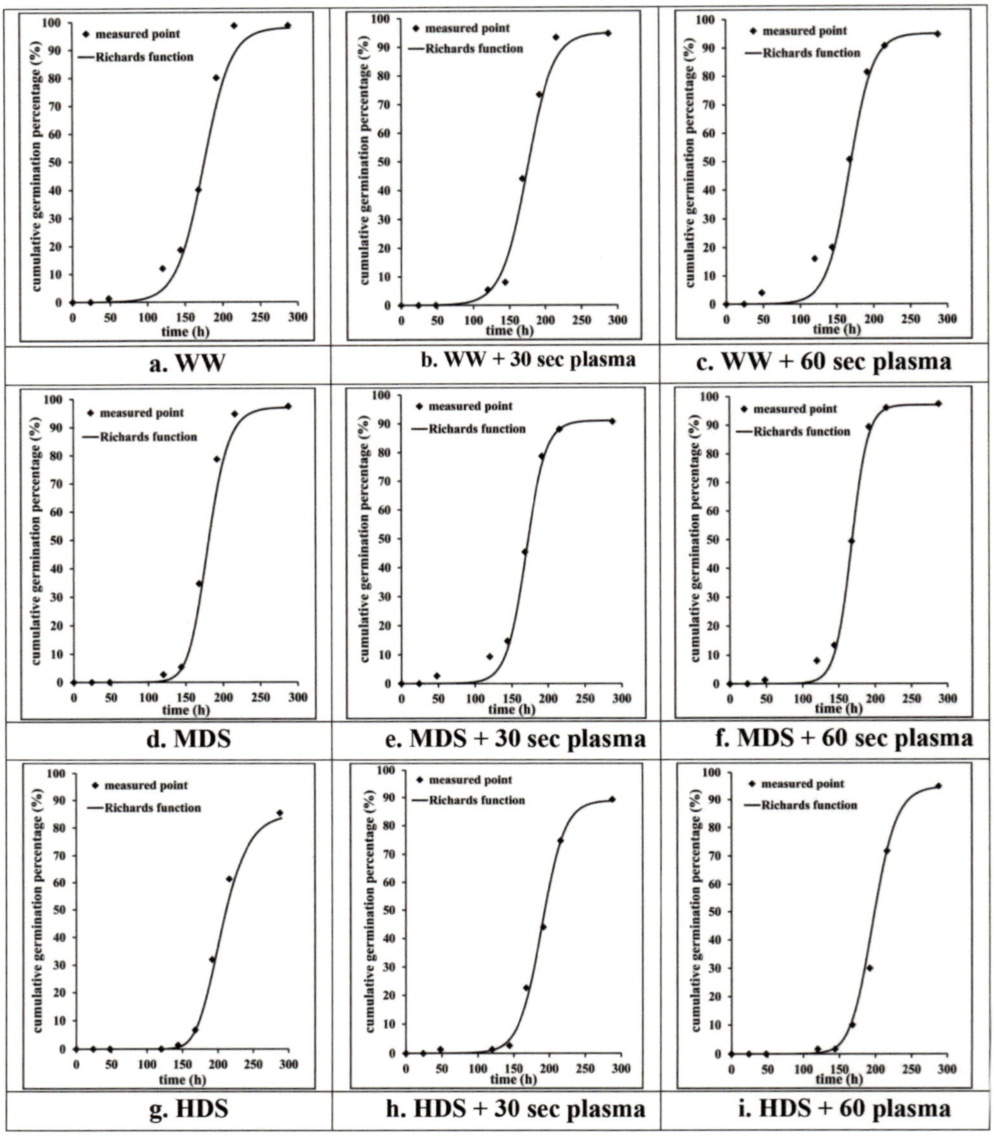


**Figure 6.** Germination curves calculated using Richards' fitting function [25] for pepper seeds (at 21°C). (a) WW, (b) WW + 30 s plasma, (c) WW + 60 s plasma, (d) MDS, (e) MDS + 30 s plasma, (f) MDS + 60 s plasma, (g) HDS, (h) HDS + 30 s plasma, and (i) HDS + 60 plasma.

seeds (see **Table 1**). PT significantly increased the germination rate (*Vi*) of tomato seeds. It increased by 11% for 30 s PT and by 7% for 60 s compared to the medium drought-stressed non-treated seeds, and it increased by 16% for 30 s PT and 11% for 60 s PT compared to the harsh drought-stressed non-treated seeds.

For pepper seeds in HDS, the cold plasma treatment increased the germination rate by 4% for 30 s PT and 10% for 60 s PT seeds when compared to harsh drought-stressed non-treated seeds. However, in the case of medium drought stress, the final percentage of germination rate was almost the same in 60 s plasma-treated and non-treated samples (see **Table 1**).

| Cultivar | Treatment | *Vi* (%) | *Me* (h) | *Qu* (h) | *Sk* (%) |
|---|---|---|---|---|---|
| Tomato | WW | 92±3a | 105±5 | 12.5 | 0.3 |
| | WW + 30 s PT | 89±3a | 105±2 | 15 | 0.23 |
| | WW + 60 s PT | 89±3a | 100±2 | 17.5 | 0.2 |
| | MDS | 84±4b | 125±6 | 20 | 0.3 |
| | MDS + 30 s PT | 95±2 | 110±4 | 14.5 | 0.23 |
| | MDS + 60 s PT | 91±2ab | 115±5 | 20 | 0.3 |
| | HDS | 81±3b | 145±5 | 26 | 0.13 |
| | HDS + 30 s PT | 97±2a | 140±5 | 20 | 0.17 |
| | HDS + 60 s PT | 92±1a | 135±5 | 20 | 0.11 |
| Pepper | WW | 98±1a | 175±5 | 20 | 0 |
| | WW + 30 s PT | 95±2a | 175±5 | 20 | 0 |
| | WW + 60 s PT | 95±2a | 168±4 | 17.5 | 0 |
| | MDS | 97±2a | 180±5 | 15 | 0. 15 |
| | MDS + 30 s PT | 91±2a | 170±5 | 15 | 0 |
| | MDS + 60 s PT | 97±2a | 168±4 | 12.5 | 0 |
| | HDS | 85±3b | 205±5 | 20 | 0.3 |
| | HDS + 30 s PT | 89±2ab | 190±5 | 17.5 | 0 |
| | HDS + 60 s PT | 95±2a | 197±10 | 17.5 | 0.12 |

Note: WW, well-watered; WW + plasma, well-watered + plasma; MDS, medium drought stress; MDS + plasma, medium drought stress + plasma; HDS, harsh drought stress; HDS + plasma, harsh drought stress + plasma

*Vi* is the viability, *Me*is the time, *Qu* is the dispersion, and *Sk* is the skewness, details in text.

The values of each experiment (bold line in the table) were significantly marked by different statistical letters at $P \leq 0.05$ according to Student's t-test.

**Table 1.** Effect of water availability conditions, temperature (21°C), and cold plasma treatment on seed germination in tomato and pepper seeds.

Under free water supply conditions (WW), the germination percentage of PT seeds is slightly and insignificantly decreased in both cultivars.

The positive influence of plasma treatment on germination of tomato seeds at 27°C temperature was also shown (see **Table 2**). The cold plasma treatment significantly increased germination rate (*Vi*) under MDS conditions by 8% for 30 s PT and by 12% for 60 s PT compared to the MDS non-treated seeds. In HDS conditions, germination increased by 6% for 30 s PT and by 11% for 60 s PT when compared to the HDS non-treated seeds. In the case of pepper seeds, the 27°C temperature significantly decreased the germination percentage in all water regimes (see **Table 2**). Interestingly, at 27°C, while in WW conditions, germination rates of non-PT pepper seeds were much higher than the rates of treated seeds, when grown under MDS and HDS, the rates of germination of PT seeds were higher than those of WW-grown pepper seeds.

In order to elucidate the data describing the kinetics of germination, Richards' curves were fitted to a number of experiments [23, 27, 28]. Fitting experimental data by Richards' curves is shown in **Figures 5** and **6**. The solution of Richards' differential equation worked out for the growth of modeling results in Richards' curve, which is an extension of the logistic or sigmoid functions, which are the *S*-shaped curves describing the kinetics of germination. The Richards' function $Y_t$ demonstrating a variable inflection point was calculated according to Eq. (1):

$$Y_t = \frac{a}{[1 + b \cdot d \cdot \exp(-c \cdot t))]^{\frac{1}{d}}} \tag{1}$$

where $Y_t$ is the germination percentage; *a*, *b*, *c* and *d* are the fitting parameters; and *t* is the time.

Fitting of experimental data by Eq. (1) supplied the best values of the fitting parameters summarized in **Table 1**, in which *Me* (median) denotes the time of 50% germination and characterizes the rate of this process. The quartile deviation of germination time *Qu* describes the deviation range of Richards' curve relative to *Me*, and *Sk* (skewness) represents the asymmetry of Richards' curve relative to the inflection point (mode) (see **Figure 7**). For calculation of these quantities, the useful formulae developed by Hara were implemented [28].

| Cultivar | Treatment | *Vi* (%) at 27°C |
|---|---|---|
| Tomato | WW | 84±3b |
| | WW + 30 s PT | 83±5b |
| | WW + 60 s PT | 97±2a |
| | MDS | 81±2b |
| | MDS + 30 s PT | 89±3ab |
| | MDS + 60 s PT | 93±4a |
| | HDS | 85±3b |
| | HDS + 30 s PT | 91±2ab |
| | HDS + 60 s PT | 96±2a |
| Pepper | WW | 64±7a |
| | WW + 30 s PT | 20±4b |
| | WW + 60 s PT | 28±6b |
| | MDS | 53±19a |
| | MDS + 30 s PT | 59±11a |
| | MDS + 60 s PT | 61±11a |
| | HDS | 31±12a |
| | HDS + 30 s PT | 44±11a |
| | HDS + 60 s PT | 45±9a |

Note: WW, well-watered; WW + plasma, well-watered + plasma; MDS, medium drought stress; MDS + plasma, medium drought stress+ plasma; HDS, harsh drought stress; HDS + plasma, harsh drought stress + plasma.

The values of each experiment (bold line in the table) were significantly marked by different statistical letters at $P \leq 0.05$ according to Student's t-test.

**Table 2.** Effect of water availability conditions, temperature (27°C), and cold plasma treatment on seed germination on tomato and pepper seeds.

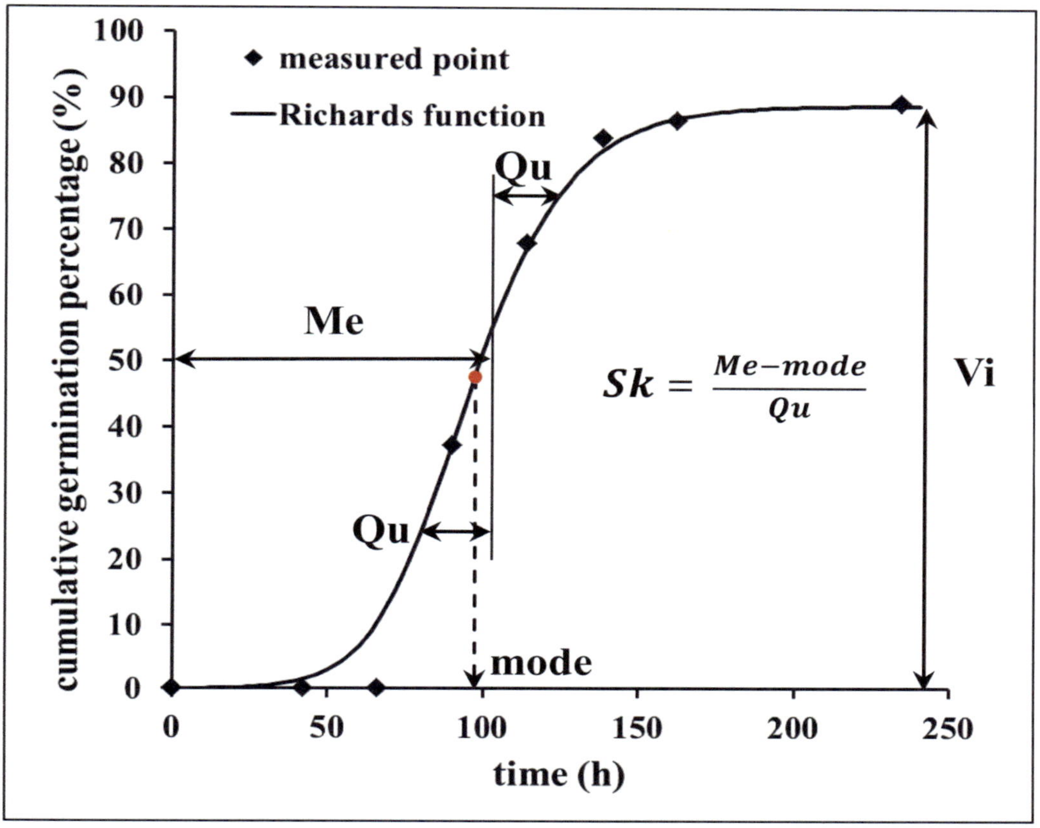


**Figure 7.** Richards' germination curve (see Eq. (1)). The quantitative parameters of germination are shown.

As it is recognized from the data supplied in **Table 1** for tomato seeds, the value of *Me* is distinctly lower (by *ca.* 5–15 and 5–10 hours) after 30 and 60 s of plasma treatment in the cases of MDS and HRS, respectively. Consider that germination is accelerated as *Me* decreases [28].

As it is seen from **Table 1**, for pepper seeds the value of *Me* is distinctly lower (by *ca.* 10–15 and 8–12 hours) after 30 and 60 s of PT in the cases of MDS and HRS, respectively. Parameters *Qu* and *Sk* were not affected markedly by the cold plasma treatment, as is seen from **Table 1**.

## 4. Discussion

In the present study, we demonstrated that the cold radiofrequency air plasma treatment has a crucial impact on the water imbibition and seed germination of pepper and tomato seeds. This effect is reasonably attributed to the hydrophilization of the surface of seeds by the plasma treatment [22, 23, 29]. The imbibition and germination in turn were adversely affected by water availability conditions of the seedbed. This effect was well expected, and it was already addressed by investigators [30–32]. However, at the conditions of the lower water availability, both the germination rate and the kinetics of germination were essentially positively affected by the cold radiofrequency air plasma treatment. The MDS conditions used in our work represent the naturally occurring conditions typical for seedbeds, where no free water is usually present, but water is given abundantly and then drained. HDS conditions represent lower water availability, causing water stress during germination. It should be emphasized that seed germination and early seedling growth are critical stages for plant establishment, and plants are more sensitive to drought stress during these stages [33]. Our results support the findings reported recently by investigators, which studied the influence of the cold plasma treatment on the oilseed rape (*Brassica napus* L.) seed germination under drought stress [31]. Ling et al. reported that, under drought stress, cold plasma treatment significantly improved the germination rate by 4.4–6.25% for various species of the oilseed rape [34]. Seedling growth characteristics, including shoot and root dry weights, shoot and root lengths, and lateral root number, were significantly improved after the cold plasma treatment. It is noteworthy that in our experiments the similar improvement of the germination rate was obtained and confined in the range of 1–12% for pepper and tomato seeds at 21°C, under various conditions of restricted water availability. Indeed, for tomato seeds it was shown that there is a direct relationship between uptake of water and germination rate (this conclusion is supported by recent results reported by other groups [35]), while for pepper seeds, no such correlation was found. This means that other possible effects of PT are responsible for the improvements in germination rates of pepper under plasma treatment (such as the regulation of energy metabolism [36]).

In addition, it was also shown that temperature plays a key role in the effect of plasma on seed germination of pepper. Our experiment demonstrates that at a high temperature of 27°C, in WW conditions, germination rates of non-PT pepper seeds were much higher than the rates of treated seeds, grown under MDS and HDS. The rates of germination of PT seeds were higher than those of WW-grown pepper seeds but considerably lower than those at 21°C. The reason for the drastic decrease in germination rates of PT seeds under WW conditions at higher temperatures remains unclear and would be further investigated. For tomato seeds the differences between germination rates at 27°C and those at 21°C were not so dramatic. These preliminary results demonstrate the importance of determining the right combination of germination temperature and PT duration for each cultivar tested. They may explain some of contradictory effects in investigations of the impact of plasma treatment on germination rates and germination speeds found reported by other researchers.

## 5. Conclusions

The influence of the cold radiofrequency air plasma treatment on the imbibition and germination of tomato and pepper seeds exerted to conditions appearing in natural seedbeds has been investigated. Plasma treatment markedly hydrophilized surface of seeds and increased the water imbibition (absorption). Various conditions of water supply to germinated seeds were studied. Drought stress essentially affected both germination rates and the kinetics of seed germination.

The intensity of harmful effects due to the water stress markedly depended on the species. In our study, a drought stress had a negative effect on the seed imbibition, germination rate, and the speed of germination of *S. lycopersicum* and of *C. annuum* seeds. This observation clarified the importance of plasma treatment, improving the water absorption by seeds. It was established that plasma treatment essentially enhanced the germination rate and the speed of germination in the case of drought stress for both studied cultivars. It was also shown that there is a direct relationship between uptake of water and germination rate of tomato seeds. Under conditions of nonrestricted water supply, the plasma treatment did not influence the germination rate.

## Acknowledgements

The work was supported by the Israel Ministry of Industry, Trade and Labor OCS Kamin Fund.

## Ethics

Does not apply.

## Data accessibility

All the data are contained within the manuscript.

## Competing interests

We have no competing interests.

## Author contributions

E.B. and E.D. conceived the research; Y.S. carried out plasma treatment of seeds, measured the water imbibition and germination rate, and analyzed the data; G.W. analyzed the data; and B.C.L. studied wetting properties of seeds. All authors reviewed the manuscript.

## Funding

This research received no specific grant from any funding agency in the public, commercial, or not-for-profit sectors.

## Author details

Yekaterina Shapira[1], Edward Bormashenko[1]*, Gene Whyman[2], Bat-Chen Lubin[3] and Elyashiv Drori[3]

*Address all correspondence to: edward@ariel.ac.il

1 Biotechnology and Materials Department, Engineering Faculty, Chemical Engineering, Ariel University, Ariel, Israel

2 Physics Department, Faculty of Natural Sciences, Ariel University, Ariel, Israel

3 Agriculture Research Department, The Samaria and Jordan Rift Regional R&D Center, Science Park, Ariel, Israel

## References

[1] Volin JC, Denes FS, Young RA, Park SMT. Modification of seed germination performance through cold plasma chemistry technology. Crop Science. 2000;**40**:1706-1718. DOI: 10.2135/cropsci2000.4061706x

[2] Filatova I, Azharonok E, Gorodetskaya L, Melnikova O, Shedikova A. Plasma-radiowave stimulation of plant seeds germination and inactivation of pathogenic microorganisms. Proceedings of International Plasma Chemistry Society. 2009;**19**:627

[3] Schnabel U, Niquet R, Krohmann U, Winter J, Schlüter O, Weltmann K D, Ejlbeck J. Decontamination of microbiologically contaminated specimen by direct and indirect plasma treatment. Plasma Processes and Polymers. 2012;**9**:569-575. DOI: 10.1002/ppap.201100088

[4] Ji SY, Choi JH, Pengkit A, Im JS, Kim JS, Kim Park Y, Hong TJ, Jung SK, Choi EH, Park G. Effects of high voltage nanosecond pulsed plasma and micro DBD plasma on seed germination, growth development and physiological activities in spinach. Archives of Biochemistry and Biophysics. 2016;**605**:117. DOI: 10.1016/j.abb.2016.02.028

[5] Selcuk M, Oksuz L, Basaran P. Decontamination of grains and legumes infected with *Aspergillus* spp. and *Penicillium* spp. by cold plasma treatment. Bioresource Technology. 2008;**99**:5104-5109. DOI: 10.1016/j.biortech.2007.09.076

[6] Nishioka T, Takai Y, Mishima T, Kawaradani M, Tanimoto H, Okada K, Misawa T, Kusakari S. Low-Pressure plasma application for the inactivation of the seed-borne pathogen *Xanthomonas campestris*. Biocontrol Science. 2016;**21**:37-43. DOI: 10.4265/bio.21.37

[7] Sera B, Spatenka P, Sery M, Vrchotova N, Hruskova I. Influence of plasma treatment on wheat and oat germination and early growth. IEEE Transactions on Plasma Science. 2010;**38**:2963-2968. DOI: 10.1109/TPS.2010.2060728

[8] Sera B, Sery M, Stranak V, Tichy M. Does cold plasma affect breaking dormancy and seed germination? A study on seeds of Lamb's Quarters (*Chenopodium album agg.*). Plasma Science and Technology. 2009;**11**:750-754. DOI: https://doi.org/10.1088/1009-0630/11/6/22

[9] Dobrin D, Magureanu M, Mandache NB, Ionita MD. The effect of non-thermal plasma treatment on wheat germination and early growth. IFSET Innovative Food and Science. 2015;**29**:255-260. DOI: http://dx.doi.org/10.1016/j.ifset.2015.02.006

[10] Ji SH, Kim J, Panngom K, Hong YY. Assessment of the effects of nitrogen plasma and plasma-generated nitric oxide on early development of *Coriandum sativum*. Plasma Processes and Polymer. 2015;**12**:1164-1173. DOI: 10.1002/ppap.201500021

[11] Dhayal M, Lee SY, Park SU. Using low-pressure plasma for *Carthamus tinctorium* L. seed surface modification. Vacuum. 2006;**80**:499-506. DOI: 10.1016/j.vacuum.2005.06.008

[12] Stolárik T, Henselová M, Martinka M, Novák O, Zahoranová A, Černak M. Effect of low-temperature plasma on the structure of seeds, growth and metabolism of endogenous phytohormones in Pea (*Pisum sativum L.*) Plasma Chemistry and Plasma Processing. 2015;**35**:659-676. DOI: 10.1007/s11090-015-9627-8

[13] Meiqiang Y, Mingjing H, Buzhou M, Tengcai M. Stimulating effects of seed treatment by magnetized plasma on tomato growth and yield. Plasma Science and Technology. 2005;**7**:3143-3147. DOI: 10.1088/1009-0630/7/6/017

[14] Živković S, Puač N, Giba Z, Grubišić D, Petrović Z. The stimulatory effect of non-equilibrium (low temperature) air plasma pretreatment on light-induced germination of *Paulownia tomentosa* seeds. Seed Science and Technology. 2004;**32**:693-701. DOI: 10.15258/sst.2004.32.3.05

[15] Kitazaki S, Koga K, Shiratanim M, Hayashi N. Growth enhancement of radish sprouts induced by low pressure $O_2$ radio frequency discharge plasma irradiation. Journal of Applied Physics. 2012;**51**:1S. DOI: 10.1143/JJAP.51.01AE01

[16] Filatova I, Azharonoki V, Kadyrov M, Beljavsky V, Gvozdov A, Shik A, Antonuk A. The effect of plasma treatment of seeds of some grain and legumes on their sowing quality and productivity. Romanian Journal of Physics. 2011;**56**:139-143. DOI: 10.1038/srep05859

[17] Jiafeng J, Xin H, Ling L, Jiangang L, Hanliang S, Qilai X, Renhong Y. Effect of cold plasma treatment on seed germination and growth of wheat. Plasma Science and Technology. 2014;**16**:54-58. DOI: https://doi.org/10.1088/1009-0630/16/1/12

[18] Shao C, Wang D, Tang X, Zhao L, Li Y. Stimulating effects of magnetized arc plasma of different intensities on the germination of old spinach seeds. Mathematical and Computer Modelling. 2013;**58**:814-818. DOI: http://dx.doi.org/10.1016/j.mcm.2012.12.022

[19] Mihai AL, Dobrin DM, Magurenau M, Popa ME. Positive effect of non-thermal plasma treatment in Radish seeds. Romanian Reports in Physics. 2014;**66**:1110-1117

[20] Randeniya LK, de Groot GJJB. Non-thermal plasma treatment of agricultural seeds for stimulation of germination, removal of surface contamination and other benefits: A review. Plasma Processes and Polymers. 2015;**12**:608-623. DOI: 10.1002/ppap.201500042

[21] Ling L, Jiafeng J, Jiangang L, Minchong S, Xin H, Hanliang S, Yuanhua D. Effects of cold plasma treatment on seed germination and seedling growth of soybean. Scientific Reports. 2014;**4**:5859. DOI: 10.1038/srep05859

[22] Bormashenko E, Grynyov R, Bormashenko Y, Drori E. Cold radiofrequency plasma treatment modifies wettability and germination speed of plant seeds. Scientific Reports. 2012;**2**:741. DOI: 10.1038/srep00741

[23] Bormashenko E, Shapira Y, Grynyov R, Whyman G, Bormashenko Y, Drori E. Interaction of cold radiofrequency plasma with seeds of beans (*Phaseolus vulgaris*). Journal of Experimental Botany. 2015;**66**:4013-4021. DOI: https://doi.org/10.1093/jxb/erv206

[24] Odrášková M, Ráhel J, Zahoranová A, Tiňo R, Černák M. Plasma activation of wood surface by diffuse coplanar surface barrier discharge. Plasma Chemistry and Plasma Physics. 2008;**28**:203-211. DOI: 10.1007/s11090-007-9117-8

[25] Černák M, Černáková L', Hudec I, Kováčik D, Zahoranová A. Diffuse coplanar surface barrier discharge and its applications for in-line processing of low-added-value materials. European Physics Journal Applied Physics. 2009;**47**:22806. DOI: 10.1051/epjap/2009131

[26] Lieberman MA, Lichtenberg AJ. Principles of Plasma Discharges and Materials Processing. 2nd ed. Hoboken, NJ, USA: John Wiley & Sons; 2005

[27] Richards FJ. A flexible growth function for empirical use. Journal of Experimental Botany. 1959;**10**:290-300. DOI: https://doi.org/10.1093/jxb/10.2.290

[28] Hara Y. Calculation of population parameters using Richards function and application of indices of growth and seed vigor to rice plants. Plant Production Science. 1999;**2**:129-135. DOI: http://dx.doi.org/10.1626/pps.2.129

[29] Clodomiro AJ, Jussier de OV, Dinnara LS da S, Mikelly de LF, Nadjamara B de LD. Water uptake mechanism and germination of *Erythrina velutina* seeds treated with atmospheric plasma. Scientific Reports. 2016;**6**:33722. DOI: 10.1038/srep33722

[30] Bitarafan Z, Rad AHS. Water stress effect on spring rapeseed cultivars yield and yield components in winter planting. International Journal of Physical Science. 2012;**7**:2755-2767. DOI: 10.5897/IJPS12.221

[31] Müller T, Lüttschwager D, Lentzsch P. Recovery from drought stress at the shooting stage in oilseed rape (*Brassica napus*). Journal of Agronomy and Crop Science. 2010;**196**:81-89. DOI: 10.1111/j.1439-037X.2009.00391.x

[32] Sadeghian SY, Yavari N. Effect of water-deficit stress on germination and early seedling growth in sugar beet. Journal of Agronomy and Crop Science. 2004;**190**:138-144. DOI: 10.1111/j.1439-037X.2004.00087.x

[33] Li FL, Bao WK, Wu N. Morphological, anatomical and physiological responses of *Campylotropis polyantha* (French.) Schindl. seedlings to progressive water stress. Scientia Horticulturae-Amsterdam. 2011;**127**:436-443. DOI: 10.1016/j.scienta.2010.10.017

[34] Ling L, Jiangang L, Minchong S, Chunlei Z, Yuanhua D. Cold plasma treatment enhances oilseed rape seed germination under drought stress. Scientific Reports. 2015;**5**:13033. DOI: 10.1038/srep13033

[35] Meng Y, Qu G, Wang T, Sun Q, Liang D, Hu Sh. Enhancement of germination and seedling growth of wheat seed using dielectric barrier discharge plasma with various gas sources. Plasma Chemistry and Plasma Processing. 2017;**37**:1105-1119. DOI: 10.1007/s11090-017-9799-5

[36] Zhang JJ, Jo JO, Huynh DL, Mongre RK, Ghosh M, Singh AK, Lee SB, Mok YS, Hyuk P, Leong DK. Growth-inducing effects of argon plasma on soybean sprouts via the regulation of demethylation levels of energy metabolism-related genes. Scientific Reports. 2017;**7**:41917. DOI: 10.1038/srep41917

Chapter 8

# Metabolic Processes During Seed Germination

Awatif S. Ali and Alaaeldin A. Elozeiri

Additional information is available at the end of the chapter

http://dx.doi.org/10.5772/intechopen.70653

**Abstract**

Seed germination is crucial stage in plant development and can be considered as a determinant for plant productivity. Physiological and biochemical changes followed by morphological changes during germination are strongly related to seedling survival rate and vegetative growth which consequently affect yield and quality. This study is aimed to focus on proceeding of the most vital metabolic processes namely reserve mobilization, phytohormonal regulation, glyoxylate cycle and respiration process under either stressful or non-stressful conditions that may be led to suggest and conduct the more successful experimental improvements. Seed imbibition triggered the activation of various metabolic processes such as synthesis of hydrolytic enzymes which resulted in hydrolysis of reserve food into simple available form for embryo uptake. Abiotic stresses potentially affect seed germination and seedling establishment through various factors, such as a reduction in water availability, changes in the mobilization of stored reserves, hormonal balance alteration and affecting the structural organization of proteins. Recent strategies for improving seed quality involved classical genetic, molecular biology and invigoration treatments known as priming treatments. $H_2O_2$ accumulation and associated oxidative damages together with a decline in antioxidant mechanisms can be regarded as a source of stress that may suppress germination. Seed priming was aimed primarily to control seed hydration by lowering external water potential, or shortening the hydration period.

**Keywords:** reserve mobilization, proteolysis, glyoxylate cycle, phytic acid, seed priming, stress tolerance mechanisms

## 1. Introduction

Seed germination is vital stage in plant development and can be considered as a determinant for plant productivity. It begins by water imbibition, mobilization of food reserve, protein synthesis and consequence radicle protrusion [1]. To sustain a good seedling development, seed stores a food reserve mainly as proteins, lipids and carbohydrates [2]. Protein and oil bodies are the major reserve in oilseed which represent a source for each of energy, carbon, and nitrogen during seedling establishment [3]. Because the physiology of reserve mobilization during germination and post-germination events is still poorly understood, extensive studies must be performed to know the metabolic mechanisms of reserve food mobilization providing insights into the ability to use such seeds as planting material [4]. Enzymatic hydrolysis of protein, lipid and carbohydrate, and transportation of metabolites is dependent mainly on water availability [5].

Physiological and biochemical changes followed by morphological changes during germination are strongly related to seedling survival rate and vegetative growth which affect yield and quality. Food reserve of starch and protein are mainly stored in the endosperm. In general, germination process can be distinguished into three phases: phase I, rapid water imbibition by seed; phase II, reactivation of metabolism; and phase III, radicle protrusion [6]. The most critical phase is phase II whereas, the essence physiological and biochemical processes such as hydrolysis, macromolecules biosynthesis, respiration, subcellular structures, and cell elongation are reactivated resulting in initiation of germination [7].

Water imbibition by reserve substances in germinating wheat seed stimulates the embryo to produce phytohormones mainly gibberellic acid (GA) which can diffuse to aleurone layer and initiate a signaling cascade resulting in the synthesis of $\alpha$-amylases and other hydrolytic enzymes. Then, hydrolytic enzymes secrete into the endosperm and hydrolyzed food reserve [8, 9]. Germination is considered a response includes bidirectional interactions between the embryo and endosperm since the endosperm can secrete signals to control embryo growth [10]. Previous studies were investigated the activity of some key enzymes in glycolysis, pentose phosphate pathway (PPP), the tricarboxylic acid cycle (TCA cycle), and amino acid metabolism during germination [11].

Seed germination is particularly vulnerable to environmental stress encountered conditions, specifically salt and water which are widespread problem around the world [12]. High salt and drought tolerance seeds might be showed rapid germination resulting in a good seedling establishment and hence expected to maintain high yield productivity [13]. Water and salt stress conditions affect seed germination with reducing germination rate and delay in the initiation of germination [14]. Under water stress, enzymes activity such as $\alpha$-amylase in *Cicer arietinum* cotyledons [15] or $\alpha$- and $\beta$-amylase in *Medicago sativa* germinating seeds [16] were reduced. In contrast, water stress conditions led to an increase in the activity of $\alpha$-amylase in *Hordeum vulgare* seedlings [17], $\beta$-amylase in *Cucumis sativus* cotyledons [18], cytosolic glyceraldehyde-3-phosphate dehydrogenase in *Craterostigma plantagineum* plants [19] and protease in *Oryza sativa* seedlings [20]. Salt stress causes ion toxicity, osmotic stress and reactive oxygen species (ROS) stress [21]. ROS reacts with cell macromolecules [22] and lipids [23], and disrupt diverse physiological and biochemical processes, such as hormonal imbalance and reduced use of reserves [24]. Plants develop ROS-scavenging mechanisms include enzymatic and non-enzymatic antioxidant systems [25] that protect plants against oxidative damage. Therefore, improvement the activity of antioxidant enzymes in plants organs is necessary for increasing plant's salt tolerance. Species and varieties/cultivars varied in their ability for salt tolerance mechanism. Comparing with adult plant, the mechanisms of stress tolerance in germinating phase are poorly interpreted and might be related to a series of factors that are inherent to the species and environment [26, 27].

Phytohormones have essential role in inducing plant acclimatization to change in environmental conditions by mediating growth, development, source/sink transitions, and nutrient allocation [28]. Phytohormones are considered the most important endogenous substances for modulating physiological and molecular responses [28]. They include auxin (IAA), cytokinins (CKs), abscisic acid (ABA), ethylene (ET), gibberellins (GAs), salicylic acid (SA), brassinosteroids (BRs), and jasmonates (JAs). The strigolactone (SL) are relatively new phytohormones.

Genetically and physiological studies have been demonstrated the effective roles of the plant hormones ABA and GAs in regulation of dormancy and germination [29]. To counteract the adverse effects of abiotic stress, seed priming methods have been applied to improve germination, uniformity, improve seedling establishment and stimulate vegetative growth in more field crops [30, 31]. Wheat seeds were priming to increase germination characteristics and stress tolerance. As seeds imbibe water, metabolic processes initiate with an increase in respiration rate [7]. Early developmental stages of seedling require fueling energy before it becomes autotrophic [32].

Seeds store mineral nutrients as sucrose or amino acids which are synthesized into starch or proteins during development to be used in early seedling emergence. Phosphorus is taken up by plants as phosphate and translocate to developed seeds where it is stored in phytic acid form mainly (about 75%).

## 2. The role of hydrolytic enzymes in seed germination

On seed hydration, separate intercellular bodies of seed stored carbohydrates, proteins, lipid and phosphate act as energy source and carbon skeleton [33]. Seed imbibition triggered many metabolic processes such as activation or freshly synthesis of hydrolytic enzymes which resulted in hydrolysis of stored starch, lipid, protein hemicellulose, polyphosphates and other storage materials into simple available form for embryo uptake. Also, consumption of an elevated level of oxygen may be induced activation/hydration of mitochondrial enzymes, involved in the Krebs cycle and electron transport chain [34, 35].

### 2.1. Hydrolysis of storage seed proteins

Proteolytic enzymes have the main role in using stored protein in metabolism of germinating seeds which proceed through many stages [36]. According to Gepstin and Ilan [37], proteolytic activity in germinating beans increased during the first 7 days which partially dependent on the embryonic axis. Proteases and peptidases have been detected in many seeds during germination whereas; plant protease and amylase inhibitors which are proteinaceous in nature are being disappeared [38]. Antitryptic and antichymotryptic activities were observed to be markedly reduced in the endosperm of finger millet on germination which might be attributed to the proteolytic activity in hydrolysis of the inhibitory proteins [39]. Hydrolysis of stored proteins produced free amino acids, which support protein synthesis in endosperm and embryo and so proceeding of germination process [40]. Schlereth et al. [41] recorded an initial little decrease in free amino acids at the beginning of vetch seeds imbibition which is attributed to leakage from the axis, but remain without change during late germination stage.

A disulfide proteome technique was developed by Yano et al. [42] to visualize redox changes in proteins. This technique was used to analyze rice bran resulting in identification of embryo-specific protein 2 (ESP2), dienelactone hydrolase, putative globulin, and globulin-1S-like protein as putative target of thioredoxin, which support the hypothesis that thioredoxin activates cysteine protease with a concurrent unfolding of its substrate during germination [43].

In buckwheat seeds, the main storage protein constituent about 16% of total seed protein is the 13S globulin with molecular mass of about 300 kDa and consists of acid and basic subunits with molecular masses ranging from 57.5 to 23.5 kDa [44]. During seed germination, 13S globulin is hydrolyzed by proteolytic enzymes through stages and the products are used by the growing seedling. The first stage of the 13S globulin degradation resulted from a limited proteolysis activity of metalloproteinase with the cleavage of about 1.5% of peptide bonds. This stage proceeds during the first 3 days of germination. It takes place during the first 3 days of germination [45]. Metalloproteinase activity is controlled by a proteinaceous inhibitor (Mr—10 kDa), present in dry buckwheat seeds in a complex with the enzyme which dissociated by bivalent cations liberated from phytin hydrolysis process. Phytin is present in buckwheat seeds in sufficient amount in the form of globoids disposed in protein bodies [46].

During the second stage of 13S globulin degradation; the products of metalloproteinase protein activity hydrolyzed into small peptides and amino acids at acid pH (5.6) by cysteine proteinase and carboxypeptidase which appear in germinating seeds [47]. It was clear that cysteine proteinase is able to hydrolyze only the modified 13S globulin but not the native. The role of carboxypeptidase is to facilitate the flow of storage protein hydrolysis and works in cooperation with cysteine proteinase. At latest stage when pH becomes more acidic (5.0) in the vacuoles, aspartic proteinase which is present in dry seeds is involved into the course of hydrolysis protein bodies.

### 2.2. Hydrolysis of storage seed starch

Carbohydrates represent the most storage food constituent in cereal grains, whereas it contains about 70–80% starch, about 15% protein, less than 5% lipids, minerals and vitamins. In cereals, most hydrolysis enzymes are produced in the aleurone or scutellum in response to germination signals. Several modified seed systems were used to detect the induction process and identify potential factors controlling enzyme induction in absence of the embryo [48].

Chrispeels and Varner [49] observed that isolated aleurone failed to synthesize $\alpha$-amylase in a manner quantitatively similar to distal half seeds led to correction by adding calcium to the medium. The role of calcium might be expected to involve amylase stability, and to have a much more complex involvement in regulating enzyme activities [50]. Because of *de novo* amylase synthesis during seed germination to stimulate the stored starch mobilization for providing young plant till photosynthesis will be initiate, amylase has been showed high activity [51]. Parys et al. [52] showed that the amylase activity is regulated by the concentration of reducing sugars in vivo in both cotyledons and axis. At the time, the amylase activity in the cotyledons increased gradually and reached a maximum on the 5th day of germination process, while the starch decreased and soluble sugars increased [53].

Many studies which concerned with studying the essentiality of α-amylase activity during seed germination under drought stress and could be summarized as follows; the promotion of drought stressed germinating seeds is a result of high α-amylase activity directly but, it might be related to adaptive strategy to water deficit since its activity is required for solutes accumulation and decrease osmotic potential [54, 55]. In addition, α-amylase synthesis inhibition might be not a mechanism by which drought prevents the germination of *Agropyron desertorum* seeds [54]. GAs can alleviate the drought stress-caused inhibition of seed germination through regulation of α-amylase [19].

### 2.3. Hydrolysis of storage seed lipids

Generally oilseeds composed of two parts, the kernel which is main part and the seed covering that enclosed the kernel and called the husk or tegument. The kernel comprised two parts which are the embryo and the endosperm. Lipase activity is investigated during seed germination where it is maximum value [56, 57]. Triacylglycerols is stored in oleosomes and comprise in range from 20 to 50% of dry. As germination proceeds, triacylglycerols are hydrolyzed to produce energy which required for the synthesis of sugars, amino acids (mainly asparagine, aspartate, glutamine and glutamate) and carbon chains required for embryonic growth [58].

Lipid level and lipase activity were studied in various germinating seeds. It was showed that β-oxidation takes place 4 days after germination of Castor been seeds [59]. The major hydrolytic enzymes concerned with the lipid metabolism during germination are the lipases which catalyze the hydrolysis of ester carboxylate bonds and releasing fatty acids and organic alcohols [60, 61] and the reverse reaction (esterification) or even various transesterification reactions [62]. The ability of lipases to catalyze these reactions with great efficiency, stability and versatility makes these enzymes highly attractive from a commercial point of view.

Villeneuve [63] and others classified lipases specificities into three main groups; the 1st group is **substrate specificity** in which glycerol esters represent the natural substrates, the 2nd group is called **regioselective** and involves the subgroups *non-specific lipases* that hydrolyze the triacylglycerols into fatty acids and glycerol in a random way with production of mono- and diacylglycerols as intermediate products (**Figure 1**); *specific 1.3 lipases* which catalyze the hydrolysis at $C_1$ and $C_3$ glycerol bonds in triacylglycerols with liberating of fatty acids and unstable intermediates 2-monoacylglycerols and 1.2-or 2.3-diacylglycerols and *specific or selective type fatty acid* that hydrolyze the ester bond of a specific fatty acid or a specific group of fatty acids at any position of triacylglycerol. The 3rd group **enantioselective** could identify enantiomers in a racemic mixture. The enantio specificities of lipases depend on the type of substrate [64].

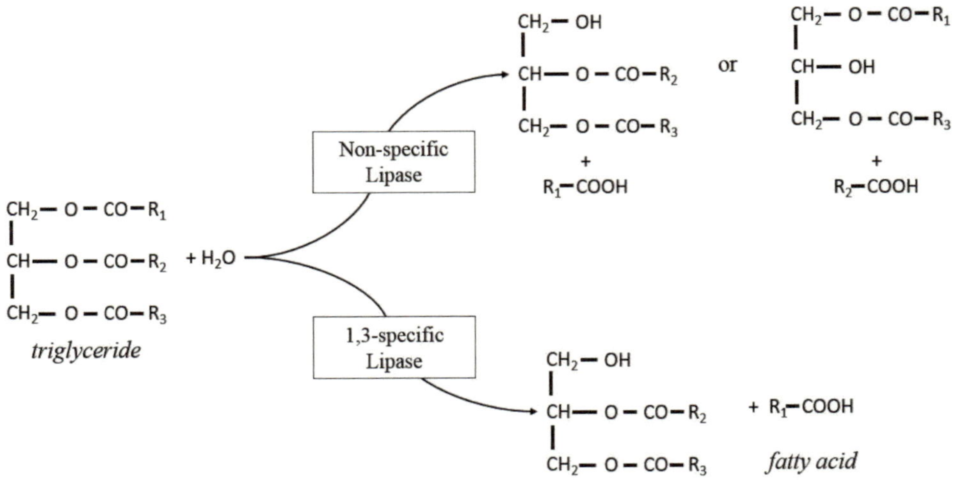


**Figure 1.** Regioselective: non-specific and 1,3 specific lipases catalyze the hydrolysis of triglycerides in different manners with the production of fatty acids.

The induction of lipase activity during germination might be dependent on factors from embryo [65]. Early study of Shoshi and Reevers [66] showed the presence of two lipases in the endosperm of Castor been seed, acid lipase in dry seed and alkaline lipase during germination. On the other hand, storage tissues of all the oilseeds except Castor bean contained only lipase activity which increased during germination [67].

Because of sucrose is the substrate for lipid biosynthesis in developing seed and the end product of lipid degradation, it might be primarily considered as regulatory factor in studying the mechanisms of lipid metabolism [58, 68]. In addition, asparagine and nitrate are considered regulatory factors in lipid metabolism of lupine [69]. In lupin germinating seeds, the level of asparagine can reach 30% of dry matter, and it is a main transport form of nitrogen from source to sink tissues [70]. Borek et al. [71] reported that asparagine controls the metabolism of carbohydrate as it caused a significant decrease in soluble sugars and increase in starch in organs of germinating lupin seed. In contrast, nitrate is not a favorable source of nitrogen in protein metabolism in lupin seeds [72] and rather does not influence the carbohydrate metabolism [71]. Nitrate similarly as N sucrose, is regarded as a factor which can regulate plant metabolism by changes in the expression of some genes [73].

Storage lipid mobilization in germinating seeds begins with hydrolysis of triacylglycerols in oleosomes by lipases into free fatty acids and glycerol. Then fatty acids undergo β-oxidation in peroxisomes. Next, glyoxylate cycle will proceed partially in the peroxisome and partially in the cytoplasm. Three of the five enzymes of the glyoxylate cycle (citrate synthase, isocitrate lyase and malate synthase) are located in peroxisomes, while two other enzymes (aconitase and malate dehydrogenase) operate in the cytoplasm [74]. Succinate transported from peroxisome to mitochondria and here is converted to malate via the Krebs cycle. Malate in turn, after transport to the cytoplasm, is converted to oxaloacetate. Finally, gluconeogenesis and the synthesis of sugars are the processes which are a form of carbon transport especially in germinating seeds proceed [58, 75].

### 2.4. Hydrolysis of phytic acid during seed germination

The greatest storage form of total phosphorus (about 50–80%) is phytic acid ($C_6H_{18}O_{24}P_6$) and also known as inositol hexophosphate (IP6) in legumes and cereals seeds [76]. Phytic is regarded as antinutrient because it has the ability to form complexes with proteins and bind with cations (especially Fe, Ca, K, Mn, Mg, Zn) via ionic association to form a mixed salt called phytin or phytate with the reduction of their digestive availability [77]. On the other hand, phytate may play an important role as an antioxidant by forming iron complex that cause a decrease in free radical generation and the peroxidation of membranes, and may also act as an anticarcinogen, providing protection against colon cancer [78]. Because of it was regarded as antioxidant, anticarcinogen or vitamin like substance, it is essential to measure and manipulate phytate content in food grains such as beans [79, 80].

One of the major breeding objectives is the development of crop cultivars with low seed phytin content. It was found that the increase in *myo*-inositol and reduced amounts of *myo*-inositol phosphate intermediates in the seeds of maize mutants with a phenotype of reduced phytic acid had a little effect on plant growth and development [81]. These findings might suggest that a high level of stored phytate is not necessary for seed viability and germination or seedlings growth.

Phytin is mainly stored in protein bodies in seeds called globoids in the aleurone layer and scutellum cells of most grains. Phytic acid has a strong ability to chelate multivalent metal ions, specially zinc, calcium, iron and as with protein residue. Seed phytate content depend mainly on the environmental mainly plant phosphorus fertilization [82]. It has been shown the important genetic variability in the phytate content of beans and it appears to be a trait controlled by several genes [83]. Also, a correlation between phytate and protein contents was found [84], so the protein content of grains can be considered another factor that regulates phytate content.

Phytin in germinating seeds is hydrolyzed by an acid phosphatase enzyme called phytase [85], with releasing of phosphate, cations, and inositol which are utilized by the seedlings. It was found little changes in extractible $P_i$ in hazel seeds during chilling accompanied with IP6 mobilization that might be suggested the rapid conversion of $P_i$ into organic form [86]. These results were discussed as evidence of active metabolism in germinating seed [87]. In agreement, phytase is strongly and competitively inhibited by $P_i$, while the decrease in phytase activity coincided with maximal IP6 turnover [88]. It was found that about 87% of IP6 is digested during the first 6 days of germination [89]. In this respect, Ogawa et al. [90] postulated that the early axiferous IP6 digestion is essential for metabolic activity of the resting tissue via supplying $P_i$ and minerals for physiological and metabolic requirements, for example, enzymes of starch metabolism. In addition, IP6 related compounds such as pyrophosphate-containing inositol phosphates (PP-IP) play a potential role in providing $P_i$ for ATP synthesis during the early stages of germination before complete dependence on aerobic mitochondrial respiration the mainly source of ATP production [91].

In stressed seeds, many vital processes such as germination, growth, respiration and other related processes are affected which consequently can trigger other effects on metabolic activities particularly the enzymes of phosphate metabolism that play an important role in germination and seed development [92]. Phosphate metabolism is one of negatively affected processes under different stressful conditions [93]. Under stressful conditions, the restriction of growth and phosphorus availability resulting in enhancement the activity of phosphatases to produce $P_i$ by hydrolysis the insoluble phosphate form that modulate mechanism of free phosphate uptake. In agreement, Olmos and Hellin [94] reported that acid phosphatases activity increased to sustain $P_i$ level which enables it to be co-transported with $H^+$ down a proton motive force gradient.

## 3. Effect of abiotic stress on metabolic activities during seed germination

Abiotic stresses including salt, drought, heavy metals, pollutants, heat, etc., potentially affect seed germination and seedling growth. Depending on the stress intensity and genetic background, germination is delayed or suppressed. Plants have developed unique strategies including a tight regulation of germination ensuring species survival [95]. It was well known that stress exposure would produce early signals such as change in intracellular $Ca^{2+}$, secondary signaling molecules such as inositol phosphate and ROS as well as activation of kinase cascades.

Seed imbibition triggers many biochemical and cellular processes associated with germination involve the reactivation of metabolism, the resumption of cellular respiration and the biogenesis of mitochondria, the translation and/or degradation of stored mRNAs, DNA repair, the transcription and translation of new mRNAs, and the onset of reserve mobilization [7, 96]. These processes are followed by ROS (mostly $H_2O_2$) accumulation as a result of a pronounced increase in the intracellular and extracellular production during early stages [97, 98].

ROS function as cellular messengers or toxic molecules on seed hydration [99]. ROS caused seed damage accompanied with a loss of seed vigor and as a repercussion of aging [100]. The highly activity of respiration during germination results in superoxide anion production during electron leakage from the mitochondrial electron transport chain followed by dismutation to $H_2O_2$. Other sources of ROS are NADPH oxidases of the plasma membrane, extracellular peroxidases, β-oxidation pathway in glyoxysomes [97]. $H_2O_2$ is along-lived ROS that can diffuse easily through membranes and that can reach targets far from production sites, and is recognized as an important signaling molecule [101]. $H_2O_2$ is considered as strong oxidizing agent, it could interact with most biomolecules resulting in oxidative stress that causes cellular damage. It causes lipid peroxidation which in turn affects polyunsaturated fatty acids (PUFAs) found in membranes or reserve lipids. Also, $H_2O_2$ cause oxidation of nucleic acids (DNA, RNA) and proteins [97]. Induction of DNA oxidation by $H_2O_2$ resulted in the accumulation of 7, 8-dihydro-8-oxoguanine (8-oxo-dG), which has been shown to cause the accumulation of double- strand breaks in genome and deleterious effects on cell viability [102]. DNA oxidation by ROS is considered a main source of DNA damage during seed storage and germination.

Kong and Lin [103] have shown that mRNA is much more sensitive to oxidative damage than DNA, mainly due to its cellular localization, single stranded structure and lack of repair mechanisms. Guanine is the most frequently oxidized base in RNA leads to the accumulation of 8-hydroxyguanosine (8-OHG). Oxidative damage to mRNA results in the inhibition of protein synthesis and in protein degradation [104]. Oxidation of protein by ROS result in alteration of protein functions due to enzymatic and binding properties modifications [105]. $H_2O_2$ accumulation and associated oxidative damages together with a decline in antioxidant mechanisms can be regarded as a source of stress that may suppress germination. On the other hand, Barba-Espin et al. [106] reported that the selective oxidation of proteins and mRNAs can act as a positive regulator of seed germination.

Using of calcium sensors called $Ca^{2+}$ binding proteins revealed an increase in intracellular calcium concentration under abiotic-stress conditions [107]. This is accompanied with enhancement of calcium-dependent protein kinases (CDPKs), calcium/calmodulin-dependent protein kinases (CCaMKs) or phosphatases which stimulate the phosphorylation/or dephosphorylation of specific transcription factors, resulting in an increase of stress-responsive genes expression [108]. However, activated $Ca^{2+}$ sensors regulate stress-responsive genes either by binding to cis-elements in the promoters or by interacting with DNA-binding proteins of genes that led to gene activation or suppression.

Stressed-germinating wheat seeds develop a powerful regulator mechanism in response to stresses which is calreticulin-like protein (M16 and M13) and abundant $Ca^{2+}$-binding protein predominantly located in the endoplasmic reticulum (ER) of higher plants [109]. Its expression trend was mainly up-regulated, especially in the last period of germination which hints that wheat seed may encounter stress in late germination [110]. Another regulator mechanism with peptidyl-prolyl cis-trans isomerase activity which involved in signal transduction, cell apoptosis, and protein folding called cyclophilin (M51) was detected in stressed germinating wheat seeds [111]. Because of the cellular structure is not complete in early germination, M51 increased slowly in first three germination stages but increased sharply in the last stage [109].

One of the most effective factors on seed imbibition and germination is the temperature. It affects water uptake and reactivation of metabolic processes [7]. Many physiological, biochemical and molecular disturbance will occur with temperature deviation away from optimal to sustain cellular homeostasis [112].

## 4. The role of phytohormones during germination

Plants are characterized by producing various types of growth regulators that differed in their chemical structure and physiological action. They include auxins, cytokinins (CK), gibberellins (GA), abscisic acid (ABA), ethylene (ET), salicylic acid (SA), jasmonates (JA), brassinosteroids (BR) and strigolactones. Each of ABA, SA, JA and ET is found to play an essential role in mediating plant defense response against stresses [113]. During the early phase of seed germination, a decrease in JA and SA contents and an increased level of auxins were recorded in *Arabidopsis* seeds [114]. Both JA and SA were shown to act as negative regulators of seed germination [115]. Auxins are considered to be regulators of the seed germination process in a crosstalk with GAs, ABA, and ET [116]. The brassinosteroids signal could stimulate germination by decreasing the sensitivity to ABA [117].

A variety of cellular processes in plants are under control of phytohormones which play key roles and coordinate various signal transduction pathways during abiotic-stress response [118]. Seed imbibitions resulted in an activation of GA biosynthesis and response pathways with the production of the bioactive GAs. Then, GAs stimulated the genes encoding for enzymes such as endo-β-1,3 glucanase [119], β-1,4 mannan endohydrolase [120] which hydrolyze the endosperm and alleviate the inhibitory effects of ABA on embryo growth potential [121]. These results are indicated the antagonistic relation between each of ABA and GA which interpret the presence of high GA and low ABA levels in seeds under favorable environmental conditions and a reverse ration under unfavorable conditions. Thus, the cross-talk relation between seed dormancy and germination is balanced by GA-ABA ration, a key mechanism for cope early abiotic-stress conditions.

ABA inhibits water uptake by preventing cell wall loosening of the embryo and thereby reduces embryo growth potential [122]. GAs are involved in direct enhancement the growth of the embryo during late phase [123]. GAs repressive the ABA effect during the early and the late phases of germination through stimulation of genes expression encoding cell wall loosening that result in remodeling enzymes such as α-expansins in early phase of germination. Light and cold act together to break dormancy of imbibed seeds and to promote seed germination by increasing GAs levels. A rapid decrease of ABA endogenous content during Phase II is one of many factors that influence the successful completion of germination [124]. Highly leakage of cellular solutes due to initial imbibition indicates cellular membranes damage caused by rehydration. In addition, drying and rapid seed dehydration processes influence DNA integrity [125]. Seeds have developed a number of repair mechanisms during seed germination, including the repair of membranes, as well as proteins and DNA [126].

Under stress conditions, phytohormones play a crucial role via responsive protein mediated stress. It was found C1-(cysteine rich protein family) domain containing proteins that play a part in plant hormone-mediated stress responses [127]. In addition, 72 responsive proteins mediated stress are identified in *Arabidopsis* that contained all three unique signature domains. Many hydrolytic enzymes biosynthesis and activity are influenced by $GA_3$ in wheat and barley. Catalase and ascorbate peroxidase activity showed a significant improvement in wheat SA- and GA-primed wheat seeds compared to the unprimed [128, 129].

## 5. Priming strategy to improve seed germination under stressful or non-stressful conditions

Under various conditions, the potential of seeds for rapid uniform emergence and development under various conditions is determined mostly by seed vigor trait [130]. Recent strategies for improvement seed quality involved classical genetic, molecular biology and invigoration treatments known as priming treatments. Seed priming was aimed primarily to control seed hydration by lowering external water potential, or shortening the hydration period, because of most seeds are partially hydrated after priming process and reach a pre-germinate stage without radicle protrusion [131]. It was reported that primed seeds showed improved germination rate and uniformity under both optimal and adverse environments in wheat [132]. The cellular mechanism of priming as it relates to improved stress tolerance in germinating seeds is still required more study.

Currently seed priming techniques include osmopriming (soaking seeds in osmotic solutions as PEG or in salt solutions), hydropriming (soaking seeds in predetermined amounts of distilled water or limiting imbibition periods), and hormone priming (seed are treated with plant growth regulators) which are more commonly studied in laboratory conditions, and thermopriming (it is a physical treatment achieved by pre-sowing of seeds at different temperature that improve germination vigor under adverse environmental conditions) and matric priming (mixing seeds with organic or inorganic solid materials and water in definite proportions and in some cases adding chemical or biological agents) [130, 133]. Hydropriming and osmopriming with large-sized priming molecules cannot permeate cell wall/membrane so water influx would be the only external factor affecting priming. The determination of suitable priming technique is dependent mainly on plant species, seed morphology and physiology. On the other hand, salts and hormone priming affect not only the seed hydration but also other germination-related processes due to absorption of exogenous ions/hormones, consequently confusing the effects of imbibition *versus* that of ions/hormones.

Improvement germination performance of primed seeds may be considered a result of advanced metabolism processes [134] including enhancement each of the efficiency of respiration [135] and antioxidant activity [136], initiation of repairing processes [137] and alteration phytohormonal balance [138]. Also, improvement of germination performance may be linked to higher expressions of gene sand proteins involved in water transport, cell wall modification, cytoskeletal organization, and cell division and increases in protein synthesis potential, post-translational processing capacity, and targeted proteolysis have been linked to the advanced germination of primed seeds [139].

Seed germination process is regulated by a network of transcription factors that have both confused and separate functions. In order to maintain or break the period of arrested germination and to complete germination under stress conditions, different metabolic pathways including phytohormones biosynthesis and signal transduction pathways, chromatin modifications, and microRNA post transcriptional regulation, are involved [140].

Many effects on metabolic processes, germination performance and seedling establishment due to seed priming with $H_2O_2$ were observed although seed soaking followed by dehydration have an important role in controlling gene expression and biosynthesis of proteins [141].

Seed priming with auxin, cytokinin, GA, and ethylene (ET) resulted in improvement of germination of pigeon pea seeds under both control and Cd-stress conditions [142]. ABA pretreated seeds showed a reduction in germination that may be attributed to metabolic deviation, limiting the available energy and changes in metabolomics or may be attributed to modulate the endogenous ABA level [143]. On contrary, $GA_3$ seed treatment has not affect seed germination substantially. It is documented that $GA_3$ have a stimulatory effects on germination and associated enzymes [144]. Also, auxin namely IAA is documented to regulate seed dormancy and plant shade avoidance syndrome that adversely affects seedling development and crop yield [145]. Cytokinin pretreatment may act as auxins in promoting seed germination by antagonizing the inhibitory effect of ABA on germination process. However, it was found that cytokinin antagonize the inhibitory effect of ABA on post-germinating growth of *Arabidopsis* through the stimulation of ABI5 protein degradation [146].

Recently published data support the existence of interactions between ROS and phytohormone signaling networks that modulate gene expression and cellular redox status [147]. Interaction between phytohormones and $H_2O_2$ can be antagonistic or synergistic. Signaling processes trigger interactions are not developed only between particular phytohormones but also between phytohormones and other signaling molecules such as NO [148], $H_2S$ [149], ·OH [150] and $H_2O_2$ [151], which is believed to play a central role in signaling processes during plant development and stress responses [152]. GA treatment enhanced ROS production namely superoxide and $H_2O_2$ in radish plants [153] and *Arabidopsis* [154]. On the other hand, exogenous application of $H_2O_2$ does not influence ABA biosynthesis and signaling but it has a more pronounced effect on GA signaling, resulting in the modulation of hormonal balance and in subsequent germination initiation [154]. It was showed that $H_2O_2$ diminished the inhibitory effects of ABA on endosperm damage. Müller et al. [155] showed that $H_2O_2$ abolishes inhibitory effects of ABA on endosperm rupture. As suggested previously by Lariguet et al. [154], $H_2O_2$ regulates the expression of gene encoding enzyme hydrolyzing the testa and endosperm, which facilitate *Arabidopsis* germination by releasing the embryo from the control of the seed envelope.

## 6. The respiratory reactivation during seed germination

The initial liberation of seed stored food at the beginning of germination is mainly by anaerobic respiration. Anaerobic respiration is catalyzed by the activity of enzymes which are not required aerobic conditions such as dehydrogenases [156]. Dehydrogenase facilitating the transport of electrons from substrates to oxygen through electron transport chain using nicotinamide adenine dinucleotide ($NAD^+$), nicotinamide adenine dinucleotide phosphate ($NADP^+$) or riboflavin as cofactor [157]. Activities of dehydrogenases have been shown to involve the activities of alcohol dehydrogenase, lactate dehydrogenase and succinate dehydrogenase [158] which mediated the conversion of storage lipid and carbohydrates through the anaerobic respiration. Succinate dehyrogenase, a complex enzyme tightly bound to the inner mitochondrial membrane oxidizes succinate to fumarate [159]. Lactate dehydrogenase catalyzes the reversible oxidation of lactate to pyruvate using $NAD^+$ as a co-enzyme. Anaerobic respiration was recorded to take place during resting stages of seeds and the initial stages of seed germination [160]. It was showed that the reactivity of dehydrogenases covered the first 3 days of cowpea seed germinations [161].

The increase in respiratory rate in germinating seeds is associated with the increase in glycolytic activity. The intermediates of glycolysis are transferred to the OPPP pathway which feeds its products back into glycolysis, so the activity of this pathway is also important in determining the flux through glycolysis [162]. During germination, seeds use sugars and other molecules as a substrate for respiration. α-amylase and β-amylase are involved in degradation of endosperm starch. Starch hydrolysis into glucose is catalyzed by action of α- and β-amylases, debranching enzyme and α-glucosidases (maltase) [163]. So, importance of amylases is related to their ability to provide growing embryo with respiratory substrates for producing energy and carbon source until the established seedling can photosynthesize. In addition, embryo growth from quiescent stage to active phase depending mainly on the utilization of stored ATP and storage lipid breakdown products [164].

Seed germination represents a good period for mitochondria development study. Results obtained from previous transcriptome studies recorded a substantial increase in mitochondrial transcripts encoding proteins and protein content accompanied with changes in their functions during early 3 h of seed imbibitions [165]. During the first 48 h of seed imbibitions, 56 differentially expressed proteins were detected which include the outer membrane channel TOM40 and the inner membrane TIM17/22/23 families, compared to dry seed.

The interpretation of suggestion that import pathway capacity is absolutely dependent on the presence of oxygen (aerobic respiration) is related to the significant decrease in capacity of the general import pathway in mitochondria under anaerobic conditions, compared to under aerobic conditions. In supporting for this suggestion, three proteins from the TIM17/22/23 family were found to be 6–14 folds up-regulated under anaerobic conditions [166] and a decline in proteins involving import apparatus was detected in the mature mitochondria that might be suggested that the accumulation of these import proteins in the dry seed could operate functions after 2 h imbibition, and then serve as donors of TCA cycle and electron transport chain components [167].

## 7. The role of glyoxylate cycle in oilseed germination

Glyoxylate cycle has been known to play a crucial role in lipid degradation in oilseeds, whereas stored lipid is converted into glucose the main respiratory substrate during germination and hence seedling establishment [168]. Seed imbibition triggers highly increase in oxygen consumption which reflects the enhancement of oxidation of produced carbohydrates from the glyoxylate cycle [169]. Alongside to glyoxylate cycle, the OPPP operates where a number of enzymes and intermediates participate the two pathways [170]. It functions to provide the cell with NADPH for biosynthetic reactions and appears to be important in the regulation of germination [171].

The action of the two glyoxylate cycle enzymes isocitrate lyase (ICL) and malate synthase (MS) that by pass the decarboxylation steps of the TCA cycle are essential in oilseed germination. Whereas, two moles of acetyl-CoA are introduced with each turn of the cycle, resulting in the synthesis of one mole of the four-carbon compound succinate that are transported from the glyoxysome into the mitochondrion and converted into malate via TCA cycle. This malate is then exported to cytosol in exchange for succinate and is converted to oxalacetate. PEP-CK catalyzes the conversion of oxaloacetate to phosphoenolpyruvate and this fuels the synthesis of soluble carbohydrates necessary to germination [169].

## 8. Conclusion

Under stressful conditions, oxidative damage to mRNA results in the inhibition of protein synthesis and in protein degradation which caused disturbance in protein functions due to enzymatic and binding properties modification. Consequently; seed germination may delay or suppress. The priming techniques improve stress acclimation mechanisms during germination but the cellular mechanism of priming is still requires more studying. In response to abiotic stresses, activity of acid phosphatases increased to match a definite level of inorganic phosphate which can be co-transported with $H^+$ down proton motive force gradient. The signaling interactions among multiple phytohormones are rather common in controlling various growth and developmental processes. Hormonal signaling coordination may be regulated through controlling biosynthesis of certain phytohormone, by modifying the available pool of hormone molecules or by elaborate regulation of the signaling process. However; seed pretreatment with each of GAs, auxins or cytokinin promote seed germination not only through stimulation of hydrolyzing enzymes but also by antagonizing the inhibitory effect of ABA on germination process. Phytohormone signal crosstalk will present valuable new avenues for genetic improvement of crop plants needed to meet the future food production targets in the face of global climate change. Surprising; seed priming with $H_2O_2$ resulted in improvement germination process and seedling establishment. This may be resulted from its effect on GA signaling and modulation of hormonal balance that promote initiation of seed germination. In addition; $H_2O_2$ diminished the inhibitory effects of ABA on endosperm damage through expression of gene encoding enzyme hydrolyzing the testa and endosperm with the releasing of embryo.

## Authors details

Awatif S. Ali[1]* and Alaaeldin A. Elozeiri[2]

Address all correspondence to: awatifali95@yahoo.com

1 Botany Department, Faculty of Science, Kafr Elsheikh University, Kafr Elsheikh, Egypt

2 Environmental Engineering Department, Zewail City of Science and Technology, Cairo, Egypt

## References

[1] Hasanuzzaman M, Nahar K, Alam MM, Roychowdhury R, Fujita M. Physiological, biochemical, and molecular mechanisms of heat stress tolerance in plants. International Journal of Molecular Sciences. 2013;**14**:9643-9684

[2] Borgheretti F, Ferreira AG. Germinação do básico ao aplicado. Porto Alegre: Artmed; 2000. 222 p

[3] Zienkiewicz Z, Zienkiewicz AK, Rejon JD, Alche JD, Castro AJ, Rodringuez-Garcia MI. Olive seed protein bodies store degrading enzymes involved in mobilization of oil bodies. Journal of Experimental Botany. 2014; **65**:103-115. DOI: 10.1093/jxb/ert355

[4] Gonçalves JFC, et al. Aspectos fisiológicos bioquímicos de plantas da Amazônia. Projeto Jacaranda Fase II: Pesquisas Florestais na Amazônia Central. Manaus: INPA; 2003. p. 89-101

[5] Bewley JD, Black M. Physiology of Development and Germination. 2nd ed. New York: Plenum Press; 1994 445 p

[6] Bewley JD. Seed germination and dormancy. The Plant Cell. 1997;**9**:1055-1066

[7] Bewley JD, Bradford K, Hilhorst H, Nonogaki H. Seeds: Physiology of Development, Germination and Dormancy. 3rd ed. New York: Springer; 2013

[8] Jacobsen JV, Chandler PM. Gibberellin and abscisic acid in germinating cereals. In: Davies PJ, editor. Plant Hormones: Physiology, Biochemistry and Molecular Biology. Boston, MA: Kluwer; 1995. p. 164-193

[9] Bethke PC, Schuurink R, Jones RL. Hormonal signaling in cereal aleurone. Journal of Experimental Botany. 1997;**48**:1337-1356. DOI: 10.1093/jxb/48.7.1337

[10] Lee SJ, Kang JY, Park HJ, Kim MD, Bae MS, Choi HI, et al. DREB2C interacts with ABF2, a bZIP protein regulating abscisic acid-responsive gene expression, and its overexpression affects abscisic acid sensitivity. Plant Physiology. 2010;**153**(2):716-727

[11] Firenzuoli AM, Vanni P, Ramponi G, Baccari V. Changes in enzyme levels during germination of seeds of *Triticum durum*. Plant Physiology. 1968;**43**:260-264. DOI: 10.1104/pp.43.2.260

[12] Carter LM, Chesson JH. Two USDA researchers develop a moisture seeking attachment for crop seeders that is designed to help grower's plant seed in soil sufficiently moist for germination. Seed World. 1996;**134**:14-15

[13] Munns R. Comparative physiology of salt and water stress. Plant, Cell & Environment. 2002;**25**:239-250

[14] Singh J, Sastry EVD, Singh V. Effect of salinity on tomato (*Lycopersicon esculentum* mill.) during seed germination stage. Physiology and Molecular Biology of Plants. 2012;**18**:45-50

[15] Kaur S, Gupta AK, Kaur N. Effect of $GA_3$, kinetin and indole acetic acid on carbohydrate metabolism in chickpea seedlings germinating under water stress. Plant Growth Regulation. 2000;**30**:61-70

[16] Zeid IM, Shedeed ZA. Response of alfalfa to putrescine treatment under drought stress. Biologia Plantarum. 2006;**50**:635-640

[17] Jacobsen JV, Hanson AD, Chandlor PC. Water stress enhances expression of $\alpha$-amylase gene in barley leaves. Plant Physiology. 1986;**80**:350-359

[18] Todaka D, Matsushima H, Morphashi Y. Water stress enhances $\alpha$-amylase activity in cucumber cotyledons. Journal of Experimental Botany. 2000;**51**:739-745

[19] Velasco R, Salamini F, Bartels D. Dehydration and ABA increase mRNA levels and enzyme activity of cytosolic GAPDH in the resurrection plant *Craterostigma plantagineum*. Plant Molecular Biology. 1994;**26**:541-546

[20] Pandey R, Agarwal RM, Jeevaratnam K, Sharma GL. Osmotic stress-induced alterations in rice (*Oryza sativa* L.) and recovery on stress release. Plant Growth Regulation. 2004;**42**:79-87

[21] Baatour O, Kaddour R, Wannes WA, Lachaâl M, Marzouk B. Salt effects on the growth, mineral nutrition, essential oil yield and composition of marjoram (*Origanum majorana*). Acta Physiologiae Plantarum. 2010;**32**:45-51

[22] Job C, Rajjou L, Lovigny Y, Belghazi M, Job D. Patterns of protein oxidation n *Arabidopsis* seeds and during germination. Plant Physiology. 2005;**138**:790-802. DOI: 10.1104/pp.105.062778

[23] Garg N, Manchanda G. ROS generation in plants: Boon or bane? Plant Biosystems. 2009;**143**:81-96

[24] Yacoubi R, Job C, Belghazi M, Job D. Proteomic analysis of the enhancement of seed vigour in osmoprimed alfalfa seeds germinated under salinity stress. Seed Science Research. 2013;**23**(2):99-110

[25] Yin H, Chen Q, Yi M. Effects of short-term heat stress on oxidative damage and responses of antioxidant system in *Lilium longiflorum*. Plant Growth Regulators. 2008;**54**:45-54

[26] Zhang Q, Li JJ, Zhang WJ, Yan SN, Wang R, Zhao JF, Li YJ, Qi ZG, Sun ZX, Zhu ZG. The putative auxin efflux carrier OsPIN3t is involved in the drought stress response and drought tolerance. The Plant Journal. 2012;**72**:805-816

[27] Das AB. Bio prospecting and genetic engineering of mangrove genes to enhance salinity tolerance in crop plants. In: Jain SM, Gupta SD, editors. Biotechnology of Neglected and Underutilized Crops. New York: Springer; 2013. p. 385-456

[28] Fahad S, Nie L, Chen Y, Wu C, Xiong D, Saud S, Hongyan L, Cui K, Huang L. Crop plant hormones and environmental stress. Sustain Agriculture Reviews. 2015;**15**:371-400

[29] Koornneef M, Bentsink L, Hilhorst H. Seed dormancy and germination. Current Opinion in Plant Biology. 2002;**5**:33-36

[30] Ansari O, Choghazardi HR, Sharif Zadeh F, Nazarli H. Seed reserve utilization and seedling growth of treated seeds of mountain ray (*Secale montanum*) as affected by drought stress. Cercetări Agronomiceîn Moldova. 2012;**2**(150):43-48

[31] Patade VY, Maya K, Zakwan A. Seed priming mediated germination improvement and tolerance to subsequent exposure to cold and salt stress in capsicum. Research Journal of Seed Science. 2011;**4**(3):125-136

[32] Mayer AM, Shain Y. Control of seed germination. Annual Review of Plant Physiology. 1974;**25**:167-193

[33] Bewley JD, Black M. Seeds: Physiology of Development and Germination. New York: Plenum Press; 1985

[34] Mayer AM, Poljakoff-Mayber A. The Germination of Seeds. 4th ed. Oxford: Pergamon Press; 1989

[35] Salisbury FB, Ross CW. Plant Physiology. 4th ed. Belmont, California, USA: Wadsworth Publishing Company; 1991

[36] Shutov AD, Vaintraub IA. Degradation of storage proteins in germinating seeds. Phytochemistry. 1987;**26**:1557-1566

[37] Gepstin S, Han I. Evidence for the involvement of cytokinin in the regulation of proteolytic activity in cotyledons of germinating beans. Plant and Cell Physiology. 1980;**21**(1):57-63

[38] Shivaraj B, Pattabiraman TN. Natural plant enzyme inhibitors, part VIII. Indian Journal of Biochemistry & Biophysics. 1980;**17**:181-185

[39] Veerabhadrappa PS, Manjunath NH, Virupaksha TK. Proteinase inhibitors of finger millet (*Eleusine coracana* gaertn.). Journal of the Science of Food and Agriculture. 1978;**29**(4):353-358

[40] Tully RE, Beevers H. Protease and peptidases of castor bean endosperm. Enzyme characterization and changes during germination. Plant Physiology. 1978;**62**:726-750

[41] Schlereth A, Becker C, Horstmann C, Tiedemann J, Müntz K. Comparison of globulin mobilization and cysteine proteinases in embryonic axes and cotyledons during germination and seedling growth of vetch (*Vicia sativa* L.). Journal of Experimental Botany. 2000;**51**:1423-1433

[42] Yano H, Wong JH, Lee YM, Cho MJ, Buchanan BB. A strategy for the identification of proteins targeted by thioredoxin. Proceedings of the National Academy of Sciences USA. 2001;**98**:4794-4799. DOI: 10.1073/pnas.071041998

[43] Yano H, Masaharu KM. Disulfide proteome yields a detailed understanding of redox regulations: A model study of thioredoxin-linked reactions in seed germination. Proteomics. 2006;**6**:294-300. DOI: 10.1002/pmic.200402033

[44] Dunaevsky YE, Belozersky MA. Proteolysis of the main storage protein of buckwheat seeds at the early stage of germination. Physiologia Plantarum. 1989;**75**:424-428

[45] Belozersky MA, Dunaevsky YE, Voskoboynikova NE. Isolation and properties of a metalloproteinase from buckwheat seeds. The Biochemical Journal. 1990;**272**:677-682

[46] Elpidina EN, Dunaevsky YE, Belozersky MA. Protein bodies from buckwheat seed cotyledons: Isolation and characteristics. Journal of Experimental Botany. 1990;**41**:969-977

[47] Dunaevsky YE, Belozersky MA. The role of cysteine proteinase and carboxypeptidase in the breakdown of storage proteins in buckwheat seeds. Planta. 1989;**179**:316-322

[48] Paleg LG. Physiological effects of gibberellic acid II. On starch hydrolysis enzymes of barley endosperm. Plant Physiology. 1960;**35**:902-906

[49] Chrispeels MJ, Varner JE. Gibberellic acid enhanced synthesis and release of $\alpha$-amylase and ribonuclease by isolated barley aleurone layers. Plant Physiology. 1967;**42**:398-406

[50] Jones RL, Jacobsen JV. Regulation of synthesis and transport of secreted proteins in cereal aleurone. International Review of Cytology. 1991;**126**:49-88

[51] Filner P, Varner JE. A simple and unequivocal test for *de novo* synthesis of enzymes: Density labeling of barley α-amylase with $H_2O18$. Proceedings of the National Academy of Sciences USA. 1967;**58**:1520-1526

[52] Parys E, Romanowska E, Poskuta J. Amylase activities in attached and excised cotyledons and in embryonic axes of *Pisum sativum* L. Plant and Cell Physiology. 1983;**21**:181-188

[53] Wang Y, Wang DM, Liang HG. Effect on plumular axis on the amylase activity in cotyledons of germinating pea seeds. Acta Phytophysiology Sinica. 1988;**14**(3):244-249

[54] Kępczyński J. Ethylene-dependent action of gibberellins in seed germination of *Amaranthus caudatus*. Physiologia Plantarum. 1986;**67**:584-587

[55] Wilson AM. Amylase synthesis and stability in crested wheatgrass seeds at low water potentials. Plant Physiology. 1971;**48**:541-546

[56] Hellyer SA, Chandler IC, Bosley JA. Can the fatty acid selectivity of plant lipases be predicted from the composition of the seed triglyceride? Biochemica Biophysica Acta. 1999;**1440**(2-3):215

[57] Paques FW, Macedo GA. Lipases de Látex Vegetais: Propriedades e Aplicações Industriais: A Review. Química Nova. 2006;**29**(1):93

[58] Quettier AL, Eastmond PJ. Storage oil hydrolysis during early seedling growth. Plant Physiology and Biochemistry. 2009;**47**:485

[59] Hutton D, Stumpf PK. Fat metabolism in higher plant. Characterisation of β-oxidation system from maturing and germinating caster bean seeds. Plant Physiology. 1969;**44**: 508-516

[60] Pereira EP, Zanin GM, Castro HF. Immobilization and catalytic properties of lipase on chitosan for hydrolysis and etherification reactions. Brazilian Journal of Chemical Engineering. 2003;**20**(4):343

[61] Leal MCM, Cammarota MC, Freire DMG, Sant'Anna JGL. Hydrolytic enzymes as coadjuvants in the anaerobic treatment of dairy waste waters. Brazilian Journal of Chemical Engineering. 2002;**19**(2):175

[62] Freire GDM, Castilho FL. Lipases em Biocatálise. In: Bon et al. (org), editor. Enzimas em biotecnologia: Produção, Aplicação e Mercado. Rio de Janeiro: Interciência; 2008

[63] Villeneuve P. Plant lipases and their applications in oils and fats modification. European Journal of Lipid Science and Technology. 2003;**105**(6):308

[64] Castro HF, Anderson WA. Fine chemicals by biotransformation using lipases. Quimica Nova. 1995;**18**(6):544-554

[65] Tavener RJA, Laidman DL. The induction of triglyceride metabolism in the germinating wheat grain. Phytochemistry. 1972;**11**:981-987

[66] Shoshii M, Reevers H. Lipase activity in castor been endosperm during germination. Plant Physiology. 1974;**54**:23-28

[67] Anthony HC, Huang AHC, Moreau RA. Lipases in the storage tissues of peanut and other oil seeds during germination. Planta. 1978;**141**:111-116

[68] Baud S, Dubreucq B, Miquel M, Rochat C, Lepiniec L. Storage reserve accumulation in *Arabidopsis*: Metabolic and developmental control of seed filling. *Arabidopsis* Book American Society of Plant Biology. 2008;**6**: e0113. DOI: 10.1199/tab.0113

[69] Ratajczak W. Asparagine metabolism in developing seeds of *Lupinus luteus* L. Biochemie und Physiologie der Pflanzen. 1986;**181**:17-22

[70] Lehmann T, Ratajczak L. The pivotal role of glutamate dehydrogenase (GDH) in the mobilization of N and C from storage material to asparagine in germinating seeds of yellow lupine. Journal of Plant Physiology. 2008;**165**:149-158. DOI: 10.1016/j.jplph.2006.12.010

[71] Borek S, Galor A, Paluch E. Asparagine enhances starch accumulation in developing and germinating lupin seeds. Journal of Plant Growth Regulation. 2013;**32**:471-482

[72] Ratajczak W, Gwóźdz′ EA, Mia˛dowicz M. Effects of nitrogen nutrition on storage protein composition yellow lupin cotyledons cultured in vitro. Acta Physiologiae Plantarum. 1996;**18**:295-304

[73] Vidal EA, Gutiérrez RA. A systems view of nitrogen nutrient and metabolite responses in Arabidopsis. Current Opinion in Plant Biology. 2008;**11**:521-529

[74] Pracharoenwattana I, Smith SM. When is a peroxisome not a peroxisome? Trends in Plant Science. 2008;**13**:522-525

[75] Borek S, Ratajczak L. Storage lipids as a source of carbon skeletons for asparagine synthesis in germinating seeds of yellow lupine (*Lupinus luteus* L.). Journal of Plant Physiology. 2010;**167**:717-724

[76] Jacela JY, DeRouehey Tokach MD, Goodband RD, Nelssen JL, Renter D, Dritz SS. Feed additives for swine: Fact sheets-prebiotics and probiotis and phytogenics. Journal of Swine Health and Production. 2010;**18**:87-91

[77] Lott JNA, Greenwood JS, Batten GD. Mechanisms and regulation of mineral nutrient storage during seed development. In: Kigel J, Galili G, editors. Seed Development and Germination. New York: Marcel Dekker Inc; 1995. p. 215-235

[78] Thompson LU, Zhang L. Phytic acid and minerals: Effect on early markers of risk for mammary and colon carcinogenesis. Carcinogenesis. 1991;**12**:2041-2045

[79] Coelho SM, Taylor AR, Ryan KP, Sousa-Pinto I, Brown MT, Brownlee C. Spatiotemporal patterning of reactive oxygen production and $Ca^{2+}$ wave propagation in *Fucus rhizoid* cells. The Plant Cell. 2002;**14**:2369-2381

[80] Okazaki Y, Katayama T. Reassessment of the nutritional function of phytic acid, with special reference to myo-inositol function. Journal of Japan Society of Nutrition and Food Sciences. 2005;**58**:151-156

[81] Shi H, Bressan R, Hasegawa PM, Zhu J-K. In: Broadlay M, White P, editors. Sodium in Plant Nutritional Genomics. London: Blackwell Publishing; 2005. p. 127-149

[82] Buerkert A, Haake C, Ruckwied M, Marschner H. Phosphorus application affects the nutritional quality of millet grain in the sahel. Field Crops Research. 1998;**57**:223-235

[83] Santos JCP. Estado nutricional do feijoeiro (*Phaseolus vulgaris* L.) e teores de nutrientes e fitatos nos grãos. Piracicaba: Universidade de São Paulo. Tese de doutorado; 1998

[84] Raboy V, Noaman MM, Taylor GA, Pickett SG. Grain phytic acid and protein are highly correlated in winter wheat. Crop Science. 1991;**31**:631-635

[85] Hubel F, Beck E. Maize root phytase. Purification, characterization, and localization of enzyme activity and its putative substrate. Plant Physiology. 1996;**112**:1429-1436

[86] Mukherji S, Dey B, Paul AK, Sircar SM. Changes in phosphorus fractions and phytase activity of rice seeds during germination. Physiologia Plantarum. 1971;**25**:94-97

[87] Silva LG, Trugo LC. Characterization of phytase activity in lupin seed. Journal of Food Biochemistry. 1996;**20**:329-340

[88] Andriotis VME, Ross JD. Isolation and characterization of phytase from dormant *Corylus avellana* seeds. Phytochemistry. 2003;**64**:689-699

[89] Azarkovich MI, Dmitrieva MI, Sobolev AM. Mobilization of protein and phytin in aleurone grains of germinating castor beans. Russian Journal of Plant Physiology. 1999;**46**:349-356

[90] Ogawa M, Tanaka K, Kasai Z. Accumulation of phosphorus, magnesium, and potassium in developing rice grains: Followed by electron microprobe X-ray analysis focusing on the aleurone layer. Plant and Cell Physiology. 1979;**20**:19-27

[91] Raboy V. Myo-Inositol-1,2,3,4,5,6-hexakisphosphate. Phytochemistry. 2003;**64**:1033-1043

[92] Fincher GB. Molecular and cellular biology associated with endosperm mobilization in germinating cereal grains. Annual Review of Plant Physiology. 1989;**40**:305-346

[93] Mihoub A, Chaoui A, El-Ferjani E. Biochemical changes associated with cadmium and copper stress in germinating pea seeds (*Pisum sativum* L.). Competes Rendus Biologies. 2005;**328**:33-41

[94] Olmos E, Hellin E. Cytochemical localization of ATpase plasma membrane and acid phosphatase by cerium based in a salt-adapted cell line of *Pisum sativum*. Journal of Experimental Botany. 1997;**48**:1529-1535

[95] Bai B, Sikron S, Gendler T, Kazachkova Y, Barak S, Grafi G, Khozin-Goldberg I, Fait A. Ecotypic variability in the metabolic response of seeds to diurnal hydration–dehydration cycles and its relationship to seed vigor. Plant and Cell Physiology. 2011;**53**(1):38-52

[96] Nonogaki H, Bassel GW, Bewley JD. Germination–stillamystery. Plant Science. 2010; **179**:574-581. DOI: 10.4161/psb.25504

[97] El-Maarouf-Bouteau H, Bailly C. Oxidative signaling in seed germination and dormancy. Plant Signaling & Behavior. 2008;**3**:175-182. DOI: 10.4161/psb.3.3.5539

[98] Kubala S, Wojtyla Ł, Quinet M, Lechowska K, Lutts S, Garnczarska M. Enhanced expression of the proline synthesis gene P5CSA in relations to seed osmopriming improvement of *Brassica napus* germination under salinity stress. Journal of Plant Physiology. 2015;**183**:1-12. DOI: 10.1016/j.jplph.2015.04.009

[99] Bailly C, El-Maarouf-Bouteau H, Corbineau F. From intra cellular signaling networks to cell death: The dual role of reactive oxygen species in seed physiology. *Comptes Rendus Biologies*. 2008;**331**:806-814. DOI: 10.1016/j.crvi.2008.07.022

[100] Kumar SPJ, Prasad SR, Banerjee R, Thammineni C. Seed birth to death: Dual functions of reactive oxygen species in seed physiology. Annals of Botany. 2015;**116**:663-668. DOI: 10.1093/aob/mcv098

[101] Møller IM, Jensen PE, Hansson A. Oxidative modifications to cellular components in plants. Annual Review of Plant Biology. 2007;**58**:459-481. DOI: 10.1146/annurev.arplant.58.032806.103946

[102] Pommier Y, Redon C, Rao VA, Seiler JA, Sordet O, Takemura H, et al. Repair of and checkpoint response to topoisomerase I-mediated DNA damage. Mutation Research. 2003;**532**:173-203. DOI: 10.1016/j.mrfmmm.08.016

[103] Kong QM, Lin CG. Oxidative damage to RNA: Mechanisms, consequences and diseases. Cellular and Molecular Life Sciences. 2010;**67**:1817-1829. DOI: 10.1007/s00018-010-0277-y

[104] Chmielowska-Bkak J, Izbiańska K, Deckert J. Products of lipid, protein and mRNA oxidation as signals and regulators of gene expression. Frontiers in Plant Science. 2015;**6**:405. DOI: 10.3389/fpls.2015.00405

[105] Davies MJ. The oxidative environment and protein damage. Biochimica et Biophysica Acta. 2005;**1703**:93-109. DOI: 10.1016/j.bbapap.2004.08.007

[106] Barba-Espin G et al. Role of thioproline on seed germination: Interaction ROS-ABA and effects on antioxidative metabolism. Plant Physiology and Biochemistry. 2011;**59**:30-36. DOI: 10.1016/j.plaphy.2011.12.002.3

[107] Kudla J, Batistic O, Hashimoto K. Calcium signals: The lead currency of plant information processing. The Plant Cell. 2010;**22**:541-563

[108] Reddy AS, Ali GS, Celesnik H, Day IS. Coping with stresses: Roles of calcium- and calcium/calmodulin-regulated gene expression. The Plant Cell. 2011;**23**(6):2010-2032

[109] Finnie C, Melchior S, Roepstorff P, Svensson B. Proteome analysis of grain filling and seed maturation in barley. Plant Physiology. 2002;**129**:1308-1319

[110] Komatsu S, Yang G. Over-expression of calcium-dependent protein kinase 13 and calreticulin interacting protein 1 confers cold tolerance on rice plants. Molecular Genetics and Genomics. 2007;**277**:713-723

[111] Galat A. Variations of sequences and amino acid composition of proteins that sustain their biological function: An analysis of the cyclophilin family of proteins. Archives of Biochemistry and Biophysics. 1999;**371**(2):149-162

[112] Fitter AH, Hay RKM. Environmental Physiology of Plants. New York: Academic Press; 1981

[113] Wani SH, Kumar V, Shriram V, Kumar S. Phytohormones and their metabolic engineering for abiotic stress tolerance in crop plants. The Crop Journal. 2016;**4**:162-176

[114] Preston J, Tatematsu K, Kanno Y, Hobo T, Kimura M, Jikumaru Y, et al. Temporal expression patterns of hormone metabolism genes during imbibition of *Arabidopsis thaliana* seeds: A comparative study on dormant and non dormant accessions. Plant and Cell Physiology. 2009;**50**:1786-1800

[115] Dave A, ML H'n, He Z, Andriotis VM, Vaistij FE, Larson TR, et al. 12-Oxo-phytodienoic acid accumulation during seed development represses seed germination in *Arabidopsis*. The Plant Cell. 2011;**23**:583-599

[116] Carrera E, Holman T, Medhurst A, Dietrich D, Footitt S, Theodoulou FL, et al. Seed after-ripening is a discrete developmental pathway associated with specific gene networks in *Arabidopsis*. The Plant Journal. 2008;**53**:214-224

[117] Steber CM, McCourt P. A role for brassinosteroids in germination in *Arabidopsis*. Plant Physiology. 2001;**125**:763-769

[118] Vob U, Bishopp A, Farcot A, Bennett MJ. Modelling hormonal response and development. Trends in Plant Science. 2014;**19**:311-319

[119] Leubner-Metzger G, Fründt C, Vögeli-Lange R, Meins FJ. β-1, 3-glucanases in the endosperm of tobacco during germination. Plant Physiology. 1995;**109**:751-759

[120] Nonogaki H, Gee OH, Bradford KJ. A germination-specific endo-β-mannanase gene is expressed in the micropylar endosperm cap of tomato seeds. Plant Physiology. 2000;**123**:1235-1245

[121] Koornneef M, Karssen CM. Seed dormancy and germination. In: *Arabidopsis* Book/Cold Spring Harbor Monograph Archive. 1994;**27**:313-334

[122] Schopfer P, Plachy C. Control of seed germination by abscisic acid. Plant Physiology. 1985;**77**(3):676-686

[123] Debeaujon I, Koornneef M. Gibberellin requirement for *Arabidopsis* seed germination is determined both by testa characteristics and embryonic abscisic acid. Plant Physiology. 2000;**122**:415-424

[124] Weitbrecht K, Müller K, Leubner-Metzger G. First off the mark: Early seed germination. Journal of Experimental Botany. 2011;**62**:3289-3309

[125] Matilla A, Gallardo M, Puga-Hermida MI. Structural, physiological and molecular aspects of heterogeneity in seeds: A review. Seed Science Research. 2005;**15**:63-76

[126] Oge L, Bourdais G, Bove J, Collet B, Godin B, Granier F, et al. Protein repair-L isoaspartyl methyltransferase 1 is involved in both seed longevity and germination vigor in *Arabidopsis*. The Plant Cell. 2008;**20**:3022-3037

[127] Bhaskar RV, Mohanty B, Verma V, Wijaya E, Kumar PP. A hormone-responsive C1-domain-containing protein At5g17960 mediates stress response in *Arabidopsis thaliana*. PLoS. 2015;**10**(1):115-118. DOI: 10.1371/journal.pone.0115418

[128] Hammerton RW, Ho T-HD. Hormonal regulation of the development of protease and carboxypeptidase activities in barley aleurone layers. Plant Physiology. 1986;**80**:692-697

[129] Stuart IM, Loi L, Fincher GB. Development of (1-3, 1-4)-fl-glucan endohydrolase isoenzymes in isolated scutella and aleurone layers of barley *(Hordeum vulgare)*. Plant Physiology. 1986;**80**:310-314

[130] Paparella S, Araújo SS, Rossi G, Wijayasinghe M, Carbonera D, Balestrazzi A. Seed priming: State of the art and new perspectives. Plant Cell Reports. 2015;**34**:1281-1293. DOI: 10.1007/s00299-015-1784-y

[131] Hilhorst HWM, Finch-Savage WE, Buitink J, Bolingue W, Leubner-Metzger G. Dormancy in plant seeds. In: Lubzens E, Cerdà J, Clarck M, editors. Dormancy and Resistance in Harsh Environments. Heideberg: Springer-Verlag; 2010. p. 43-67

[132] Zhuo J, Wang W, Lu Y, Sen W, Wang X. Osmopriming-regulated changes of plasma membrane composition and function were inhibited by phenylarsin oxide in soybean seeds. Journal of Integrative Plant Biology. 2009;**9**:858-867

[133] Jisha KC, Vijayakumari KJT, Puthur JT. Seed priming for abiotic stress tolerance: An overview. Acta Physiologiae Plantarum. 2013;**3**:1381-1396. DOI: 10.1007/s11738-012-1186-5

[134] Soeda Y, Konings MCJM, Vorst O, van Houwelingen AMML, Stoopen GM, Maliepaard CA, Koddle J, Bino RJ, Groot SPC, van der Geest AHM. Gene expression programs during *Brassica oleracea* seed maturation, osmopriming, and germination are indicators of progression of the germination process and the stress tolerance level. Plant Physiology 2005;**137**:354-368

[135] Sun J, Hutchins DA, Feng Y, Seubert EL, Caron DA, Fu F-X. Effects of changing $pCO_2$ and phosphate availability on domoic acid production and physiology of the marine harmful bloom diatom *Pseudo-nitzschia multiseries*. Limnology and Oceanography. 2011;**56**:829-840. DOI: 10.4319/lo.2011.56.3.0829

[136] Chen K, Arora R. Dynamics of the antioxidant system during seed osmopriming, post-priming germination, and seedling establishment in spinach *(Spinacia oleracea)*. Plant Science. 2011;**180**:212-220. DOI: 10.1016/j.plantsci.2010.08.007

[137] Balestrazzi A, Macovei C, Carbonera D. Seed imbibition in *Medicago truncatula* Gaertn. Expression profiles of DNA repair genes in relation to PEG-mediated stress. Journal of Plant Physiology. 2011;**168**:706-713. DOI: 10.1016/j.jplph.2010.10.008

[138] El-Araby MM, Moustafa SMA, Ismail AI, Hegazi AZA. Hormone and phenol levels during germination and osmopriming of tomato seeds, and associated variations in protein patterns and anatomical seed features. Acta Agronomica Hungarica. 2006;**54**:441-458. DOI: 10.1556/AAgr.54.2006.4.7

[139] Kubala S, Garnczarska M, Wojtyla Ł, Clippe A, Kosmala A, Zmiénko A, et al. Deciphering priming-induced improvement of rape seed (*Brassica napus* L.) germination through an integrated transcriptomic and proteomic approach. Plant Science. 2015;**231**:94-113. DOI: 10.1016/j.plantsci.2014.11.008

[140] Hilhorst HWM. Definition and hypotheses of seed dormancy. In: Bradford KJ, Nonogaki H, editors. Seed Development, Dormancy and Germination. Annual Plant Reviews. Vol. 27, Chap 4. Sheffield, UK: Blackwell Publishing; 2007. p. 50-71

[141] Sneideris LC, Gavassi MA, Campos ML, D'Amico-Damião V, Carvalho RF. Effects of hormonal priming on seed germination of pigeon pea under cadmium stress. Anais da Academia Brasileira de Ciências. 2015;**87**:1847-1852. DOI: 10.1590/0001-3765201520140332

[142] Sneideris AC, Gavassi MA, Campiao VDA, Carvalho RF. Effects of hormonal priming on seed germination of pigeon pea under cadmium stress. Annals of the Brazilian Academy of Sciences. 2015;**87**(3):1847-1852. DOI: 10.1590/0001-3765201520140332

[143] Fincher GB. Molecular and cellular biology associated with endosperm mobilization in germinating cereal grains. Annual Review of Plant Physiology and Plant Molecular Biology. 1989;**40**:305-346

[144] Liu W-Z, Kong D-D, Gu X-X, Gao HB, Wang J-Z, Xia M, et al. Cytokinins can act as suppressors of nitric oxide in *Arabidopsis*. Proceedings of the National Academy of Sciences USA. 2013;**110**:1548-1553. DOI: 10.1073/pnas.1213235110

[145] Procko C, Crenshaw CM, Ljung K, Noel JP, Chory J. Cotyledon generated auxin is required for shade-induced hypocotyl growth in *Brassica rapa*. Plant Physiology. 2014;**165**:1285-1301. DOI: 10.1104/pp.114.241844

[146] Guan C, Wang X, Feng J, Hong S, Liang Y, Ren B, et al. Cytokinin antagonizes abscisic acid-mediated inhibition of cotyledon greening by promoting the degradation of abscisic acid insensitive 5 protein in *Arabidopsis*. Plant Physiology. 2014;**164**:1515-1526. DOI: 10.1104/pp.113.234740

[147] Xia X-J, Zhou Y-H, Shi K, Zhou J, Foyer CH, Hu J-Q. Inter play between reactive oxygen species and hormones in the control of plant development and stress tolerance. Journal of Experimental Botany. 2015;**66**:2839-2856. DOI: 10.1093/jxb/erv089

[148] Sanz L, Albertos P, Mateos I, Sánchez-Vicente I, Lechón T, Fernández Marcos M, Lorenzo O. Nitric oxide (NO) and phytohormones crosstalk during early plant development. Journal of Experimental Botany. 2015;**66**:2857-2868

[149] Jin Z, Pei Y. Physiological implications of hydrogen sulfide in plants: Pleasant exploration behind its unpleasant odour. Oxidative Medicine and Cellular Longevity. 2015;**397502**. DOI: 10.1155/2015/397502

[150] Richards SL, Wilkins KA, Swarbreck SM, Anderson AA, Habib N, Smith AG, et al. The hydroxyl radical in plants: From seed to seed. Journal of Experimental Botany. 2015;**66**:37-46. DOI: 10.1093/jxb/eru 398

[151] Diaz-Vivancos P, Barba-Espín G, Hernández JA. Elucidating hormonal/ROS networks during seed germination: Insights and perspectives. Plant Cell Reports. 2013;**32**:1491-1502. DOI: 10.1007/s00299-013-1473-7

[152] Petrov VD, Van Breusegem F. Hydrogen peroxide–a central hub for information flow in plant cells. AoB Plants. 2012;**2012**:pls014. DOI: 10.1093/aobpla/pls014

[153] Schopfer P, Plachy C, Frahry G. Release of reactive oxygen intermediates (superoxide radicals, hydrogen peroxide, and hydroxyl radicals) and peroxidase in germinating radish seeds controlled by light, gibberellin, and abscisic acid. Plant Physiology. 2001;**125**:1591-1602. DOI: 10.1104/pp.125.4.1591

[154] Lariguet P, Ranocha P, DeMeyer M, Barbier O, Penel C, Dunand C. Identification of a hydrogen peroxide signaling pathway in the control of light-dependent germination in *Arabidopsis*. Planta. 2013;**238**:381-395. DOI: 10.1007/s00425-013-1901-5

[155] Müller K, Hess B, Leubner-Metzger G. A role for reactive oxygen species in endosperm weakening. In: Adkins S, Shmore SA, Navie S, editors. Seeds: Biology, Development and Ecology. Wallingford: CAB International; 2007. p. 287-295

[156] Turner JF, Turner DH. The regulation of carbohydrate metabolism. Annual Review of Plant Physiology. 1975;**26**:159-186

[157] Robert KM, David AB, Katheleen MB, Peter JK, Victor WR, Weil P. A Harper's illustrated biochemistry. Biologic Oxidation. 2009;**12**:99-100

[158] Oaikhena EE, Ajibade GA, Appah J, Bello M. Dehydrogenase enzyme activities in germinating cowpea (*Vigna unguiculata* (L) Walp). Journal of Biology, Agriculture and Healthcare. 2013;**3**:32-36

[159] Devlin T. Text book of biochemistry with clinical correlations. Bioenergetics mitochondria and oxidative. Metabolism. 2011;**14**:554-555

[160] Jones JD, Burneth P, Zollman P. The glyoxylate cycle does it function in the dormant or active seed. Comparative Biochemistry and Physiology Part B: Biochemistrty and Molecular Biology. 1999;**124**(2):177-179

[161] Ebukanson GJ, Bassey ME. About Seed Plants. Kaduna: Baraka Press and Publishers Ltd; 1992. p. 36-39

[162] Podestá FE, Plaxton WC. Regulation of cytosolic carbon metabolism in germinating Ricinus *communis* cotyledons. II. Properties of phosphoenolpyruvate carboxylase and cytosolic pyruvate kinase associated with the regulation of glycolysis and nitrogen assimilation. Planta. 1994;**194**:381-387

[163] Andriotis VME, Saalbach G, Waugh R, Field RA, Smith AM. Maltase involved in starch metabolism in barley endosperm is encoded by a single gene. PLoS One. 2016;**11**(3):e0151642. DOI: 10.1371/journal.pone.0151642

[164] Nandi S, Das G, Sen-Mendi S. β-amylase activity as an index for germination potential in rice. Annals of Botany. 1995;**75**:463-467

[165] Howell KA, Narsair R, Carroll A, Ivanova A, Lohse M, Usadel B, et al. Mapping metabolic and transcript temporal switches during germination in rice high- lights specific transcription factors and the role of RNA in stability in the germination process. Plant Physiology. 2009;**149**:961-980. DOI: 10.1104/pp.108.129874

[166] Howell KA, Cheng K, Murcha MW, Jenkin LE, Millar AH, Whelan J. Oxygen initiation of respiration and mitochondrial biogenesis in rice. The Journal of Biological Chemistry. 2007;**282**:15619-15619

[167] Howell KA, Millar AH, Whelan J. Ordered assembly of mitochondria during rice germination begins with promitochondrial structures rich in component of the protein import apparatus. Plant Molecular Biology. 2006;**60**:201-223

[168] Eastmond PJ, Germain V, Lange PR, Bryce JH, Smith SM, Graham IA. Post germinative growth and lipid catabolism in oilseeds lacking the gloxylate cycle. Proceedings of the National Academy of Sciences of the United States of America. 2000;**97**:5669-5674

[169] Muscolo A, Sidari M, Mallamaci C, Attinà E. Changes in germination and glyoxylate and respiratory enzymes of *Pinus pinea* seeds under various abiotic stresses. Journal of Plant Interactions. 2007;**2**(4):273-279. DOI: 10.1080/17429140701713795

[170] ap Rees T. Integration of pathways of synthesis and degradation of hexose phosphates. In: Preiss J, editor. The Biochemistry of Plants. London: Academic Press. 1980; pp. 1-42.

[171] Perino C, Come D. Physiological and metabolic study of the germination phases in apple embryo. Seed Science and Technology. 1991;**19**:1-14

Chapter 9

# Free Radicals and Antioxidant System in Seed Biology

Fadime Eryılmaz Pehlivan

Additional information is available at the end of the chapter

http://dx.doi.org/10.5772/intechopen.70837

**Abstract**

Reactive oxygen species (ROS) are involved in various development stages of seed biology. During seed desiccation, germination and aging, oxidative stress may increase in higher levels, leading to cellular damage and seed deterioration. Plant cells have antioxidant system, detoxifying enzymes and antioxidant compounds, that scavenge ROS, participating in seed survival. This antioxidant system has various roles in desiccation and germination of developing seeds, seed storability, and seed aging. On the other hand, ROS are accepted as molecules involving in cellular signaling, and having regulatory functions in seed development. ROS are also found to have roles in gene expression in early embryogenesis, dormancy and germination. Abscisic acid is a plant hormone and a signaling molecule in seed development and that is reported to have relationships with ROS. The objective of this article is to review the roles of ROS and the importance of antioxidant system in orthodox seeds, and to emphasize the dual effects of ROS in seed biology.

**Keywords:** free radicals, reactive oxygen species (ROS), antioxidant, seed biology, ROS signaling

## 1. Introduction

ROS are usually thought as hazardous molecules, attacking to biomolecules, leading to membrane and DNA injuries, and deleterious effects in seed germination processes [1–5]. Seed aging is a process that the roles of ROS are well documented [6–9]. ROS also have damages in desiccation of seeds by dehydration [10, 11]. Although ROS have been considered as detrimental to seeds up to now [12–16], recent advances in plant physiology signaling pathways have led to reconsider their role [17–29]. ROS accumulation can therefore be also beneficial for seed germination and seedling growth by regulating cellular growth, providing a protection against pathogens, and controlling the cellular redox status [30–33]. ROS are also proved to act as a positive signal in seed dormancy release [30–33]. The dual function of ROS in plants depends on the levels of antioxidant compounds, and enzyme activities release [34–38]. By this way, plants can eliminate potentially harmful ROS that is produced under stress conditions, or control ROS concentrations in order to regulate various signaling pathways [34–38]. This dual function of ROS is a very interesting subject in seed physiology. Even though there is a huge progress in this field, and the dual functions of ROS are quite well documented in the literature, it should also be regarded from a different point of view. The involvement of ROS in seed filling processes is not well documented, and the mobility of ROS in seeds has not yet been documented, thus, more data is needed on roles of ROS in seed germination and development physiology. Under light of the increasing progress made in the understanding of mechanisms driven by ROS, the role of ROS in seed biology may need to be revisited. To date, many distinct roles for ROS, apart from their toxic effects, have been identified.

## 2. ROS and antioxidant system

Oxygen is an essential element for the life of aerobic organism but it may become toxic at higher concentrations. Oxygen molecule in its ground state is relatively unreactive; but its partial reduction gives rise to reactive oxygen species (ROS). ROS are highly reactive oxygen molecules consisting of free radicals. Free radicals are an atom or molecule having an unpaired electron which is extremely reactive, starting chain reactions that generate many more free radicals, that are capable of attacking the healthy cells, causing them to lose their structure and function [1–5]. Types of ROS include the hydroxyl radical, the superoxide anion radical, hydrogen peroxide, singlet oxygen, nitric oxide radical, hypochlorite radical, and various lipid peroxides [1–5] (**Table 1**). Reduction of oxygen leads to the formation of the superoxide radical ($O_2^{\bullet-}$), which is a molecule with an uncoupled electron and can react with other molecules to stabilize its energy. Hydrogen peroxide ($H_2O_2$) result from the nonenzymatic reduction of $O_2^{\bullet-}$ in the presence of $H^+$ ions, or from the action of catalase on $O_2^{\bullet-}$. $H_2O_2$ has a strong oxidizing capacity, and its life span is longer than that of superoxide. $H_2O_2$ can also diffuse through membranes and therefore reach target molecules at some distance from its production site [1–5].

| Free radicals | Nonradicals |
|---|---|
| Superoxide, $O_2^{\bullet-}$ | Hydrogen peroxide, $H_2O_2$ |
| Hydroperoxyl, $HO_2$ | Ozone, $O_3$ |
| Peroxyl, ROO· | Singlet oxygen, $^1O_2$ or $^1\Delta g$ |
| Hydroxyl, ·OH | Hypochlorous acid, HOCl |
| Alkoxyl, RO· | Peroxynitrite, $ONOO^-$ |

**Table 1.** Main reactive oxygen species (ROS) [1].

| Enzymatic | Nonenzymatic or low molecular weight |
|---|---|
| Catalase | Glutathione (GSH) |
| Superoxide dismutase (SOD) | Ascorbic acid (vitamin C) |
| Ascorbate-glutathione cycle enzymes | Tocopherols (vitamin E) |
| Peroxidases | Polyphenols (flavonoids) |
| NADP-dehydrogenases | |
| Peroxiredoxin (Prx) | |

**Table 2.** Main plant antioxidants [2].

The Haber-Weiss and Fenton reactions involve superoxide radicals and $H_2O_2$. In the presence of iron or other transition metals, $O_2^{\bullet -}$ and $H_2O_2$ lead to the formation of the hydroxyl radical, $OH^{\bullet}$, the most aggressive form of ROS, including the radical derivatives of oxygen ($O_2^{\bullet -}$, $OH^{\bullet}$), and also the peroxyl, alkoxyl or hydroperoxyl radicals, which are named as free radicals. Free radicals contain one or more unpaired electrons, but they also include nonradical derivatives of oxygen such as $H_2O_2$ and singlet oxygen [2, 5]. These free radicals are highly toxic and electrically charged molecules, i.e., they have an unpaired electron which causes them to seek out and capture electrons from other substances in order to neutralize themselves, all are capable of reacting with membrane lipids, nucleic acids, proteins and enzymes, and other small molecules, resulting in cellular damage, thus generate oxidative stress in plants [1–5].

Plants have developed a wide range of defense strategies to combat with these free radicals and deactivate their harmful effects known as antioxidants. The evolution of efficient antioxidant systems has enabled plant cells to overcome ROS toxicity and to use these reactive species as signal transducers [4, 5]. Antioxidants have diverse physiological roles in plants, acting as a scavenging and deactivating agent against oxidation, and converting the radicals to less reactive species, even at relatively small concentrations. The antioxidative system copes up with the harmful free radicals both by enzymatic (superoxide dismutase (SOD), catalase (CAT), peroxidase (POX), ascorbate peroxidase (APX), glutathione reductase (GR), polyphenol oxidase (PPO), etc.), and by nonenzymatic (ascorbic acid (vitamin C); $\alpha$-tocopherol, carotenes, flavonoids, polyphenolics, etc.) systems (**Table 2**). Under unfavorable conditions such as extreme oxidative stress, this antioxidant system scavenges the toxic radicals, and thus helps the plants to survive through such conditions [1–5].

## 3. ROS production in seeds

In plants, transport chain of electrons toward oxygen can potentially generate ROS. Seeds represent a particular case in this regard. During germination, the seed metabolic activity vary dramatically, meanwhile the sources of ROS in seeds also vary considerably [13, 39–45]. The mitochondrial respiratory chain is one of the major sources of ROS; electron leakage from the transport chain generates superoxide, and subsequently $H_2O_2$, by dismutation of the former. During germination, respiratory activity increased and production of ROS enhanced [6, 14, 46, 47]. Another source of ROS is peroxisomes. Peroxisomes divided into: glyoxysomes (oily seeds), peroxisomes of photosynthetic tissues, nodule peroxisomes (*Fabaceae* nodules) and gerontosomes (senescing tissues) [14, 27, 46–50]. In glyoxysomes, lipid reserves of oily seeds are converted into sugars during the first stages of seedling development [49–51]. During this lipid oxidation process $H_2O_2$ is produced. In peroxisomal matrix, xanthine is also oxidized into uric acid by xanthine oxidase resulting with the production of superoxide [49–51]. Catalase ($H_2O_2$ eliminating enzyme) is localized in peroxisomes [49–52]. Production of nitric oxide (NO), (a free radical and also an important cellular signaling compound in plants) also takes place in peroxisomes [48, 51–54]. NADPH oxidases of the cell membrane are another sources of ROS in plants, these enzymes transfer electrons from cytoplasmic NADPH to oxygen, producing superoxide radical and its dismutating product $H_2O_2$. NADPH oxidases are increased during plant infections [28, 29], in plant growth processes [55], and under severe abiotic stress conditions [56]. Enhanced activity of NADPH oxidase is reported in ABA induced generation of ROS under water stress [57, 58]. During biotic stress cell wall peroxidases and amine oxidases are induced leading to the formation of $H_2O_2$ in the apoplast [59]. As a result, mitochondria and peroxisomes are the major sources of ROS in nonquiescent seeds, during seed development and germination. Aquaporins and peroxiporins (transmembrane proteins) are shown to play roles in the transport of $H_2O_2$ in vegetative tissues [56, 60], but the mobility of ROS in seeds has not yet been documented. Finally, lipid oxidation can generate ROS that could be trapped in seed tissues [16, 61].

# 4. The dual effect of ROS: from toxicity to signaling

## 4.1. Toxicity of ROS

The oxidative stress may cause damage to DNA resulting in cancer and aging [62], and the presence of reactive oxygen also may initiate a chain reaction at the cellular level resulting in damage to critical cell bio-molecules [63–65]. The uncontrolled accumulation of ROS, particularly of $OH^•$ is highly toxic for the cell. These radicals are highly toxic and thus generate oxidative stress in plants. ROS can react with the majority of biomolecules, thus resulting in oxidative stress that can become irreversible and cause cellular damage [1–5]. Many harmful effects of ROS on cellular macromolecules have been identified [1–5]. All are capable of reacting with membrane lipids, nucleic acids, proteins and enzymes, and other small molecules, resulting in cellular damage [1–5]. Lipid peroxidation, which is a free-radical chain process leading to the deterioration of polyunsaturated fatty acids (PUFAs), is the best known cellular hazard among these, and has been studied intensively in food science [66]. Lipid peroxidation is initiated by free-radical attack upon a lipid, that gives starting to a chain reaction, removing a hydrogen atom from another fatty acid chain to form a lipid hydroperoxide (LOOH) in a propagation step [67]. This process is likely to degrade PUFAs present in membranes or in reserve lipids of oily seeds. Beside membranes, nucleic acids and proteins are also potential targets of ROS [67]. The hydroxyl radical, $OH^•$, can damage both nuclear and organelle DNA directly, by having ability to attack deoxyribose, purines and pyrimidines [67, 68]. Enzymes can also be inactivated easily by ROS, by degrading amino acids [69, 70]. ROS can damage transport proteins, receptors and ion channels and then lead to extensive cellular dysfunction [1–5, 69, 70].

## 5. The roles of ROS: cell signaling

Cellular antioxidant mechanisms control ROS concentrations, rather than to eliminate them completely, suggesting that some ROS may act as signaling molecules [5, 34–37, 46]. Although ROS have been considered as detrimental to seeds, advances in plant physiology evaluated them as messengers of various signal transduction pathways in plants. ROS are suggested as being beneficial for seed germination, seedling growth, protection against pathogens and controlling the cell redox status [28–40]. $H_2O_2$ is shown to be involved in the tolerance to various abiotic stresses acting as a secondary messenger [71], in cellular defense mechanisms against pathogens [72]. $H_2O_2$ has also been identified in many processes in plants, including programmed cell death (PCD) [8, 73], somatic embryogenesis [17], root gravitropism [19], and ABA-mediated stomatal closure [20, 21], response to wounding [74]. Superoxide ($O_2^{-\bullet}$) found to have roles in cell death and plant defense [24]. $H_2O_2$ also proved to have roles in protein phosphorylation through mitogen-activated protein kinase (MAP kinase) cascades [75, 76], calcium mobilization [77, 78], and regulation of gene expression [79, 80].

## 6. Control of ROS levels: detoxifying mechanisms

In plants and animals ROS are deactivated by antioxidants. These antioxidants act as an inhibitor of the process of oxidation, even at relatively small concentration and thus have diverse physiological roles [40]. Antioxidant constituents of plant materials act as radical scavengers, and convert the radicals to less reactive species [81, 82]. Plants have developed an array of defense strategies (antioxidant system) to cope up with oxidative stress. Plant cells are equipped with mechanisms allowing scavenging (in the case of oxidative stresses) or homeostasis of ROS (for cellular signaling) [83]. The antioxidative system includes both enzymatic and nonenzymatic systems. The nonenzymatic system includes ascorbic acid (vitamin C); $\alpha$-tocopherol, carotenes, etc., and enzymic system include superoxide dismutase (SOD), Superoxide dismutase, which can be mitochondrial (MnSOD), cytosolic (Cu/ZnSOD) or chloroplastic (CuZnSOD, FeSOD), dismutates superoxide radicals into $H_2O_2$ and oxygen [84]. Hydrogen peroxide is eliminated by the action of catalase (CAT), which is located in glyoxysomes and peroxisomes [51]. The ascorbate-glutathione cycle (also called the Halliwell-Asada cycle) also takes part in $H_2O_2$ scavenging. Ascorbate peroxidase (APX), monodehydroascorbate reductase (MDAR), dehydroascorbate reductase (DHAR), and glutathione reductase (GR) are involved in this cycle (**Figure 1**), and are present in chloroplasts, the cytoplasm, mitochondria, peroxisomes and the apoplast. These enzymes also participate in the regeneration of the powerful antioxidants such as reduced glutathione (GSH), ascorbic acid (vitamin C), and $\alpha$-tocopherol (vitamin E) (**Figure 1**). Glutathione an

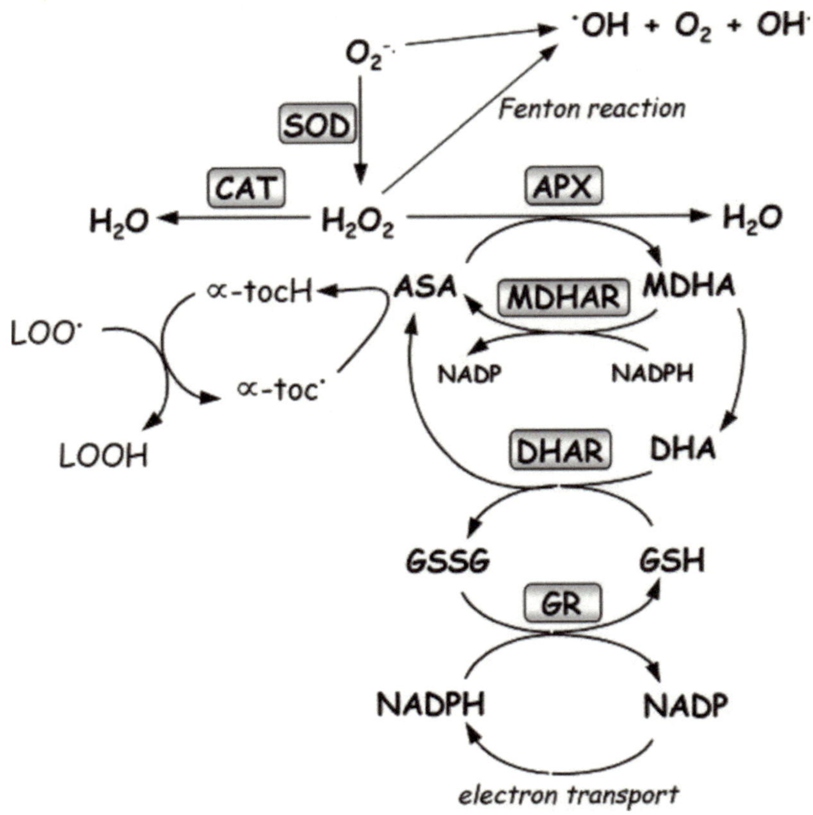


**Figure 1.** Main detoxifying mechanisms in plants. CAT, catalase; SOD, superoxide dismutase; APX, ascorbate peroxidase; MDHAR, monodehydroascorbate reductase; DHAR, dehydroascorbate reductase; GR, glutathione reductase; ASA, ascorbate; MDHA, monodehydroascorbate; DHA, dehydroascorbate; GSSG, oxidized glutathione; GSH, reduced glutathione; α-tocH, α-tocopherol; α-toc, α-tocopheryl; LOOH, lipid peroxide; LOO, lipid radical (Halliwell-Asada cycle).

important water-soluble antioxidant and is synthesized from the amino acids glycine, glutamate, and cysteine, which directly scavenges ROS such as lipid peroxides, and also plays a vital role in xenobiotic detoxification [85–88]. Research suggests that glutathione and vitamin C work interactively to quench free radicals and that they have a sparing effect upon each other [85–87]. Glutathione peroxidases (GPX) may also catalyze the reduction of $H_2O_2$ and hydroperoxides [85–88]. Polyphenol oxidase (PPO), the function of this antioxidant system is to scavenge the toxic radicals produced during oxidative stress and thus help the plants to survive through such conditions. Various compounds, such as polyphenols, flavonoids and peroxiredoxins [89] also have a strong antioxidant function.

## 7. ROS and seed development

Seed development consists of embryogenesis, reserve accumulation and maturation/drying on the mother plant, leading from a zygotic embryo to a mature, quiescent seed. During maturation

http://dx.doi.org/10.5772/intechopen.70837

seeds undergo a period of desiccation where water content is reduced and the embryo is at a state of quiescence [10]. ROS are involved in final stage of seed development, in desiccation in tolerance. A dramatic loss of water becomes during desiccation or maturation phase which requires cellular adaptative mechanisms, at this stage ROS scavenging plays a key role, for allowing seed survival [10]. Recently, LEA (late embryogenesis abundant)-related proteins which are cited as accumulating proteins during drought conditions are correlated with desiccation tolerance, but their biological functions remain unclear [90]. A group-2 LEA class of proteins has been suggested to act as free-radical scavengers [91], emphasizing the importance of ROS scavenging in dehydration tolerance mechanisms. In developing or germinating seeds, the active mitochondria are probably one of the major sources of ROS, generating superoxide, and subsequently $H_2O_2$ [14, 32]. ROS is also generated in chloroplasts in the beginning of seed development, but they rapidly become nonfunctional [15, 63]. $O_2^{\bullet-}$ and $H_2O_2$ are produced in peroxisomes, and in seeds, glyoxysomes, which is a particular type of peroxisomes involving in mobilization of lipid reserves [15, 63]. High amounts of $H_2O_2$ are produced in glyoxysomes resulting from the activity of enzymes such as glycolate oxidase. $H_2O_2$ is known to promote seed germination of cereal plants, and exogenously applied $H_2O_2$ is shown to ameliorate seed germination in many plants [7, 92]. Ascorbic acid is the most important reducing substrate for removal of $H_2O_2$, acting as an antioxidant, in plant cells. It is reported that ascorbic acid suppresses the germination of wheat seeds, recently [93]. In plant cells, ascorbate peroxidase (APX) and catalase (CAT) that are involved in scavenging $H_2O_2$ are localized at the site of $H_2O_2$ generation [93] (**Figure 2**). $H_2O_2$ is mentioned to induce expression of many genes, coding defense-related proteins, transcription factors, phosphatases, kinases and enzymes involving in ROS synthesis or degradation [37, 54, 56, 79, 80] (**Figure 2**).

Seed filling is also associated with the high potential of the $H_2O_2$ detoxification machinery, mainly due to APX and CAT activities [94]. It is suggested that cellular membranes in germinating tissues are vulnerable to damage from desiccation [10, 69]. After the loss of desiccation tolerance several products of peroxidized lipids are accumulated [45], and activated forms of oxygen are generated through xanthine oxidase [35, 48, 50]. Some studies have also suggested

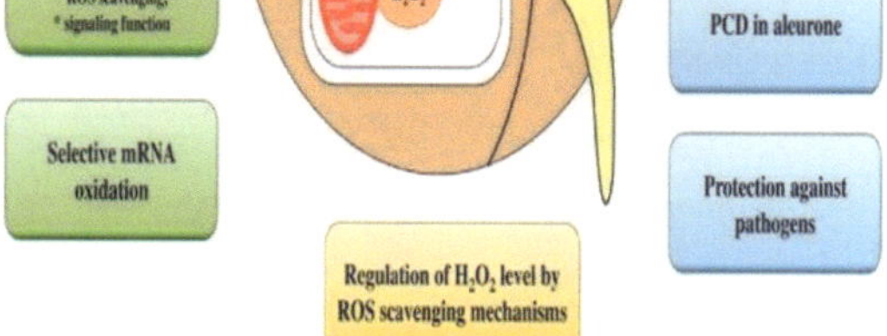


**Figure 2.** Production and functions of $H_2O_2$ in seed biology [37].

that ROS metabolism might also be important during initial embryogenesis [17, 95]. During embriogenesis, metabolic activity and mitochondrial respiration are increased, suggesting that developing embryos have the potential to generate significant amounts of ROS [17, 95]. The antioxidant ascorbate system reported to play an important role in embryogenesis and cell growth [41, 85]. Ascorbate content proposed to influence cell growth by modulating the expression of genes involved in hormonal signaling pathways [96]. Totipotency also related to antioxidant system, because of high ROS content and repressed expression of totipotency [97]. Conversely, ROS have beneficial effects in growth and development of plants. Seed germination requires release from dormancy. Treatment of dormant seeds with methylviologen (as a generator of ROS including $OH^{\bullet}$) is reported to break dormancy [98]. Hydroxyl radicals are also postulated to be involved in cell wall extension during cell growth, and auxin-induced increases in $OH^{\bullet}$ production is speculated to be involved in cell wall elongation, stiffening, and lignification depending on the concentration of auxin [55, 99]. Hydrogen peroxide is suggested to participate in lignin deposition in the cell walls in a peroxidase-catalyzed reaction [100]. The involvement of a diamine oxidase in $H_2O_2$ production has been demonstrated along with lignin deposition in the chalazal cells, in developing barley grains, in developing barley grains [100]. Production of ROS and their release in the surrounding medium are supposed to play a part in protecting the embryo against pathogens during seed imbibition [99]. Some of the selected published reviews on the dual roles of ROS in seed biology are listed in **Table 3**.

As shown above, the effects of ROS, and more particularly $H_2O_2$ on transcriptome have been widely studied [56]. However, up to date, there is no information available establishing a direct link between the changes in ROS content and gene expression during seed germination and development. Further experiments in this area, will be highly informative for getting a comprehensive view of ROS in seed biology.

| ROS molecule | Physiological trait | Reference |
|---|---|---|
| Hydrogen peroxide ($H_2O_2$) | Alleviation of seed dormancy | Wang et al. [104] |
| $H_2O_2$ | Somatic embryogenesis | Cui et al. [17] |
| Superoxide ($O_2^-$) | Plant defense response | Wisniewski et al. [24] |
| $O_2^-$ | Survival and germination seeds | Roach et al. [105] |
| $H_2O_2$ | Response to wounding | Oroczo-Cardenas et al. [74] |
| $H_2O_2$ | Seed germination-ABA levels | Barba-Espin et al. [102] |
| Hydroxyl radical ($OH^-$) | Breakdown of polysaccharides | Schweikert et al. [106] |
| $O_2^-$ | Cell growth by auxin | Schopfer et al. [55] |
| $OH^-$ | Cell wall loosening | Müller et al. [107] |
| $H_2O_2$ | Lateral root formation | Chen et al. [101] |
| $H_2O_2$ | Seed germination via pentose phosphate pathway | Barba-Espin et al. [103] |
| $H_2O_2$ | Programmed cell death | de Jong et al. [8] |
| $O_2^-$ | Cell death | Doke et al. [22] |

**Table 3.** Published reviews on the dual role of ROS in seed physiology [34].

http://dx.doi.org/10.5772/intechopen.70837

## 8. Conclusion

ROS and antioxidants play important roles in seed biology. In seed life, ROS are involved in all the stages of seed development, from embryogenesis to germination. ROS can react with the majority of biomolecules, resulting in cellular damage. In developing or germinating seeds, major amounts of ROS are generated, which are highly toxic and thus generate oxidative stress in seed cells. Plants have developed an array of defense strategies (antioxidant system) to cope up with oxidative stress. Conversely, ROS are suggested to have beneficial effects in growth and development of seeds, and are considered as part of a signaling network involving in numerous regulatory components of seed development. For example, $H_2O_2$ is known to promote seed germination of cereal plants. The antioxidant system reported to play an important role in embryogenesis and cell growth. Ascorbate content is proposed to influence cell growth by modulating the expression of genes involved in hormonal signaling pathways. The above findings show that, these dual effects of ROS in seed biology are very interesting subjects and need further examinations for determination of the roles of ROS in seed physiology. Depending on the progress that has been required in seed tissue physiology, cellular production sites of ROS and their diffusion within the cell are established. Investigations in this field encourage to enlighten the cellular mechanisms involved in acquisition of the desiccation tolerance, germination and alleviation of dormancy. Finally, ROS signaling transduction pathway in seeds, from sensing to changes in gene expression, is not fully understood yet. Therefore, there is still a domain to be examined in future studies dealing with seed biology and ROS, which concerns the direct effects of these compounds on gene expression. Analyses of gene expression using the novel methods will be of help in elucidating the mechanisms underlying the interplay of ROS with hormones and their cross-talk in seed germination and development, providing a challenge for future research in this area.

## Author details

Fadime Eryılmaz Pehlivan

Address all correspondence to: eryilmazfadime@gmail.com

Department of Biology, Botany Section, Faculty of Science, Istanbul University, Istanbul, Turkey

## References

[1] Halliwell B, Gutteridge JMC. Free Radicals in Biology and Medicine. United Kingdom, Oxford University Press; 2007

[2] Halliwell B. Reactive species and antioxidants. Redox biology is a fundamental theme of aerobic life. Plant Physiology. 2006;**141**:312-322

[3] Mittler R, Vanderauwera S, Gollery M, Van Breusegem F. Reactive oxygen gene network of plants. Trends in Plant Science. 2004;**9**:490-498 [PubMed]

[4] Apel K, Hirt H. Reactive oxygen species: Metabolism, oxidative stress, and signal transduction. Annual Review of Plant Biology. 2004;**55**:373-399 [PubMed]

[5] Foyer CH, Noctor G. Redox homeostasis and antioxidant signaling: A metabolic interface between stress perception and physiological responses. Plant Cell. 2005;**17**:1866-1875 [PMC free article] [PubMed]

[6] Sun WQ, Leopold AC. The Maillard reaction and oxidative stress during aging of soybean seeds. Physiologia Plantarum. 1995;**94**:94-104

[7] Fath A, Bethke PC, Jones RL. Enzymes that scavenge reactive oxygen species are down-regulated prior to gibberellic acid-induced programmed cell death in barley aleurone. Plant Physiology. 2001;**126**:156-166 [PMC free article] [PubMed]

[8] de Jong AJ, Yakimova ET, Kapchina VM, Woltering EJ. A critical role for ethylene in hydrogen peroxide release during programmed cell death in tomato suspension cells. Planta. 2002;**214**:537-545 [PubMed]

[9] Pellinen RI, Korhonen MS, Tauriainen AA, Palva ET, Kangasjärvi J. Hydrogen peroxide activates cell death and defense gene expression in birch. Plant Physiology. 2002;**130**:549-560 [PMC free article][PubMed]

[10] Pammenter NW, Berjak P. A review of recalcitrant seed physiology in relation to desiccation tolerance mechanisms. Seed Science Research. 1999;**9**:13-37

[11] Vertucci CW, Farrant JM. Acquisition and loss of desiccation tolerance. In: Kigel J, Galili G, editors. Seed Development and Germination. New York: Marcel Dekker; 1995. p. 237-271

[12] Bailly C. Active oxygen species and antioxidants in seed biology. Seed Science Research. 2004;**14**:93-107

[13] Bewley JD, Black M. Seeds Physiology of Development and Germination. 2nd ed. New York: Plenum Press; 1994

[14] Moller IM. Plant mitochondria and oxidative stress: Electron transport, NADPH turnover, and metabolism of reactive oxygen species. Annual Review of Plant Physiology and Plant Molecular Biology. 2001;**52**:561-591 [PubMed]

[15] Noctor G, De Paepe R, Foyer CH. Mitochondrial redox biology and homeostasis in plants. Trends in Plant Science. 2007;**12**:125-134 [PubMed]

[16] McDonald MB. Seed deterioration: Physiology, repair and assessment. Seed Science and Technology. 1999;**27**:177-237

[17] Cui K, Xing G, Liu X, Xing G, Wang Y. Effect of hydrogen peroxide on somatic embryogenesis of *Lycium barbarum* L. Plant Science. 1999;**146**:9-16

[18] Buchanan BB, Balmer Y. Redox regulation: A broadening horizon. Annual Review of Plant Biology. 2005;**56**:187-220 [PubMed]

[19] Joo JH, Bae YS, Lee JS. Role of auxin-induced reactive oxygen species in root gravitropism. Plant Physiology. 2001;**126**:1055-1060 [PMC free article] [PubMed]

[20] Pei ZM, Murata Y, Benning G, Thomine S, Klüsener B, Allen GJ, Grill E, Schroeder JI. Calcium channels activated by hydrogen peroxide mediate abscisic acid signalling in guard cells. Nature. 2000;**406**:731-734 [PubMed]

[21] Zhang X, Zhang L, Dong F, Gao J, Galbraith D, Song CP. Hydrogen peroxide is involved in abscisic acid-induced stomatal closure in *Vicia faba*. Plant Physiology. 2001;**126**:1438-1448 [PMC free article][PubMed]

[22] Doke N, Miura Y, Sanchez LM, Kawakita K. Involvement of superoxide in signal transduction: Responses to attack by pathogens, physical and chemical shocks and UV irradiation. In: Foyer CH, Mullineaux P, editors. Causes of Photooxidative Stress and Amelioration of Defense Systems in Plants. Boca Raton: CRC Press; 1994. p. 177-218

[23] Jabs T, Dietrich RA, Dangl JL. Initiation of runaway cell death in an Arabidopsis mutant by extracellular superoxide. Science. 1996;**27**:1853-1856 [PubMed]

[24] Wisniewski JP, Cornille P, Agnel JP, Montillet JL. The extensin multigene family responds differentially to superoxide or hydrogen peroxide in tomato cell cultures. FEBS Letters. 1999;**447**:264-268 [PubMed]

[25] Chance B, Boveris A, Oshino N, Loschen G. The nature of catalase intermediate and its biological function. In: King TE, Mason HS, Morrison, editors. Oxidases and Related Redox Systems. Baltimore: University Park Press; 1973. p. 350-353

[26] Puntarulo S, Sanchez RA, Boveris A. Hydrogen peroxide metabolism in soybean embryonic axes at the onset of germination. Plant Physiology. 1988;**86**:626-630 [PMC free article] [PubMed]

[27] Huang AHC, Trelease RN, Moore TS. Plant Peroxisomes. London: Academic Press; 1983

[28] Lamb C, Dixon RA. The oxidative burst in plant disease resistance. Annual Review of Plant Physiology and Plant Molecular Biology. 1997;**48**:251-275 [PubMed]

[29] Grant JJ, Loake GJ. Role of reactive oxygen intermediates and cognate redox signaling in disease resistance. Plant Physiology. 2000;**124**:21-30 [PMC free article] [PubMed]

[30] Kucera B, Cohn MA, Leubner-Metzger G. Plant hormone interactions during seed dormancy release and germination. Seed Science Research. 2005;**15**:281-307

[31] El-Maarouf-Bouteau H, Bailly C. Oxidative signaling in seed germination and dormancy. Plant Signaling & Behavior. 2008;**3**(3):175-182

[32] Fontaine O, Huault C, Pavis N, Billard JP. Dormancy breakage of *Hordeum vulgare* seeds: Effects of hydrogen peroxide and scarification on glutathione level and glutathione reductase activity. Plant Physiology and Biochemistry. 1994;**32**:677-683

[33] Hendricks SB, Taylorson RB. Breaking of seed dormancy by catalase inhibition. Proceedings of the National Academy of Sciences. 1975;**72**:306-309

[34] Kumar SPJ, Prasad SR, Banerjee R, Thammineni C. Seed birth to death: Dual functions of reactive oxygen species in seed physiology. Annals of Botany. 2015;**116**:663-668

[35] del Río LA. ROS and RNS in plant physiology: An overview. Journal of Experimental Botany. 2015;**66**(10):2827-2837

[36] You J, Chan Z. ROS regulation during abiotic stress responses in crop plants. Frontiers in Plant Science. 2015;**6**:1-15

[37] Wojtyla L, Lechowska K, Kubala S, Garnczarska M. Different modes of hydrogen peroxide action during seed germination. Frontiers in Plant Science. 2016;**7**:1-16

[38] Barba-Espín G, Hernández JA, Diaz-Vivancos P. Role of $H_2O_2$ in pea seed germination. Plant Signaling & Behavior. 2012;**7**(2):193-195

[39] Caliskan M, Cuming AC. Spatial specificity of $H_2O_2$-generating oxalate oxidase gene expression during wheat embryo germination. The Plant Journal. 1998;**15**:165-171 [PubMed]

[40] Chien CT, Lin TP. Mechanism of hydrogen peroxide in improving the germination of *Cinnamonum camphora* seed. Seed Science and Technology. 1994;**22**:231-236

[41] De Gara L, de Pinto MC, Arrigoni O. Ascorbate synthesis and ascorbate peroxidase activity during the early stage of wheat germination. Physiologia Plantarum. 1997;**100**: 894-900

[42] Hite DRC, Auh C, Scandalios JG. Catalase activity and hydrogen peroxide levels are inversely correlated in maize scutella during seed germination. Redox Report. 1999;**4**:29-34

[43] Puntarulo S, Galleano M, Sanchez RA, Boveris A. Superoxide anion and hydrogen peroxide metabolism in soybean embryonic axes during germination. Biochimica et Biophysica Acta. 1991;**1074**:277-283 [PubMed]

[44] Simontacchi M, Caro A, Fraga CG, Puntarulo S. Oxidative stress affects $\alpha$-tocopherol content in soybean embryonic axes upon imbibition and following germination. Plant Physiology. 1993;**103**:949-953

[45] Smith MT, Berjak P. Deteriorative changes associated with the loss of viability of stored desiccation tolerant and desiccation-sensitive seeds. In: Kigel J, Galili G, editors. Seed Development and Germination. New York: Marcel Dekker; 1995. p. 701-746

[46] Foyer CH, Noctor G. Redox sensing and signalling associated with reactive oxygen in chloroplasts, peroxisomes and mitochondria. Physiologia Plantarum. 2003;**119**:355-364

[47] Staniek K, Nohl H. Are mitochondria a permanent source of reactive oxygen species? Biochimica et Biophysica Acta. 2000;**1460**:268-275

[48] Corpas FJ, Barroso JB, del Rio LA. Peroxisomes as a source of reactive oxygen species and nitric oxide signal molecules in plant cells. Trends in Plant Science. 2001;**6**:145-150

[49] Reumann S. The structural properties of plant peroxisomes and their metabolic significance. Biological Chemistry. 2000;**381**:639-648

[50] del Rio LA, Pastori GM, Palma JM, Sandalio LM, Sevilla F, Corpas FJ, Jimenez A, Lopez-Huertas E, Hernandez JA. The activated oxygen role of peroxisomes in senescence. Plant Physiology. 1998;**116**:1195-1200

[51] Willekens H, Inzé D, Van Montagu M, Van Camp W. Catalases in plants. Molecular Breeding. 1995;**1**:207-228

[52] Kawano T. Roles of the reactive oxygen species-generating peroxidase reactions in plant defense and growth induction. Plant Cell Reports. 2003;**9**:829-837 [PubMed]

[53] Durner J, Klessig DF. Nitric oxide as a signal in plants. Current Opinion in Plant Biology. 1999;**2**:369-374

[54] Neill SJ, Desikan R, Hancock JT. Nitric oxide signalling in plants. New Phytologist. 2003;**159**:11-35

[55] Schopfer P, Liszkay A, Bechtold M, Frahry G, Wagner A. Evidence that hydroxyl radicals mediate auxin-induced extension growth. Planta. 2002;**214**:821-828 [PubMed]

[56] Neill S, Desikan R, Hancock J. Hydrogen peroxide signalling. Current Opinion in Plant Biology. 2002;**5**:388-395

[57] Jiang M, Zhang J. Water stress-induced abscisic acid accumulation triggers the increased generation of reactive oxygen species and up-regulates the activities of antioxidant enzymes in maize leaves. Journal of Experimental Botany. 2002;**53**:2401-2410

[58] Jiang M, Zhang J. Involvement of plasmamembrane NADPH oxidase in abscisic acid- and water stress-induced antioxidant defense in leaves of maize seedlings. Planta. 2002;**215**:1022-1030

[59] Bolwell GP, Bindschedler LV, Blee KA, Butt VS, Davies DR, Gardner SL, Gerrish C, Minibayeva F. The apoplastic oxidative burst in response to biotic stress in plants: A three-component system. Journal of Experimental Botany. 2002;**53**:1367-1376

[60] Henzler T, Steudle E. Transport and metabolic degradation of hydrogen peroxide in *Chara corallina*: Model calculations and measurements with the pressure probe suggest transport of $H_2O_2$ across water channels. Journal of Experimental Botany. 2000;**51**:2053-2066

[61] Nelson KA, Labuza TP. Relationship between water and lipid oxidation rates: Water activity and glass transition theory. In: St Angelo AJ, editor. Lipid Oxidation in Foods. Washington, DC: American Chemical Society. 1992. pp. 93-103

[62] Mittler R, Vanderauwera S, Suzuki N, Miller G, Tognetti VB, Vandepoele K, Gollery M, Shulaev V, van Breusegem F. ROS signaling: The new wave? Trends in Plant Science. 2011;**16**:300-309

[63] Sung JM. Lipid peroxidation and peroxide scavenging in soybean seeds during aging. Physiologia Plantarum. 1996;**97**:85-89

[64] Wilson DO, McDonald MB. The lipid peroxidation model of seed aging. Seed Science and Technology. 1986;**14**:269-300

[65] Bailly C, Benamar A, Corbineau F, Côme D. Changes in malondialdehyde content and in superoxide dismutase, catalase and glutathione reductase activities in sunflower seeds as related to deterioration during accelerated aging. Physiologia Plantarum. 1996;**97**:104-110

[66] Beckman KB, Ames BN. Oxidants, antioxidants, and aging. In: Scandalios JG, editor. Oxidative Stress and the Molecular Biology of Antioxidant Defenses. New York: Cold Spring Harbor Laboratory Press; 1997. p. 201-246

[67] Halliwell B, Gutteridge JMC. Free Radicals in Biology and Medicine. 3rd ed. New York: Oxford University Press; 1999

[68] Chen SX, Schopfer P. Hydroxyl-radical production in physiological reactions: A novel function of peroxidase. European Journal of Biochemistry. 1999;**260**:726-735

[69] Smirnoff N. The role of active oxygen in the response of plants to water deficit and desiccation. New Phytologist. 1993;**125**:27-58

[70] Charles SA, Halliwell B. Effect of hydrogen peroxide on spinach (*Spinacia oleracea*) chloroplast fructose bisphosphatase. Biochemical Journal. 1980;**189**:373-376

[71] Prasad TK, Anderson MD, Martin BA, Stewart CR. Evidence for chilling-induced oxidative stress in maize seedlings and a regulatory role for hydrogen peroxide. Plant Cell. 1994;**6**:65-74

[72] Levine A, Tenkanen R, Dixon R, Lamb C. $H_2O_2$ from the oxidative burst orchestrates the plant hypersensitive disease resistance response. Cell. 1994;**79**:583-593

[73] Desikan R, Reynolds A, Hancock JT, Neill SJ. Harpin and hydrogen peroxide both initiate programmed cell death but have differential effects on defence gene expression in Arabidopsis suspension cultures. Biochemical Journal. 1998;**330**:115-120

[74] Orozco-Cardenas ML, Narvaez-Vasquez J, Ryan CA. Hydrogen peroxide acts as a second messenger for the induction of defense genes in tomato plants in response to wounding, systemin, and methyl jasmonate. Plant Cell. 2001;**13**:179-191 [PMC free article] [PubMed]

[75] Kovtun Y, Chiu WL, Tena G, Sheen J. Functional analysis of oxidative stress-activated mitogen-activated protein kinase cascade in plants. Proceedings of the National Academy of Sciences. 2000;**97**:2940-2945

[76] Samuel MA, Miles GP, Ellis BE. Ozone treatment rapidly activates MAP kinase signalling in plants. Plant Journal. 2000;**22**:367-376

[77] Bowler C, Fluhr R. The role of calcium and activated oxygens as signals for controlling crosstolerance. Trends in Plant Science. 2000;**5**:241-246

[78] Murata Y, Pei ZM, Mori IC, Schroeder J. Abscisic acid activation of plasma membrane $Ca^{2+}$ channels in guard cells requires cytosolic NAD(P)H and is differentially disrupted upstream and downstream of reactive oxygen species production in abi1-1 and abi2-1 protein phosphatase 2c mutants. Plant Cell. 2001;**13**:2513-2523

[79] Desikan R, Clarke A, Hancock JT, Neill SJ. $H_2O_2$ activates a MAP kinase-like enzyme in *Arabidopsis thaliana* suspension cultures. Journal of Experimental Botany. 1999;**50**:1863-1866

[80] Desikan R, Mackerness SAH, Hancock JT, Neill SJ. Regulation of the *Arabidopsis* transcriptome by oxidative stress. Plant Physiology. 2001;**127**:159-172

[81] Mittler R. Oxidative stress, antioxidants and stress tolerance. Trends in Plant Science. 2002;**7**:405-410

[82] Wolin, MS. and Mohazzab-H, KM. Mediation of signal transduction by oxidants. In: Scandalios, JG. (Ed.) Oxidative Stress and the Molecular Biology of Antioxidant Defenses. New York, Cold Spring Harbor Laboratory Press. pp. 21-48

[83] Guan LM, Zhao J, Scandalios JG. Cis elements and trans-factors that regulate expression of the maize Cat1 antioxidant gene in response to ABA and osmotic stress: $H_2O_2$ is the likely intermediary signaling molecule for the response. Plant Journal. 2000;**22**:87-95

[84] Bowler C, Van Montagu M, Inzé D. Superoxide dismutase and stress tolerance. Annual Review of Plant Physiology and Plant Molecular Biology. 1992;**43**:83-116

[85] De Paula M, Pérez-Otaola M, Darder M, Torres M, Frutos G, Martinez-Honduvilla CJ. Function of the ascorbate–glutathione cycle in aged sunflower seeds. Physiologia Plantarum. 1996;**96**:543-550

[86] De Tullio MC, Arrigoni O. The ascorbic acid system in seeds: To protect and to serve. Seed Science Research. 2003;**13**:249-260

[87] Tommasi F, Paciolla C, de Pinto MC, De Gara L. A comparative study of glutathione and ascorbate metabolism during germination of *Pinus pinea* L. seeds. Journal of Experimental Botany. 2001;**52**:1647-1654

[88] Eshdat Y, Holland D, Faltin Z, Ben-Hayyim G. Plant glutathione peroxidases. Physiologia Plantarum. 1997;**100**:234-240

[89] Aalen RB. Peroxiredoxin antioxidants in seed physiology. Seed Science Research. 1999;**9**:285-295

[90] Buitink J, Hoekstra FA, Leprince O. Biochemistry and biophysics of tolerance systems. In: Black M, Pritchard HW, editors. Desiccation and Survival in Plants: Drying without Dying. Wallingford: CABI Publishing; 2002. p. 293-318

[91] Hara M, Terashima S, Fukaya T, Kuboi T. Enhancement of cold tolerance and inhibition of lipid peroxidation by citrus dehydrin in transgenic tobacco. Planta. 2003;**217**:290-298

[92] Fath A, Bethke P, Beligni V, Jones R. Active oxygen and cell death in cereal aleurone cells. Journal of Experimental Botany. 2002;**53**:1273-1282

[93] Andarwulan N, Fardiaz D, Wattimena GA, Shetty K. Antioxidant activity associated with lipid and phenolic mobilization during seed germination of *Pangium edule* Reinw. Journal of Agricultural and Food Chemistry. 1999;**47**:3158-3163

[94] De Gara L, de Pinto MC, Moliterni VMC, D'Egidio MG. Redox regulation and storage processes during maturation in kernels of *Triticum durum*. Journal of Experimental Botany. 2003;**54**:249-258

[95] Otegui MS, Capp R, Staehelin LA. Developing seeds of *Arabidopsis* store different minerals in two types of vacuoles and in the endoplasmic reticulum. Plant Cell. 2002;**14**: 1311-1327

[96] Kato N, Esaka M. Changes in ascorbate oxidase gene expression and ascorbate levels in cell division and cell elongation in tobacco cells. Physiologia Plantarum. 1999;**105**: 321-329

[97] Papadakis AK, Siminis CI, Roubelakis-Angelakis KA. Reduced activity of antioxidant machinery is correlated with suppression of totipotency in plant protoplasts. Plant Physiology. 2001;**126**:434-444

[98] Whitaker C, Beckett RP, Minibayeva FV, Kranner I. Alleviation of dormancy by reactive oxygen species in *Bidens pilosa* L. seeds. South African Journal of Botany. 2010;**76**:601-605

[99] Schopfer P, Plachy C, Frahry G. Release of reactive oxygen intermediates (superoxide radicals, hydrogen peroxide, and hydroxyl radicals) and peroxidase in germinating radish seeds controlled by light, gibberellin, and abscisic acid. Plant Physiology. 2001;**125**:1591-1602 [PMC free article] [PubMed]

[100] Asthir B, Duffus CM, Smith RC, Spoor W. Diamine oxidase is involved in $H_2O_2$ production in the chalazal cells during barley grain filling. Journal of Experimental Botany. 2002;**53**:677-682

[101] Chen Y-H, Chao Y-Y, Hsu YY, Kao CH. Heme oxygenase is involved in $H_2O_2$-induced lateral root formation in apocynin-treated rice. Plant Cell Reports. 2013;**32**:219-226

[102] Barba-Espin G, Diaz-Vivancos P, Clemente-Moreno MJ. Interaction between hydrogen peroxide and plant hormones during germination and the early growth of pea seedlings. Plant, Cell and Environment. 2010;**33**:981-994

[103] Barba-Espín G, Diaz-Vivancos P, Job D, Belghazi M, Job C, Hernández JA. Understanding the role of $H_2O_2$ during pea seed germination: A combined proteomic and hormone profiling approach. Plant, Cell & Environment. 2011;**34**(11):1907-1919

[104] Wang M, Heimovaara-Dijkstra S, Van Duijn B. Modulation of germination of embryos isolated from dormant and nondormant grains by manipulation of endogenous abscisic acid. Planta. 1995;**195**:586-592

[105] Roach T, Beckett RP, Minibayeva FV. Extracellular superoxide production, viability and redox poise in response to desiccation in recalcitrant *Castanea sativa* seeds. Plant, Cell and Environment. 2010;**33**:59-75

[106] Schweikert C, Liszkay A, Schopfer P. Polysaccharide degradation by Fenton reaction- or peroxidase-generated hydroxyl radicals in isolated plant cell walls. Phytochemistry. 2002;**61**:31-35 [PubMed]

[107] Müller K, Heß B, Leubner-Metzger G. A role for reactive oxygen species in endosperm weakening. In: Adkins S, Ashmore S, Navie S, editors. Seeds: Biology, Development and Ecology. Wallingford: CAB International; 2007. pp. 287-295

Section 3

# Seed Functional Morphology

Chapter 10

# Anatomical and Chemical Insights into the White Clover (*Trifolium repens* L.) Seed Coat Associated to Water Permeability

Alberto Aníbal Galussi and María Esther Moya

Additional information is available at the end of the chapter

http://dx.doi.org/10.5772/intechopen.70313

**Abstract**

White clover (*Trifolium repens* L.) seeds can exhibit physical dormancy, which produces hard seeds within a seed lot. These seeds do not germinate because they do not imbibe water due to a barrier to water entry in the seed coat. The aim of this work was to analyze the anatomical and chemical characteristics of the testa of white clover seeds with respect to water permeability levels. Seeds of the cv. NK Churrinche (2004 and 2006 harvests) were characterized via anatomical studies and determination of structural substances, polyphenols, tannins, and cutins present in the testa of seeds of different water permeability levels. Anatomically, increased testa thickness was associated with a decreased permeability level. Very slow-hydration seed coats exhibit thicker cuticle, longer macrosclereids, thicker cell wall, and the presence of wide osteosclereids than rapid-hydration seed coats; these differences are associated with a slower hydration speed and with a barrier to water entry to the seed. From the physiological and chemical points of view, the mechanism of physical dormancy of the testa would be explained by a greater amount of hydrophobic components that cement the cell wall, such as polyphenols, lignins, condensed tannins, pectic substances, and a higher proportion of cellulose and hemicellulose.

**Keywords:** hard seeds, testa, macrosclereids, cell wall, lignin, tannins

## 1. Introduction

The seed coat is the primary defense against an adverse medium, and its characteristics determine seed permeability. A hard seed coat protects the seed from mechanical stress, microorganisms, and changes in temperature and humidity [1]. There are several types of dormancy, with most of them being induced by several factors. Legume seeds have a seed coat known as testa, which is characterized by having several layers. Outermost is the epidermis, which is uniseriate and consists of palisade macrosclereids with uneven thickened walls and interior lumen, closely packed and containing different chemical substances (quinones); the hilar region consists of two palisade layers. Below the epidermis is the hypodermis or columnar cells (osteosclereids), followed by the lacunose parenchyma, which is composed of several layers of flattened cells, the aleurone layer and the endosperm, and then the embryo [2–4].

In some leguminous species, the cuticle wax present in the hard coat plays an important role in water permeability. Some studies suggested that the osteosclereids pose the main barrier to water entry, since most seeds start to imbibe water only after those cells are perforated [3, 5]. Studies of the structure and chemical components of the seed coat of some species indicated the presence of ions, such as K, Ca, and Mg, and some phenolic compounds with a role in hardening and protection [1, 6]. The anatomy of the seed coat does not vary between hard and a non-hard seeds [7, 8]; however, some differences in content of different components, porosity, and "linea lucida" or light line were found in soybean [9]. [10, 11] related lignin content to resistance to mechanical damage in soybean cultivars; however, they exhibit hydrophobic traits and can be related to impermeability [12]. Reports on extended seed longevity are mainly related to Fabaceae and Malvaceae taxa containing Malpighian cells or osteosclereids [2, 5, 8, 13, 14]. [15] assume that dormancy breakage in these seeds (except for mechanical scarification) is due to the formation of an opening in the specialized anatomical structure of the seed coat (or of the fruit) through which water enters and hydrates the embryo.

Seed coat of white clover (*Trifolium repens*) was found to have different water permeability levels, with the corresponding different dormancy levels, which influence seed physiological quality [16]. The factors hindering water entry are not clearly defined and may be more than one. A comparative study of testas of seeds with different water permeability levels may help to explain the different dormancy levels and to determine the necessary techniques useful for breaking dormancy in seed lots. The aim of this work was to analyze the anatomical and chemical characteristics of the testa of white clover cv. NK Churrinche relative to water permeability levels.

## 2. Materials and methods

### 2.1. Seeds

Seeds of white clover cv. NK Churrinche (2004 and 2006 harvests) obtained from Criadero Barenbrug Palaversich S.A., Argentina, were used for the study. Seeds were checked under magnifying glass (10×) for purity, and those with visible physical damage were discarded.

### 2.2. Morphological and anatomical analysis of the testa by scanning electron microscope (SEM)

Seeds were selected according to their water permeability level; for this, seeds were classified by the hydration rate into rapid, slow, and very slow hydration [16, 17]. Seeds that imbibed water within the first 15 min were discarded due to possible damage in the seed coat. Seeds that hydrated after 2 h of being immersed in water were classified as of rapid hydration; seeds that hydrated between 4 and 14 h of water imbibition were classified as of slow hydration; and seeds with dormancy (hard seeds) were those did not imbibe water after 22 h of immersion and were determined as of very low hydration [18]. Testas of hard seeds that were immersed in water and failed to hydrate between 3000 h (125 days) and 7488 h (312 days) were observed. For each permeability level, five testas per sample were analyzed by SEM (2004 and 2006 harvests). The testa was observed and photographed on surface and lateral views under different magnifications. In micrographs, the testa topography and cuticle appearance were observed and compared among seeds; cell layers, cell size, and cuticle thickness were determined in lateral sections. Observations were made near and around the central point [19] and in the lens area. In addition, the cell wall of macrosclereids was observed and compared among testas of different permeability levels. SEM observations of the exomorphology and anatomy of the testa were made at CRIBABB (Centro Regional de Investigaciones Básicas y Aplicadas de Bahía Blanca) and CERIDE (Centro Regional de Investigación y Desarrollo de Santa Fe), using the method mentioned in [20, 21].

### 2.3. Observation of the cell wall of macrosclereid cells in cross sections of white clover testa

Testas were hydrated during 4 days. The fixing solution consisted of 2.5% glutaraldehyde in 0.1 M (pH 7.2) sodium phosphate buffer. A tissue section (0.5 mm long) was cut from the testa with a razor blade and immediately placed in the fixing solution for 24 h. The following steps until anatomical observations were those described in [20, 21]. Observations were made under a scanning electron microscope JEOL 100 CXII at 80 kV [22] at CRIBABB. Means (μm) were obtained from observations of nine cells of three testas per permeability level (2004 and 2006 harvests). The testas of very slow-permeability seeds that were observed and microphotographed had been immersed in water for 3000, 5000, and 7488 h and were not hydrated.

### 2.4. Determination of structural polymeric substances, pectic substances, and polyphenols, tannins, and cutins

The testas were selected according to their water permeability level: (a) rapid hydration (permeable). Under these conditions seeds are completely swollen and exhibit a crack in the testa due to the size increase caused by imbibed water. To obtain the testas for analysis, hydrated seeds were placed to germinate in moistened paper towel, which was rolled up, placed in nylon bags and taken to germination chamber (25°C). Rolls were placed in a 45° angle. Then, after 5 days, testas of germinated seedlings were collected, placed in Petri dishes, and left to dry in an oven (25°C) until 3 g of testas of rapid-germination seeds was obtained: (b) very low hydration (impermeable). To obtain testas of hard seeds, the seed coat was incised at the distal end of cotyledons and soaked in water until fully imbibed (4 h). Then they were placed to germinate in moistened paper towel, following the same procedure as that used for rapid-hydration (permeable) testas, until 3 g of testas of hard seeds was obtained.

Structural components of the testa cell wall were determined following the methods described in [23]: total polyphenols using the technique of Folin-Denis; condensed tannins using the vanilla method; and cell wall percentage, cell contents, and percentage of cellulose, lignin, and cutin via the Van Soest method. Determinations were made in 0.5 g testa samples with two repetitions (approximately 3000 testas) per permeability level, using testas of both harvests [20, 21], at the Institute of Cellulose Technology, Faculty of Chemical Engineering, National University of Litoral, Santa Fe, Argentina.

## 3. Results and discussion

### 3.1. Morphological and anatomical SEM analyses of the testa

The testa of the rapid-hydration white clover seed (nondormant seeds) observed under SEM (**Figure 1**) was 30–40 μm thick, showing a thin cuticle (1μm) with slight cracks, depressions, and openings, as mentioned by [24] for the soybean testa surface. At the surface level, the ends of the macrosclereids were visible, giving an irregular and rugged appearance. Macrosclereids (**Table 1**) were approximately 30 μm long and were arranged in palisade, but not too compressed, with a visible light line. In cross section, an irregular polygonal contour was observed (three to seven sides) (**Figure 2A**). The long and short axes varied in length; cell wall width was irregular and smaller than in macrosclereids of slow-hydration testas. Osteosclereids were not visible in these seed testas, and parenchyma cells were present.

Seed coats of very slow hydration (dormant seeds) (**Figure 3**) were 45–50 μm thick and had a thick cuticle (3–5 μm), giving the surface a smooth appearance. Macrosclereids (**Table 1**) were about 40 μm long; they were arranged in a single, very compressed palisade layer and had a visible clear line. In cross section, they had an irregular, polygonal contour (three to seven sides, **Figure 2B**). The long and short axes varied in length, depending on the cell; cell wall thickness was also variable but was thicker than in permeable testa macrosclereids. Osteosclereids (**Figure 3B** and **C**) were visible, 15–18 μm in width and relatively short (5–8 μm); parenchyma cells were present. Anatomical characteristics of hard seed testas with different immersion time were similar. The anatomical characteristics of the hard seed testas of 3000 h were similar to those of 5000 h and 7488 h (data not shown). Hard seeds of this species have very deep physical dormancy [16], suggesting similar anatomical and chemical traits among seeds within that category.

### 3.2. Determination of structural polymeric substances, pectic substances, and polyphenols, tannins, and cutins

The results show (**Table 2**) that cellulose, hemicellulose, lignins, and condensed tannins were present in higher amounts in impermeable testas, whereas in permeable ones, a higher amount of cell content was observed. These findings are related to the presence of macrosclereids of greater size and thicker cell wall found in dormant seeds than in rapid-hydration seeds, as well as to the presence of osteosclereids.

Cellulose is associated with hemicellulose, with both being the most important structural substances in the cell wall. The cellulose values found in very low-hydration testas (48.6%)

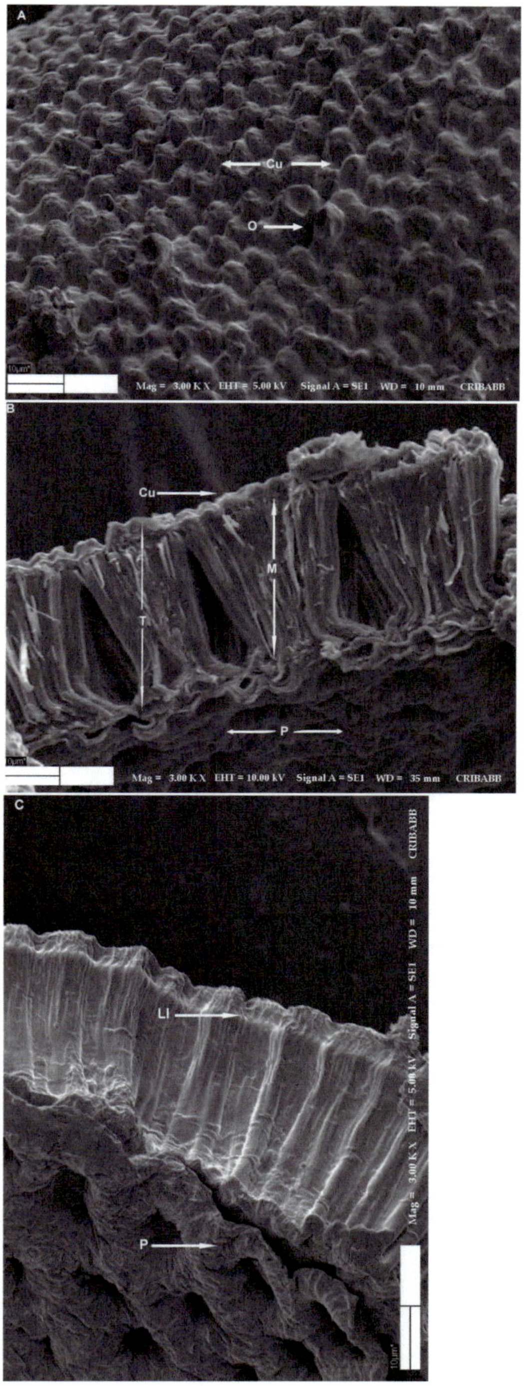


**Figure 1.** Testa of *T. repens* cv. NK Churrinche seed (2004 harvest) permeable after 2 h of water imbibition. (A) Testa surface. (Cu) Cuticle and (O) opening. (B and C) 3D view of testa. (T) Testa, (Cu) cuticle and waxes, (M) macrosclereids, (O) osteosclereids, (P) parenchyma, (Fs) slight fissure among cells, and (Ll) light line ×3000 (bar, 10μm).

| Testa | Length (μm) | Axis (cell width) (μm) | | | | | | Cell wall thickness (μm) | | |
|---|---|---|---|---|---|---|---|---|---|---|
| | | Long | | | Short | | | | | |
| | | Mean | Min | Max | Mean | Min | Max | Mean | Min | Max |
| Rapid hydration | 25-30 | 2.72 | 2.51 | 3.26 | 1.79 | 1.67 | 1.86 | 0.54 | 0.37 | 0.93 |
| Very slow hydration | 30-40 | 9.40 | 7.20 | 11.38 | 3.41 | 2.35 | 4.59 | 2.2 | 1.18 | 3.67 |

Mean values (μm) were obtained through SEM observations (×2700; ×6700 and ×10,000) of cross sections of nine cells of three teguments, 2006 harvest. (Min) minimum value; (Max) maximum value.

**Table 1.** Macrosclereid measurements in testa of seeds of white clover (*Trifolium repens*) cv. NK Churrinche as a function of testa permeability to water.

are close to those mentioned by [25, 26] for secondary wall (41 and 45%, respectively), with the amount of hemicellulose found (16.1%) being lower than that reported by those authors (30%). In alfalfa seeds, cellulose and hemicellulose in testas amounted to 39.9 and 20.7%, respectively [20, 21]. Although these components exhibit hydrophilic characteristics, in the cell wall of macrosclereids, lignification eventually occurs [2], forming hydrophobic secondary walls that provide rigidness [11, 27]. Lignin was found in greater amount in impermeable seed teguments than in permeable ones. This component is a highly insoluble polymer of

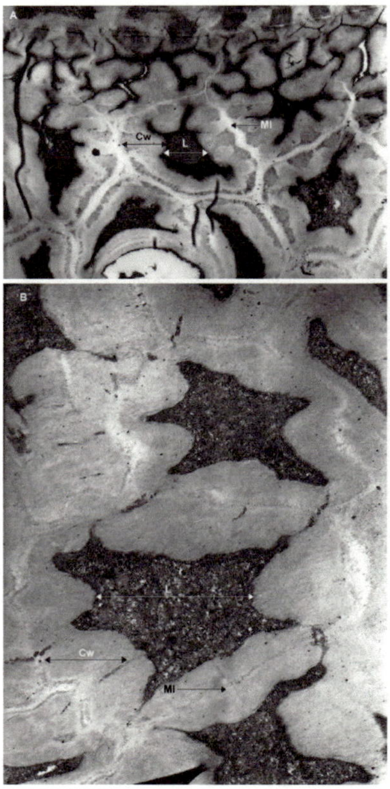


**Figure 2.** Macrosclereids of seed testa of *T. repens* cv. NK Churrinche (2004 and 2006 harvests) (A) rapid-hydration testa, permeable after 2 h of water imbibition. (B) Very low-hydration testa, impermeable after 3000 h water imbibition. (Cw) Cell wall, (Ml) middle lamella, and (L) cell lumen. (A) ×2700, 1 cm, 0.91μm; (B) ×6700, 1 cm, 1.31μm.

http://dx.doi.org/10.5772/intechopen.70313

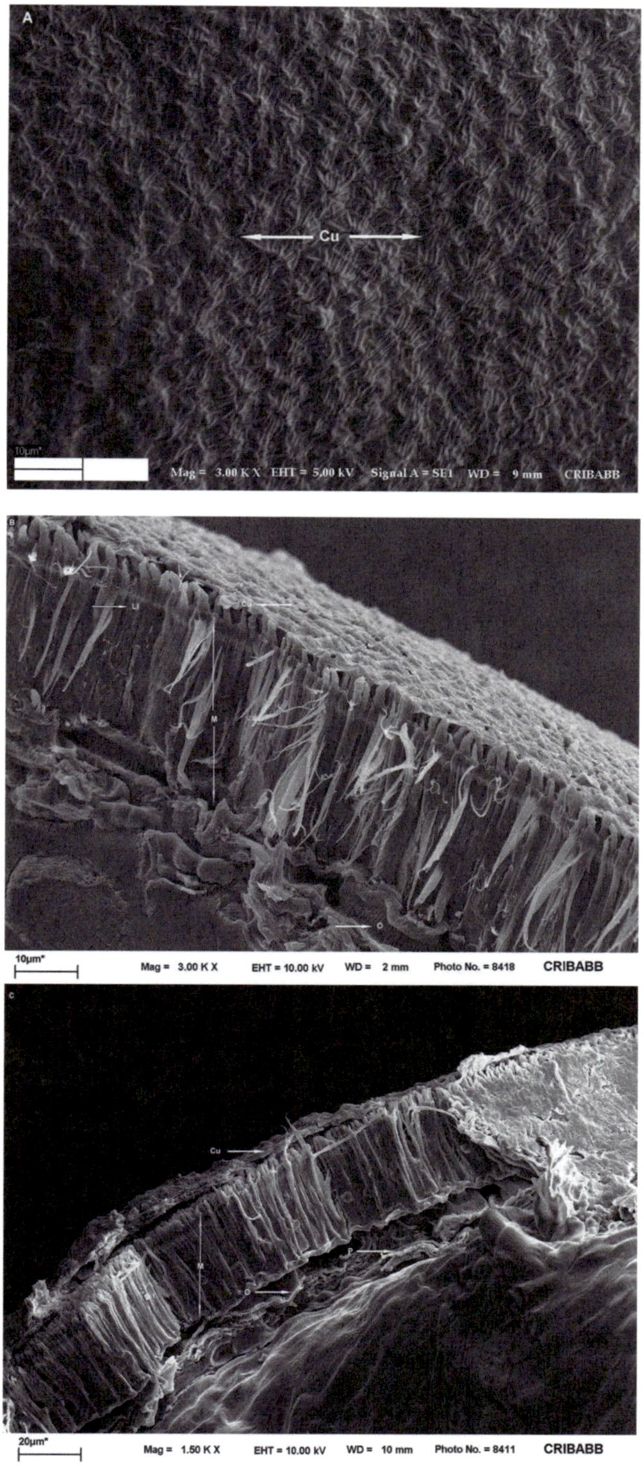


**Figure 3.** Testa of *T. repens* cv. NK Churrinche seed (2006 harvest) of very low hydration, impermeable after 3000 h of water imbibition. (A) Testa surface, (Cu) cuticle; (B and C) cross section of two testas, (Ll) light line, (M) macrosclereid, (O) osteosclereid, and (P) parenchyma. A and C: SEM ×3000 (A and B, bar 10 μm; C, bar 20 μm).

| Testa | Tp | Ct | Cw | Cc | Cell wall | | ADF | | |
|---|---|---|---|---|---|---|---|---|---|
| | | | | | ADF + sílica | hc | c | l | cu |
| Rapid hydration | 0.47 | 0.16 | 32.8 | 67.2 | 25.9 | 6.9 | 21.0 | 1.3 | 0.5 |
| Slow hydration | 0.36 | 0.14 | 69.3 | 30.1 | 56.1 | 13.8 | 45.3 | 3.1 | 9.1 |
| Very slow hydration | – | 0.29 | 77.1 | 22.9 | 61 | 16.1 | 48.6 | 5.5 | * |

Tp, Total polyphenols; Ct, Condensed tannins; Cw, Cell wall, (hc) hemicellulose, gums, mucilages, cellulose, lignin, and cysteine; Cc, cell content, glucose, fructose, sucrose, galactose, starch, and fructans (protoplasts and pectic substances); ADF, Acid detergent fiber. (c) Cellulose, (l) lignin, and (cu) cutin.

*It could not be determined by the extraction method; the presence was observed via SEM.

**Table 2.** Determination of organic compounds (%) in dry matter of seeds testa of *T. repens* cv. NK Churrinche with different permeability levels.

phenolic units that form a large network of crossed bonds. The lignin content (5.5%) found in impermeable testas is relatively high, compared to the 8% of cereals straw [28] and 16.85% of wheat straw [29]. In alfalfa, 3.34 and 7% of lignins (cell wall) were reported at leaf and preflowering stages, respectively [30, 31]. And 7.7% was integument of very slow-hydration seeds [21]. Condensed tannin is another component found in higher amount (0.29%) in hard seed testas than in permeable seeds (0.16%) and might provide the testa with astringent and feeding deterrent characteristics, as well as with defense from predators.

Regarding cell content determined in the testa (**Table 2**), a great difference was observed between rapid-hydration (67%) and very low-hydration (30.1 and 22.9%) seeds. That fraction includes diverse substances, such as pectic substances. Based on observations of the lumen in micrographs of macrosclereids, we assume that the amount of cell content determined in hard seed testas largely corresponds to pectins of the middle lamella of macrosclereids. Pectins act as cementing substances [32] and become lignified in older cells [2].

Cutin values found in rapid- and slow-hydration testas were 0.5 and 9.1%, respectively (**Table 2**). In the very slow-hydration teguments, cutin proportion could not be determined via the extraction method used; however, SEM observations showed a proportionately thicker cutin layer than in seeds of other degrees of permeability.

From an anatomical perspective, the highest level of seed physical dormancy in white clover seeds would be attributed to the combination of the effects of a thicker cuticle, thicker cell wall of macrosclereids, and greater length and width of macrosclereids than in seeds of the other permeability levels, as well as to the presence of osteosclereids.

Physiologically and chemically, the dormancy mechanism via hard coats is explained by an increased amount of hydrophobic and cementing substances, such as cutins, lignins, tannins, and pectins. Lignification of the thick cell wall would be one of the main components determining physical dormancy. White clover seeds with physical dormancy contain viable embryos and very low water content, as found in alfalfa [17]; once water permeability is produced and other appropriate factors are present, embryo imbibition occurs, originating normal seedlings [16]. The traits found in these seeds would provide physical dormancy with the capacity to preserve the embryo during the time when seeds are dormant.

## Acknowledgements

This work was supported by grants from National University of Entre Ríos, Argentina, to PID UNER 2100 of the Faculty of Agricultural Sciences.

## Author details

Alberto Aníbal Galussi* and María Esther Moya

*Address all correspondence to: agalussi@arnet.com.ar

Faculty of Agricultural Sciences, National University of Entre Ríos, Argentina

## References

[1] Yassen MY, Barringer SA, Splittstoesser WE, Constanza S. The role of seed coat in seed viability. The Botanical Review. 1994;**60**(4):426-439

[2] Esau K. Anatomía de las plantas con semillas. Buenos Aires, Argentina: Hemisferio Sur; 1987. 512 pp

[3] Bewly J, Black M. Seeds. Physiology of Development and Germination. v 2. New York: Pleum Press; 1986. pp. 182

[4] Besnier RF. Semillas Biología y Tecnología. Madrid: Mundiprensa; 1989. 637 pp

[5] Mc Kee GW, Pfeiffer RA, Mosheninn NN. Seed coat impermeability to water. Agronomy Journal. 1977;**69**:53-58

[6] Peihua T, Hua J, Zohongchen G, Zaijie J, Chenglie Y, Aiqiu L. Studies on the structure, chemical elements and functions of seed coats and their artificial imitations. In: Fu J, Khan A, editors. Advances in the Science and Technology of Seeds. Scientific Press; Beging, New York; 1992. pp. 167-175

[7] Nikolaeva MG. Factors controlling the seed dormancy pattern. In: The Physiology and Biochemistry of Seed Dormancy and Germination. Amsterdam, The Netherlands: Elsevier/North-Holland Biomedical Press; 1980. pp. 51-74

[8] Martens H, Jakobsen HB, Lyshede OB. Development of the strofiole in seeds of white clover (*Trifolium repens* L.). Seed Science Research. 1995;**5**:171-176

[9] Harris WH. Comparative ultrastructure of developing seed coats of "hard-seed" and "soft-seeded" varieties of soybeans, i (L.). Merr. Bot. Gaz. 1987;**148**(3):324-331

[10] Alvarez PJC, Krzyzanowski FC, Mandarino JMG, Franca Neto JB. Relationship between soybean seed coat lignin content and resistance to mechanical damage. Seed Science and Technology. 1997;**25**:209-214

[11] Capeleti I, Ferrarese ML, Krzyzanowski F, Ferrarese F. A new procedure for quantification of lignin in soybean (*Glycine max* (L.) Merrill) seed coat and their relationship with the resistance to mechanical damage. Seed Science and Technology. 2005;**33**:511-515

[12] Paniagua R. Citología e Histología Vegetal y Animal. 3ª ed. Madrid, España: McGraw-Hill-Interamericana de España, S.A.U.; 2004. pp. 779-814

[13] Boesewinkel FD, Bouman F. The seed: Structure. In: Johri BM, editor. Experimental Embryology of Vascular Plants. Berlin: Springer; 1982. pp. 567-608

[14] Baskin JM, Baskin CC. A classification system for seed dormancy. Seed Science Research. 2004;**14**:1-6

[15] Baskin JM, Baskin CC. Taxonomy, ecology, and evolution of physical dormancy in seeds. Plant Species Biology. 2000;**15**:139-152

[16] Galussi AA, Argüello JA, Moya ME, Zuriaga FD, Zimmermann LR. Seed dormancy mechanism as a factor influencing seed physiological quality in alfalfa (*Medicago sativa* L ) cv. Baralfa 85. Seed Science and Technolology. 2013;**41**:50-59

[17] Galussi AA, Moya ME, Zuriaga FD, Zimmermann LR, Basso R. Effect of the dormancy mechanism in the quality of seeds of *Trifolium repens* L. (white clover). Journal of Seed Technology. 2013;**35**(2):25-37

[18] International Rules for Seed Testing. International Seed Testing Association (ISTA). Bassersdorf, CH-Switzerland; 2007

[19] Aniszewski T. *Phaseolus lunatus* testa topography and biometrics of crushed cells. Seed Science and Technology. 2009;**37**:121-131

[20] Galussi AA. Estudios morfológicos y fisológicos de la dormición de semillas de alfalfa (*Medicago sativa* L.) [Tesis para optar al Grado Académico de Doctor en Ciencias Agropecuarias]. República Argentina: Universidad Nacional de Córdoba; 2010

[21] Galussi AA, Argüello JA, Cerana MM, Maximino M, Moya ME. Características anatómicas y químicas del tegumento seminal de *Medicago sativa* L. (alfalfa) cv. Baralfa 85 y su asociación con la dormición. PHYTON International Journal of Experimental Botany. 2015;**84**:163-175

[22] Sorrivas V, de Lozano, Morales A. Introducción a la Microscopía Electrónica. Bahía Blanca, Argentina: CRIBABB; 1986. 220 pp

[23] Van Soest PJ. Nutritional Ecology of the Ruminant. Ithaca, NY, USA: Cornell University Press; 1994. 476 pp

[24] Ma F, Cholewa E, Mohamed T, Peterson CA, Gijzen M. Cracks in the palisade cuticle of soybean seed coats correlate with their permeability to water. Annals of Botany. 2004;**94**:213-228

[25] Salisbury FB, Ross CW. Fisiología de las plantas. Células: agua, soluciones y superficies. Madrid, España: Thomson, Spain Paraninfo S.A.; 1992. 305 pp

[26] Taiz L, Zeiger E. Plant Physiology. 2ª ed. Massachusetts: Sinauer Associates, Inc.; 1998. 792 pp

[27] Ferrer AJ. Las células de los tejidos vegetales. Barcelona, España: Vedrá, S.L.; 1997. 392 pp

[28] FEDNA. [Internet] 2017. In: http://www.fundacionfedna.org/ingredientes_para_piensos/paja-de-cereales-trigo-y-cebada [Accessed: May 02 2017]

[29] Carroll A, Somerville C. Cellulosic biofuels. Annual Review of Plant Biology. 2009;**60**: 165-18

[30] Faner C. La pastura de alfalfa como fuente de alimentación para cerdos en crecimiento y terminación Boletín AACP [Internet]. 2007. In: www.produccion-animal.com.ar/produccion_porcina/72-alfalfa.pdf [Accessed: October 01 2007]

[31] Kephart K., Hollis G.R., Danielson, M. Forages for Swine. Perdue University. Cooperative Extension Service. West Lafayette, Indiana USA . PIH 126;1990

[32] Sandoval ZE. Técnicas aplicadas al estudio de la anatomía vegetal. Cuadernos 38. DF, México: Universidad Nacional Autónoma de México; 2005. 51 pp

Chapter 11

# Morphological Studies on Seeds of Scrophulariaceae s.l. and Their Systematic Significance

Balkrishna Ghimire, Go Eun Choi, Hayan Lee,
Kweon Heo and Mi Jin Jeong

Additional information is available at the end of the chapter

http://dx.doi.org/10.5772/intechopen.70572

**Abstract**

This study employed scanning electron microscopy and light microscopy to observe seed surface micromorphology and seed coat anatomy in the Scrophulariaceae s.l. to investigate seed characters of taxonomic importance. Seeds of 41 taxa corresponding to 13 genera of the family were carefully investigated. Seeds were minute and less than or slightly larger than 1 millimeter in length except for *Melampyrum* and *Pedicularis* species. The seed shape ranged from elliptical to broad elliptical and ovoid. In the studied species the surface sculpture was predominantly reticulate-striate, regular reticulate, sometimes colliculate, and rugose, or - rarely - ribbed, as in *Lindernia procumbens* and *Paulownia coreana*. Seed coats comprised the epidermis and the endothelium. Nevertheless, in all *Melampyrum* and some *Veronica* species the seed coat was very poorly represented and only formed by a papery layer of epidermis. According to correspondence analysis (CA) and unweighted pair group method with arithmetic mean (UPGMA) based cluster analysis the close affinities among the species of *Scrophularia* were well supported by their proximity to one another. Similarly, the proximity of *Melampyrum* species and *Pedicularis* species cannot be denied. In contrast, *Veronica* species were divided into two groups in CA plots and even three in the UPGMA tree. Regardless of the limited range taxa considered we found that similarities and differences in seed morphology between different genera could help us to understand the systematic relationships involved.

**Keywords:** Scrophulariaceae s.l., seed morphology, seed anatomy, scanning electron microscopy, light microscopy, surface sculpture, seed coat, systematics

## 1. Introduction

In traditional classification, Scrophulariaceae sensu lato (s.l.) are the largest family in the Lamiales. The members of the family can be distinguished from related families by bilaterally symmetrical and often tubular flowers, ovaries with axile placentation and numerous ovules, and many seeded capsular fruits. However, recent molecular studies have suggested that it is not the presence of a suite of uniquely derived characteristics that allows for the easy recognition of a member of Scrophulariaceae s.l. but, rather, the absence in Scrophulariaceae of the synapomorphies that characterize closely related families [1]. The taxonomic problem of the Scrophulariaceae s.l. is one aspect of the widespread difficulties that reflect the problems of describing natural groups within the order Lamiales. As with most large families, previous classification of the family includes many treatments that differ in their circumscription (for example, see [2]). The most influential classifications for the nineteenth century concept of Scrophulariaceae were those of in [3, 4], from which most contemporary classifications of the family are derived. A large-scale investigation of phylogenetic relationships in the Scrophulariaceae s.l. and related families using DNA sequence data has radically altered the circumscription of many families in Lamiales [1, 2, 5–7].

Before the revolution in molecular systematics, many studies proposed that Scrophulariaceae s.l. were monophyletic and several morphologically similar groups of taxa, which are now assigned to different families, and were placed together (for example, see [8]). For the first time in [2], the study identified two clearly separated clades consisting of members of the family and suggested that the Scrophulariaceae are polyphyletic. Subsequently, a third clade was identified consisting of parasitic members of the Scrophulariaceae and Orobanchaceae [9–11]. Later on five distinct phylogenetic lineages composed mainly of taxa previously assigned to Scrophulariaceae were recognized [5]. However, the emerging classification for the plants which are traditionally assigned to Scrophulariaceae consists of at least seven groups that bear the rank of family [12–14]. Most notable, following changes in the circumscription of families and the disintegration of Scrophulariaceae s.l., were the dramatic changes in the size of some families: the Scrophulariaceae itself was more than halved in size to just over 1800 species, and the Plantaginaceae increased to about 1900 species [15].

The morphological similarities and differences among the groups of taxa and their alignment in various families usually depend on the characters emphasized by different researchers. On the other hand, complications in discriminating between genera or groups of genera are usually the result of the available suite of usable characters, and thus, several comparative studies were carried out. In general, systematics uses morphological characteristics to carve diversity into its taxonomic subunits, and since the beginning of the discipline, plant systematics has frequently used morphological character ranges from roots, leaves, inflorescence, flowers, and fruit to seeds.

Due to its great uniformity, seed morphology has been recognized as an important source of useful phylogenetic information. A number of angiosperm taxa have already been investigated intensively in terms of their seed morphology, in combination with phenetic or phylogenetic analyses at the genus level. In the past, the variation in seed morphology has been used variously in plant systematics ranging from identification [16, 17] and taxonomic circumscription [18, 19] to phylogenetic inference [20, 21] and character-state evolution [22, 23]. Both macro- and micromorphological seed characters have been shown to be of essential systematic importance within and among the genera of traditional Scrophulariaceae, *Orobanchaceae*, and *Plantaginaceae* [24–35], in which seed morphological characters have been used widely to differentiate the different taxa or to find affinities between them.

Recent studies on seed morphology of Scrophulariaceae s.l. have focused mainly on common genera like *Veronica, Scrophularia,* and *Pedicularis*. Juan et al. [30] observed fruits and seeds of Scrophulariaceae from southwest Spain, and the systematic significance of seed morphology of *Veronica* and *Pedicularis* has been examined in some comprehensive studies [32–34]. Despite the aforementioned, no comparative studies on seed anatomy or seed coat characteristics together with surface structure have been conducted on any genera of Scrophulariaceae s.l. It is necessary to make extensive investigation of the seed morphology and anatomy of Scrophulariaceae s.l. to determine whether they can be used as additional support for disintegration of genera in the family. The objectives of this study were to (1) understand the utility of seed morphology and anatomy in Scrophulariaceae s.l. systematics, (2) discuss the proximity of studied genera based on these characters, and (3) highlight the characters that can be used to describe different genera and possible variation in infrageneric classification of *Veronica* on a similar basis.

## 2. Materials and methods

### 2.1. Specimens

More than 2500 seeds from 41 taxa and 56 accessions were investigated, corresponding to 13 genera of family Scrophulariaceae s.l. originating from seed bank and herbarium specimens at the Korea National Arboretum, Pocheon, Korea. Names of investigated species and accession numbers are presented in **Table 1**.

### 2.2. Scanning electron microscopy

Sampling seeds were directly taken from seed bank (stored in $-18^{\circ}$C) and thus no pretreatment was needed for scanning electron microscopy (SEM). Before SEM observation, the seed samples were rinsed with absolute ethyl alcohol and sputtered with gold coating in a KIC-IA COXEM ion-coater (COXEM Co., Ltd., Korea). SEM photographs were taken with the help of COXEM CX-100S scanning electron microscope at 20 kV at the Seed Test Laboratory of the Korea National Arboretum. Scale bars were added manually during image alignment.

### 2.3. Light microscopy

At least 10 to 12 seeds from each species were sectioned and investigated under light microscope. Microscopic slides were prepared using conventional microtome resin method. Mature seeds were dehydrated through alcohol series (50, 70, 80, 90, 95, and 100%). The complete dehydrated seeds were transferred in alcohol/Technovit combination (3:1, 1:1, 1:3, and 100% Technovit) and then embedded in Technovit 7100 resin. Histo-blocks were prepared from each

| Family | Genus | Species | Voucher number |
|---|---|---|---|
| Linderniaceae | *Lindernia* | *Lindernia procumbens* (Krock.) Philcox | L10041 |
| | | *L. crustacea* (L.) F.Muell. | L3029 |
| Phrymaceae | *Mazus* | *Mazus pumilus* (Burm.f.) Steenis | L10128 |
| Orobanchaceae | *Lathraea* | *Lathraea japonica* Miq. | 2014REC006 |
| | *Melampyrum* | *Melampyrum koreanum* K. J. Kim & S. M. Yun | L9710 |
| | | *M. roseum* Maxim. | L2324 |
| | | *M. roseum* var. *japonicum* Franch. & Sav. | L1707 |
| | | *M. roseum* var. *ovalifolium* Nakai ex Beauverd | L10163 |
| | | *M. setaceum* (Maxim. ex Palib.) Nakai | L8578 |
| | | *M. setaceum* var. *nakaianum* (Tuyama) T.Yamaz. | L3449 |
| | *Pedicularis* | *Pedicularis mandshurica* Maxim. | L10193 |
| | | *P. resupinata* f. *albiflora* (Nakai) W.T.Lee | L9042 |
| | | *P. resupinata* L. | 2014KNA031 |
| | | *P. resupinata* var. *umbrosa* Kom. ex Nakai | L10246 |
| | | *Pedicularis verticillata L.* | L8699 |
| | | *Phtheirospermum japonicum* (Thunb.) Kanitz | L9691 |
| | *Siphonostegia* | *Siphonostegia chinensis* Benth. | L10657 |
| Paulowniaceae | *Paulownia* | *Paulownia coreana* Uyeki | 2014 REC081 |
| Plantaginaceae | *Limnophila* | *Limnophila indica* (L.) Druce | L10453 |
| | *Veronica* | *Veronica arvensis* L. | L3304 |
| | | *V. dahurica* Steven | L10154 |
| | | *V. didyma* var. *lilacina* (*H. Hara*) T.Yamaz. | L2006 |
| | | *V. incana* L. | L10791 |
| | | *V. kiusiana* var. *diamantiaca* (Nakai) T.Yamaz. | L10532 |
| | | *V. kiusiana* var. *glabrifolia* (Kitag.) Kitag. | L9413 |
| | | *V. linariifolia* Pall. ex Link | L7983 |
| | | *V. longifolia* L. | L9903 |
| | | *V. nakaiana* Ohwi | 2014GB024 |
| | | *V. peregrina* L. | L2695 |
| | | *V. persica* Poir. | L2702 |
| | | *V. pusanensis* Y. Lee | L11087 |
| | | *V. pyrethrina* Nakai | L10564 |
| | | *V. rotunda* var. *subintegra* (Nakai) T.Yamaz. | L10130 |
| | | *V. undulata* Wall. | 2014 cc56 |
| | *Veronicastrum* | *Veronicastrum sibiricum* (L.) Pennell | L10658 |

| Family | Genus | Species | Voucher number |
|---|---|---|---|
| | *Linaria* | *Linaria japonica* Miq. | L8982 |
| Scrophulariaceae s.s. | *Scrophularia* | *Scrophularia buergeriana* Miq. | L8488 |
| | | *S. grayana* Maxim. ex Kom. | L10574 |
| | | *S. kakudensis* Franch. | L9643 |
| | | *S. koraiensis* Nakai | L9617 |
| | | *S. takesimensis* Nakai | L12516 |

**Table 1.** List of the plant species with their voucher number included in this study.

embedded materials and then sectioned using a Leica RM2255 rotary microtome (Leica Microsystems GmbH, Germany). Serial sections of 4–6 μm thickness were cut with stainless blades, fixed onto a slide glass, and dried on electric slide warmer for about 12 h. In order to stain, dried slides were immersed in 0.1% Toluidine blue O for 60–90 s, rinsed with running water, and again dried with an electric slide warmer for more than 6 h to remove water. After complete removal of the water, slides were mounted with Entellan (Merck Co., Germany) and pressed with metal blocks for a couple days to remove air bubbles. After 2 days, the prepared slides were observed under an AXIO Imager A1 light microscope (Carl Zeiss, Germany). Photomicrographs were taken with an AxioCam MRc5 attached camera system, and seed coat measurement was carried out by using AxioVision software for Windows (release 4.7, 2008). Multiple image alignment was arranged by Photoshop CS4 for Windows 2007. None of the image-alteration facilities of Photoshop were used to modify the original images.

### 2.4. Morphometry and data analysis

A total of approximately 2500 seeds were used for morphometric measurement. Digital images of whole seeds were taken with a Leica DFC420 C multifocal camera attached to a Leica MZ16 FA microscope (Leica Microsystems). The length and width of a minimum of 20 seeds from each taxon were measured from digital images using Leica LAS V3.8 software for Windows. Seed length (SL) and width were measured, length/width ratios (LWRs) were calculated, and mean values are presented. Individual seed morphological parameters and their features are given in Appendix 1.

For the analysis, 13 seed characters were treated as non-ordered and coded as unweighted consecutive numbers (**Table 2** and Appendix 2). Correspondence analysis (CA) and cluster analysis (UPGMA) were performed to reveal whether the seed features allowed the grouping of taxa using PAST version 3.14 [36]. The eigenvectors and character scores of the first four axes in CA are presented in **Table 3**, together with the percentage of total variance. The results were presented in a two-dimensional biplot of axis 1 (in *X* axis) and axis 2 (*Y*-axis). To visualize the relationship between the different taxa based on seed features, a cluster analysis-based UPGMA tree using Euclidean distance measurement was constructed. Bootstrap support values were based on 100 replicates, and values above 50 were presented above tree branches.

| |
|---|
| 1. Shape: 0, elliptical; 1 ovoid |
| 2. Seed length: 0, <1 mm; 1, 1–2 mm, 2, >2 mm |
| 3. Length/width ratio: 0, <2; 1, >2 |
| 4. Hilum position: 0, lateral; 1, basal. |
| 5. Hilum character: 0, distinctly protuberant; 1, flat to indistinctly protuberant; 2, deep |
| 6. Primary surface sculpture: 0, reticulate; 1, colliculate; 2, others |
| 7. Epidermal cell shape: 0, rectangular; 1, polygonal; 2, irregular |
| 8. Anticlinal wall: 0, shallow to indistinctly raised; 1, distinctly raised |
| 9. Periclinal wall: 0, convex; 1, flat; 2, concave |
| 10. Periclinal wall ornamentation: 0, striate; 1, papillate/granulate; 2, smooth/folded |
| 11. Seed coat anatomy: 0, seed coat distinct; 1, indistinct |
| 12. Epidermis: 0, well represent; 1, degenerate |
| 13. Endothelium: 0, present; 1, absent |

**Table 2.** Seed characters with their coding states used in analysis.

| Character number | Character code | Axis 1 | Axis 2 | Axis 3 | Axis 4 |
|---|---|---|---|---|---|
| 1 | SH | −0.23118 | **−0.76272** | −0.26612 | 0.263812 |
| 2 | SL | 0.007571 | **0.467293** | 0.281203 | −0.31876 |
| 3 | LWR | 0.654796 | **0.952965** | 0.162676 | −0.4085 |
| 4 | HP | 0.10317 | 0.025843 | −0.09821 | **−0.26307** |
| 5 | HC | 0.065757 | **−0.91809** | −0.06242 | −0.80719 |
| 6 | PSS | **0.817606** | 0.046239 | −0.7419 | 0.29745 |
| 7 | ECS | −0.09684 | −0.15219 | −0.04569 | **0.273533** |
| 8 | AW | **−0.8693** | 0.080012 | 0.101048 | 0.143169 |
| 9 | PW | **−0.88318** | 0.176183 | −0.03262 | 0.129772 |
| 10 | PWO | **0.491091** | 0.221064 | −0.31801 | −0.04532 |
| 11 | SCA | **0.995628** | −0.19168 | 0.954453 | 0.561853 |
| 12 | EP | 0.382648 | −0.61797 | **0.775164** | −0.04003 |
| 13 | EN | **0.91233** | −0.07943 | 0.531445 | 0.705178 |
| | Eigenvalue | 0.325 | 0.176 | 0.123 | 0.115 |
| | % total variance | 34.355% | 18.675% | 13.041% | 12.213% |

AW, anticlinal wall; EN, endothecium; EP, epidermis; HC, hilum character; HP, vn; LWR, length/width ratio; PSS, primary surface sculpture; PW, periclinal wall; PWO, primary wall ornamentation; SCA, seed coat anatomy; ECS, epidermal cell shape; SH, seed shape; and SL, seed length.

**Table 3.** Eigenvectors and character scores of the first four axes of a CA of the 13 seed characters.

## 3. Results

A total of 41 taxa from Scrophulariaceae s.l. were studied, belonging to six families: Scrophulariaceae sensu stricto (s.s.; 1 genus, 5 species), Plantaginaceae (4 genera, 14 species, 4 varieties), Orobanchaceae (5 genera, 9 species, 4 varieties, 1 forma), Linderniaceae (1 genus, 2 species), Phrymaceae (1 species), and Paulowniaceae (1 species). Selected scanning electron micrographs and light micrographs of seeds are presented in **Figures 1–11**. A comprehensive description of seed features by family and genus is given here.

### 3.1. *Scrophularia* (Scrophulariaceae s.s)

Five species of *Scrophularia* were investigated in this study (**Figure 1A–O**). Seeds were minute with a small of variation in size within studied species; they were ovoid to broadly elliptical in shape and usually black or sometimes brown in color (Appendix 1, online supplementary resource; **Figure 1A, B, D, E, G, H, J, K, M, N**). The hilum was terminally positioned and slightly protruding. In all species, gross surface sculpture was typical reticulate-striate, and epidermal cells were polygonal or elongated in one direction (**Figure 1C, F, I, L, O**). The periclinal wall (PW) of the testal cell was slightly concave with parallel striation, and the anticlinal wall (AW) was highly raised, straight to sinuous, or wavy and unevenly thickened.

### 3.2. *Limnophila, Linaria, Veronica,* and *Veronicastrum* (Plantaginaceae)

Altogether, 18 taxa from 4 genera of Plantaginaceae, including 11 species and 4 varieties of the genus *Veronica,* were investigated. Seeds were minute: the largest seeds of *Veronica didyma* var. *lilacina* ranged from 1.01 × 0.81 mm to 1.49 × 1.25 mm, and the smallest seeds of *V. rotunda* var. *subintegra* ranged from 0.44 × 0.32 mm to 0.81 × 0.49 mm. Seed were pale yellow or dark brown to black in color, and seed shape (SS) ranged from ovoid, broad ovoid, to sub-spherical, and mostly they were flat or plano-convex and bifacial (**Figures 2A–O**, **3A–O**, **4A–O**, and **5A–F**). The surface sculpture was predominantly reticulate-striate, cristate, and sometimes reticulate-verrucate, as in *V. persica, V. arvensis,* and *V. didyma* var. *lilacina* (**Figures 2F, I** and **4O**), reticulate-corrugate as in *Veronica undulata* (**Figure 4L**), or typical reticulate as in *V. peregrina* (**Figure 2C**). Epidermal cell shape was polygonal, elongated, and isodiametric or rarely irregular as in *V. arvensis* (**Figure 2F**). The periclinal wall of the surface cell was flat to slightly concave and striate but rarely folded, whereas the anticlinal wall was slightly to highly raised, straight to wavy, folded, and sometimes of uneven thickness.

### 3.3. *Pedicularis, Melampyrum, Lathraea, Phtheirospermum,* and *Siphonostegia* (Orobanchaceae)

A total of 14 taxa belonging to 5 genera (6 taxa from *Melampyrum,* 5 taxa from *Pedicularis,* and 1 species from each three genera) of Orobanchaceae were investigated. The seeds were larger than the taxa of Scrophulariaceae s.s. and Plantaginaceae. *Melampyrum roseum* had the largest seeds (4.31 × 1.77 mm to 5.42 × 2.2 mm), whereas *Phtheirospermum japonica* had the smallest (1.04 × 0.49 mm to 1.89 × 0.84 mm; **Figures 6A–O**, **7A–O**, and **8A–L**). Mostly seeds were elliptical and cylindrical (*Melampyrum*), ellipsoidal, navicular (*Pedicularis*), ovoid to subglobose

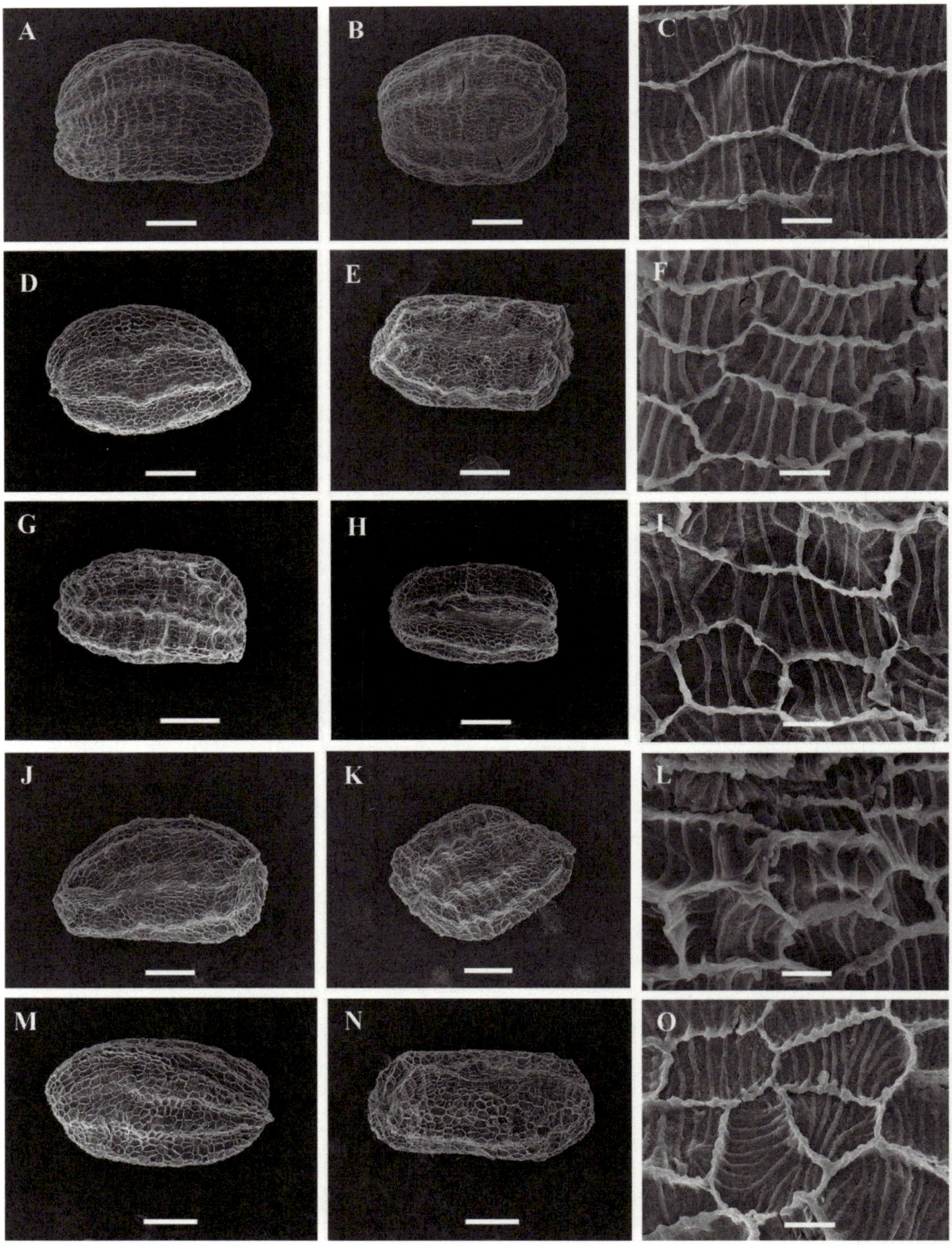


**Figure 1.** Scanning electron micrographs of seeds of Scrophulariaceae s.l. *Scrophularia kakudensis* (A, B, C), *S. koraiensis* (D, E, F), *S. takesimensis* (G, H, I), *S. buergeriana* (J, K, L), and *S. grayana* (M, N, O). Scale bars: 300 μm (A, B, D, E, G, H, J, K, M, N), 30 μm (C, O), and 20 μm (F, I, L).

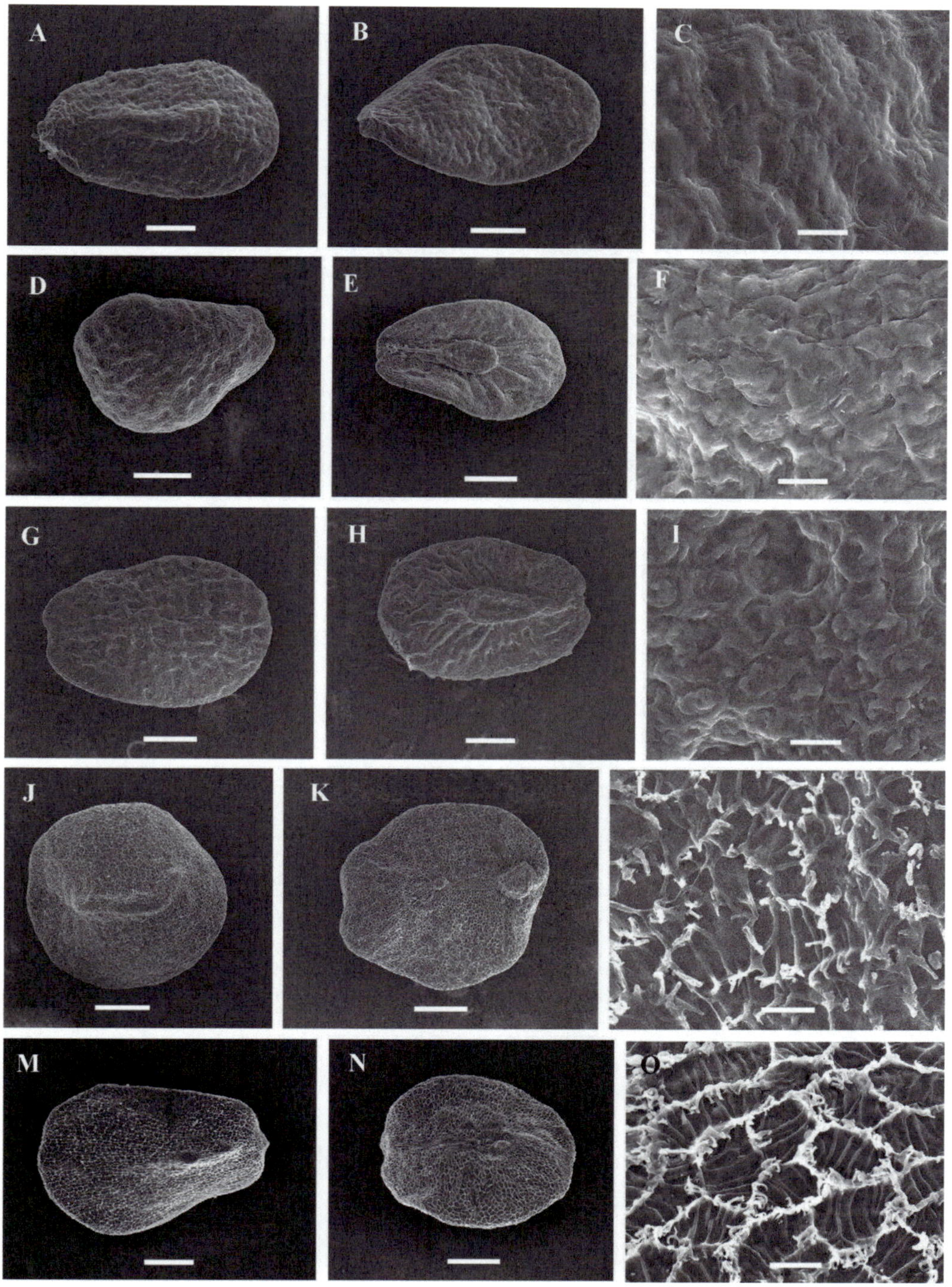


**Figure 2.** Scanning electron micrographs of seeds of Plantaginaceae. *Veronica peregrina* (A, B, C), *V. arvensis* (D, E, F), *V. persica* (G, H, I), *V. longifolia* (J, K, L), and *V. linariifolia* (M, N, O). Scale bars: 300 μm (D, E, G, H, J, K, M, N), 150 μm (A, B), and 20 μm (C, F, I, L, O).

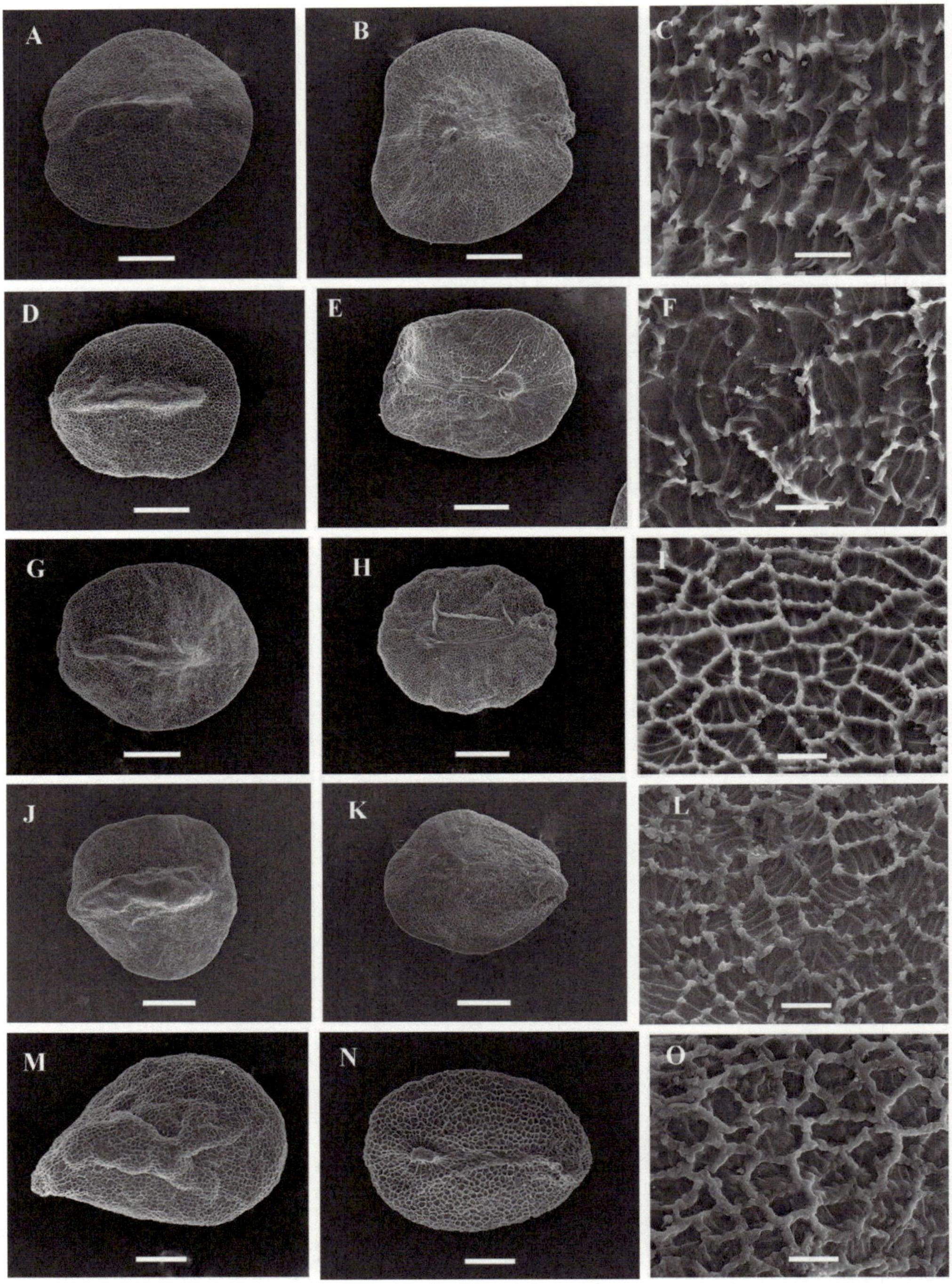


**Figure 3.** Scanning electron micrographs of seeds of Plantaginaceae. *Veronica dahurica* (A, B, C), *V. rotunda* var. *sabinteara* (D, E, F), *V. pusanensis* (G, H, I), *V. incana* (J, K, L), and *V. pyrethrina* (M, N, O). Scale bars: 300 μm (A, B, D, E, G, H, J, K), 150 μm (M, N), and 20 μm (C, F, I, L, O).

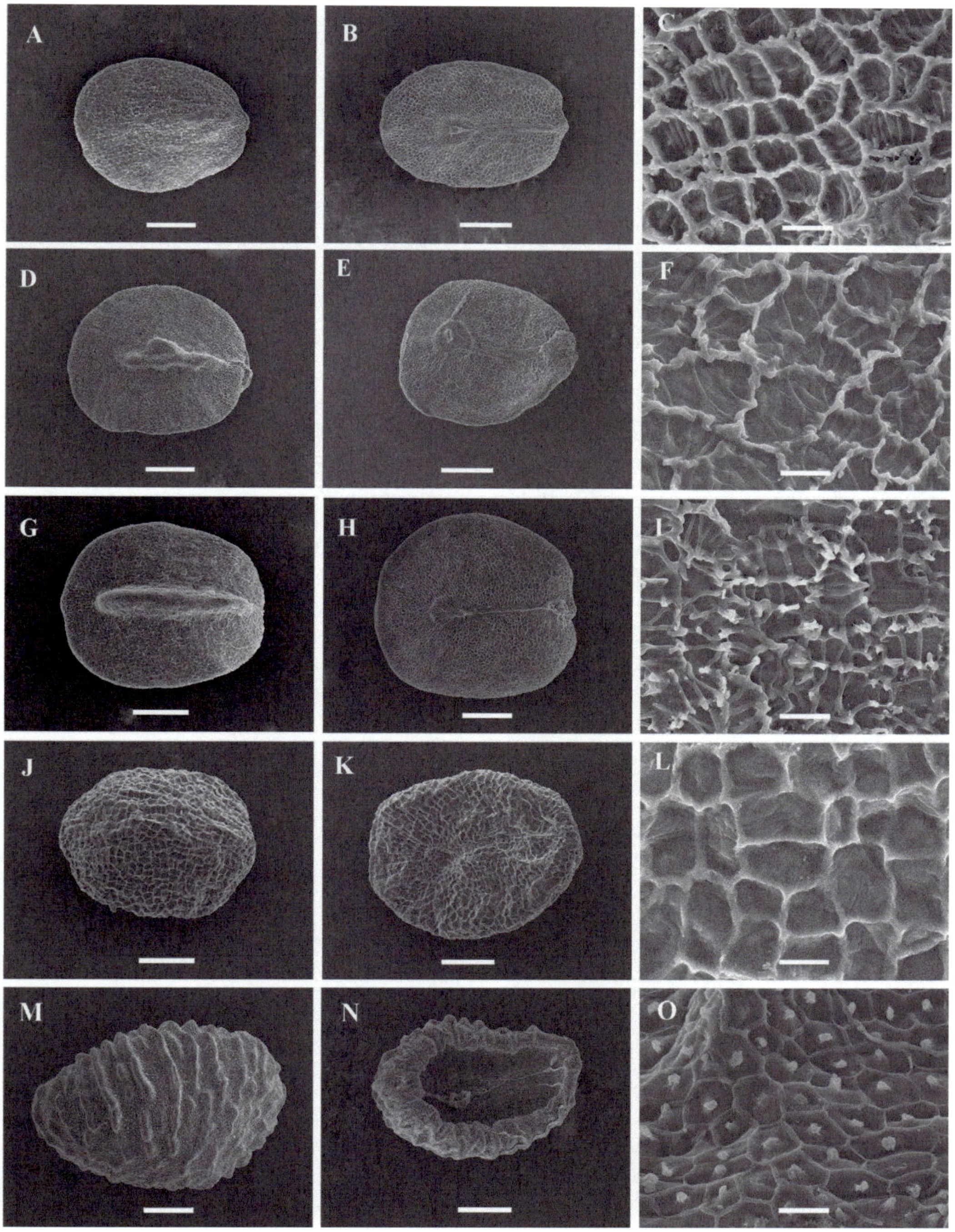


**Figure 4.** Scanning electron micrographs of seeds of Plantaginaceae. *Veronica nakiana* (A, B, C), *V. kiusiana* var. *diamentica* (D, E, F), *V. kiusiana* var. *glabrifolia* (G, H, I), *V. undulata* (J, K, L), and *V. didyma* var. *lilacina* (M, N, O). Scale bars: 300 μm (A, B, D, E, G, H, M, N), 150 μm (J, K), and 20 μm (C, F, I, L, O).

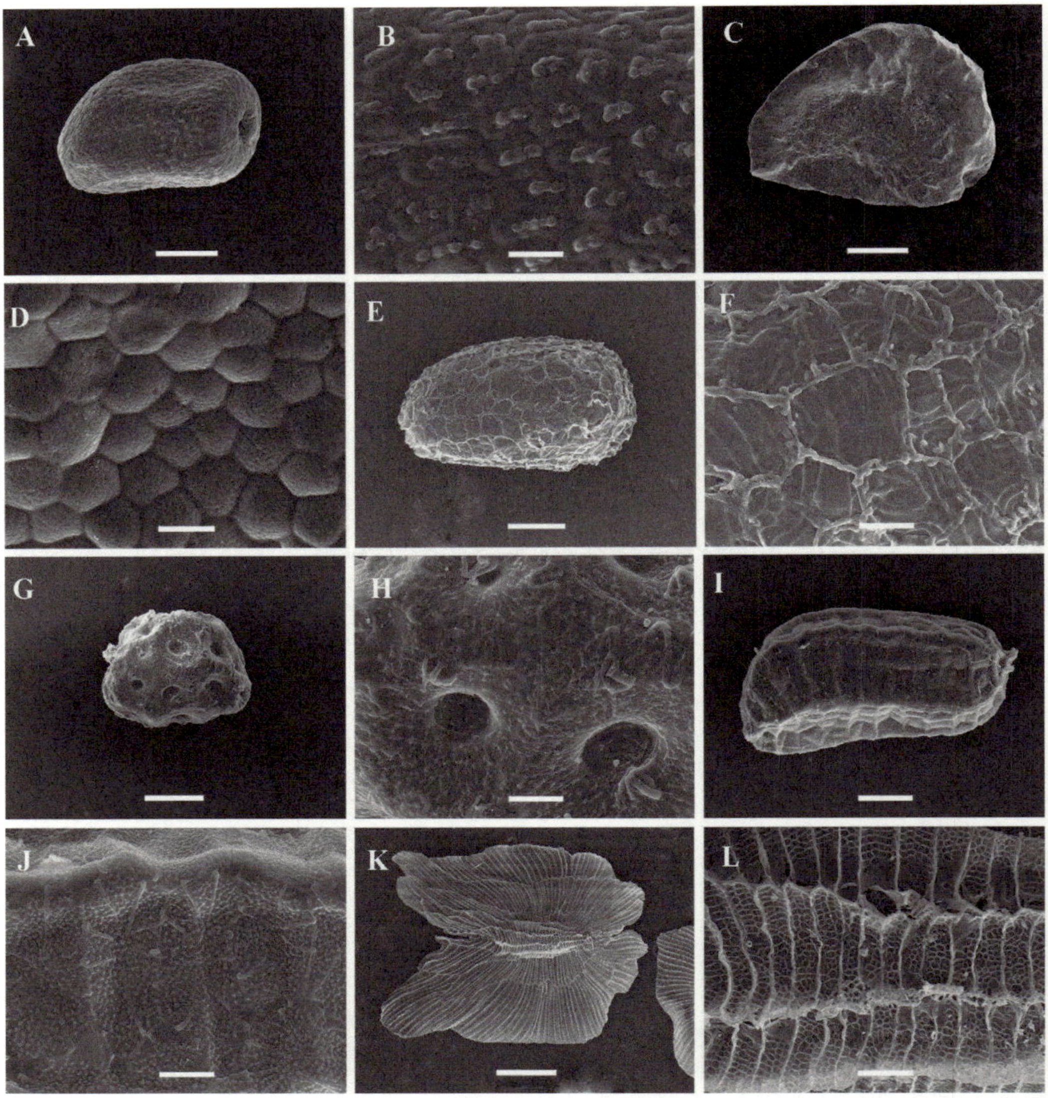

**Figure 5.** Scanning electron micrographs of seeds of Plantaginaceae, Linderniaceae, and Paulowniaceae. *Limnophila indica* (A, B), *Linaria japonica* (C, D), *Veronicastrum sibiricum* (E, F), *Lindernia crustacea* (G, H), *Lindernia procumbens* (I, J), *Paulownia coreana* (K, L). Scale bars: 1000 μm (K), 600 μm (C), 150 μm (A, G), 100 μm (L), 42 μm (F), 30 μm (B, D, H), and 15 μm (J).

(*Lathraea japonica*), ovoid (*Phtheirospermum japonica*), or elliptical beaked (*Siphonostegia chinensis*). The surface sculpture was highly variable, being rugose, colliculate, reticulate, scalariform, or papillate. The testa dermal cell shape was predominantly irregular; in some species, it was rectangular to elongated or polygonal-isodiametric. The testa periclinal wall was concave, flat concave, or slightly convex, with either smooth to finely folded or sessile papillae. The anticlinal walls were mostly raised but sometimes shallow and sometimes deep. The raised wall was smooth, finely papillate, or finely folded and unevenly thickened.

http://dx.doi.org/10.5772/intechopen.70572

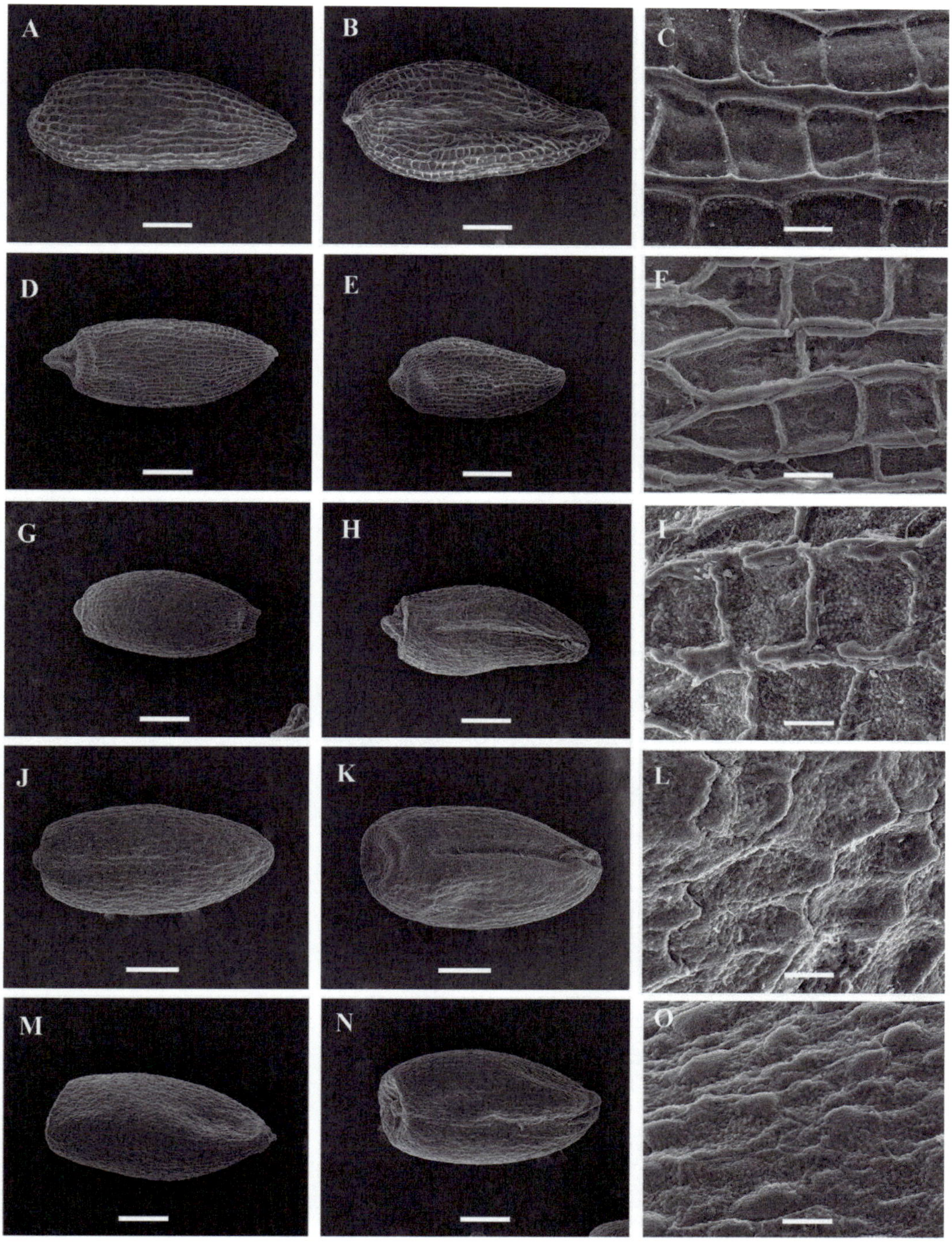


**Figure 6.** Scanning electron micrographs of seeds of Orobanchaceae. *Pedicularis mandshurica* (A, B, C), *P. resupinata* var. *umbrosa* (D, E, F), *P. verticillata* (G, H, I), *P. resupinata* (J, K, L), and *P. resupinata* for *albiflora* (M, N, O). Scale bars: 600 μm (A, B, D, E, G, H, J, K, M, N), 42 μm (C, F, I, L, O).

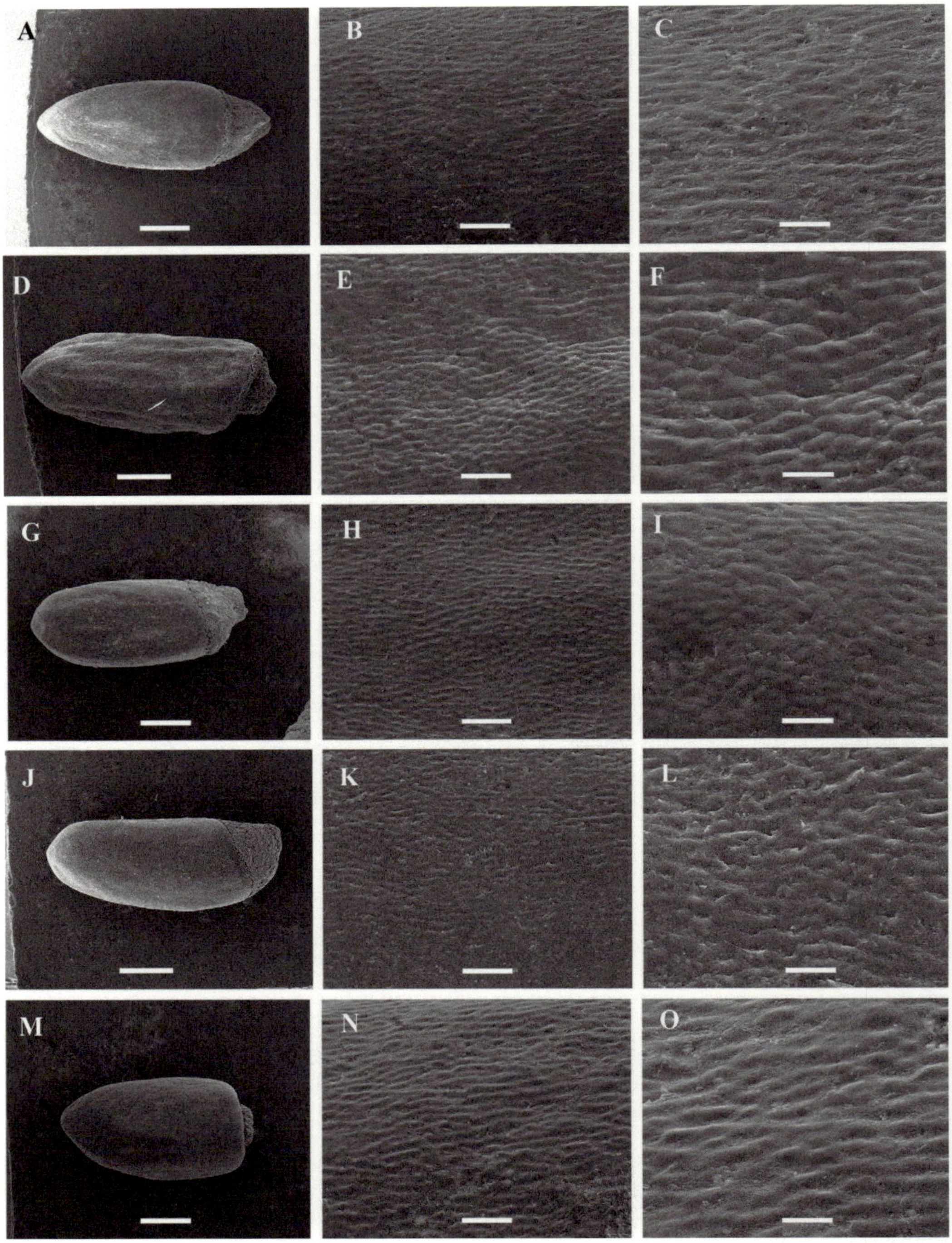


**Figure 7.** Scanning electron micrographs of seeds of Orobanchaceae. *Melampyrum koreanum* (A, B, C), *M. roseum* (D, E, F) *M. roseum* var. *japonicum* (G, H, I), *M. roseum* var. *ovalifolium* (J, K, L), and *M. setaceum* (M, N, O). Scale bars: 1000 μm (A, D, G, J, M), 100 μm (B, E, H, K, N), and 42 μm (C, F, I, L, O).

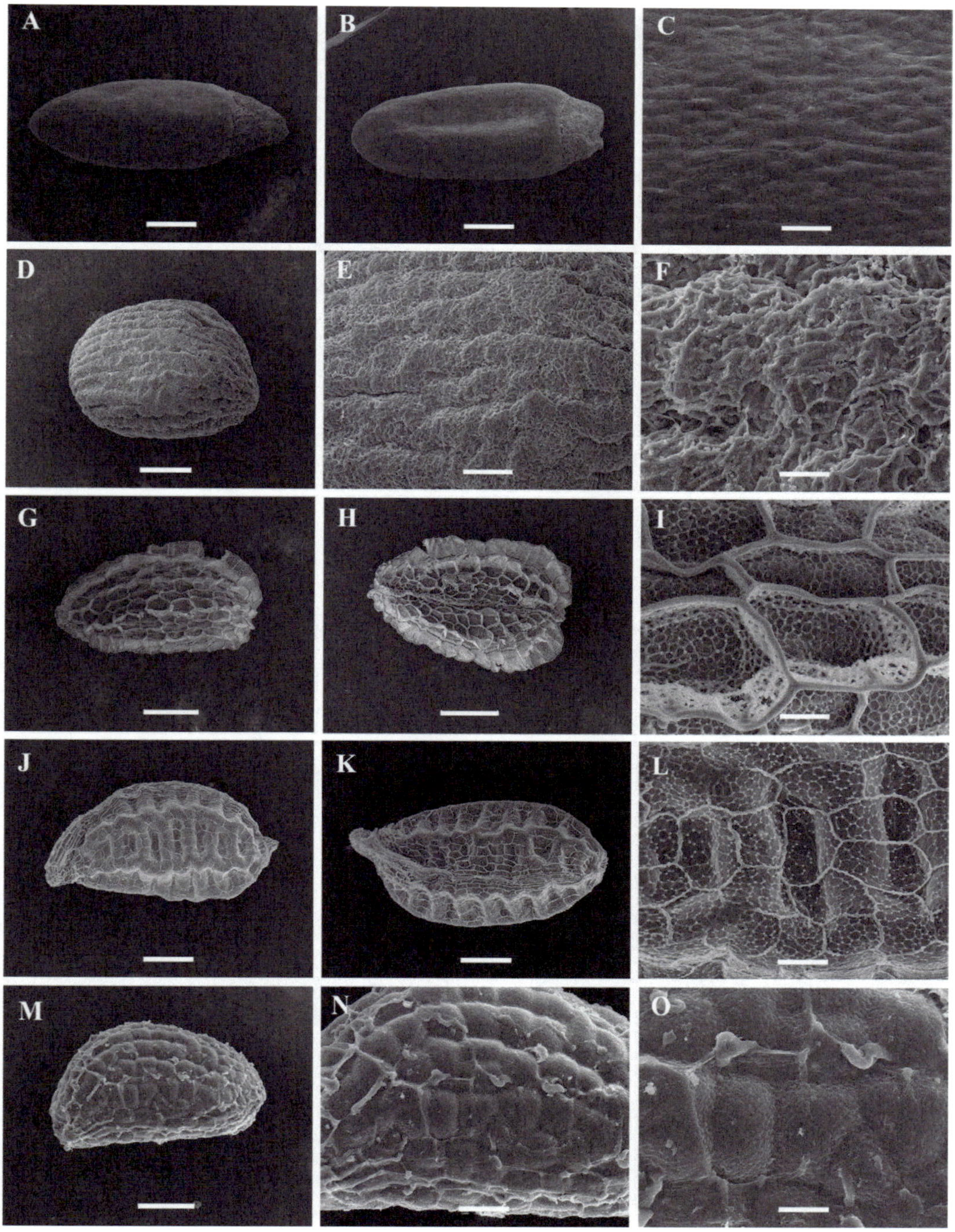


**Figure 8.** Scanning electron micrographs of seeds of Orobanchaceae and Phrymaceae. *Melampyrum setaceum* var. *nakaianum* (A, B, C), *Lathraea japonica* (D, E, F), *Phtheirospermum japonicum* (G, H, I), *Siphonostegia chinensis* (J, K, L), and *Mazus pumilus* (M, N, O). Scale bars: 1000 μm (A, B), 600 μm (D), 200 μm (E, J, K), 300 μm (G, H), 100 μm and 42 μm (C, F, I, L, M, NM), and 20 μm (O).

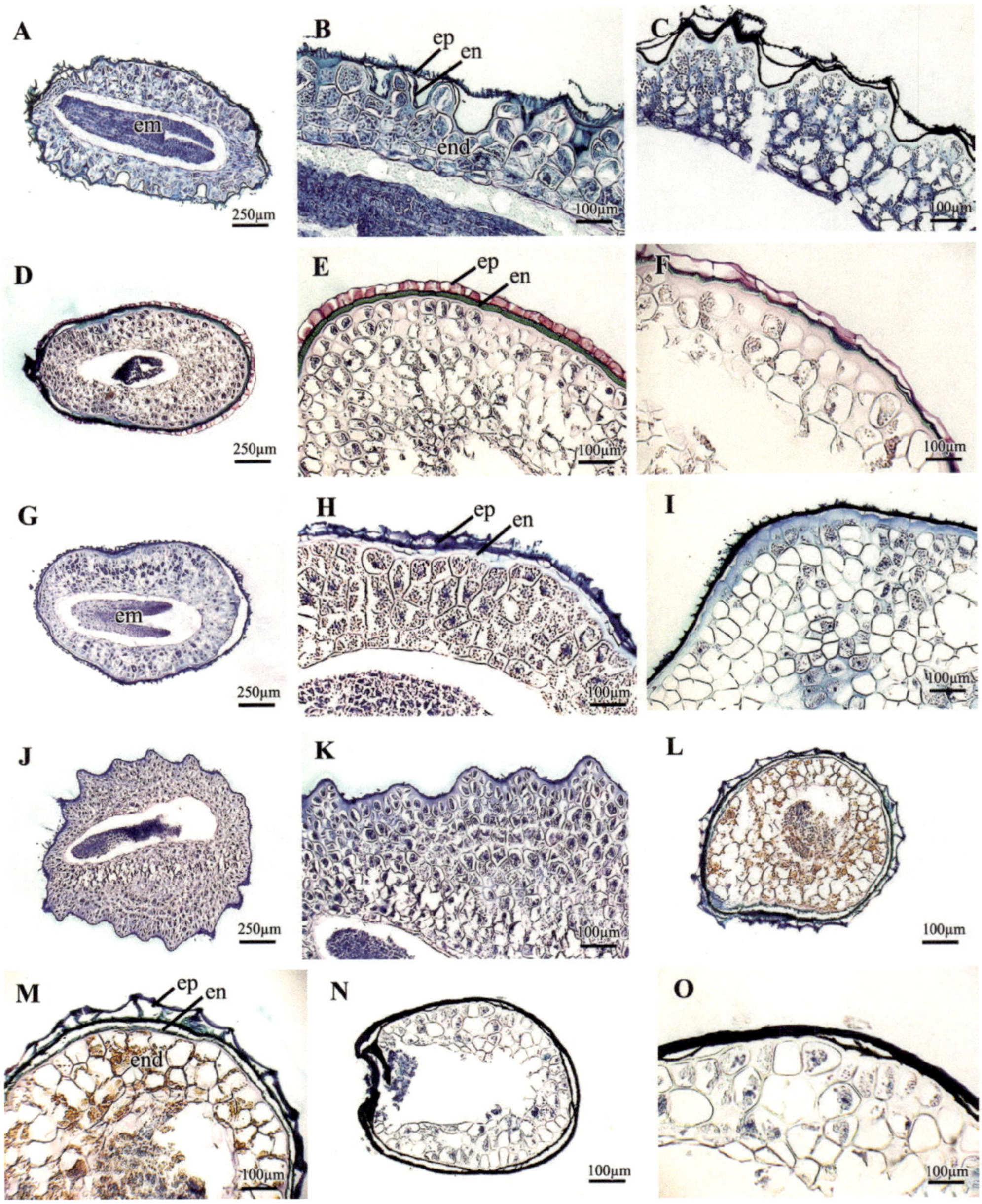


**Figure 9.** Seed anatomy of (A, B) *Scrophularia buergeriana*, (C) *S. koraiensis*, (D, E) *Veronica peregrina*, (F) *V. undulata*, (G, H) *V. incana*, (I) *V. pusanensis*, (J, K) *V. didyma* var. *lilacina*, (L, M) *Veronicastrum sibiricum*, and (N, O) *Limnophila indica*. em, embryo; en, endothelium; end, endosperm; and ep, epidermis.

### 3.4. *Lindernia* (Linderniaceae)

Two species were included and seeds were ovoid or ovoid to oblong (**Figure 5G–J**). The surface was ribbed, ridged, or rugose pitted. The epidermal cell shape was rectangular, elongated (*L. procumbens*), or irregular (*Lindernia crustacea*). The testal periclinal wall was flat,

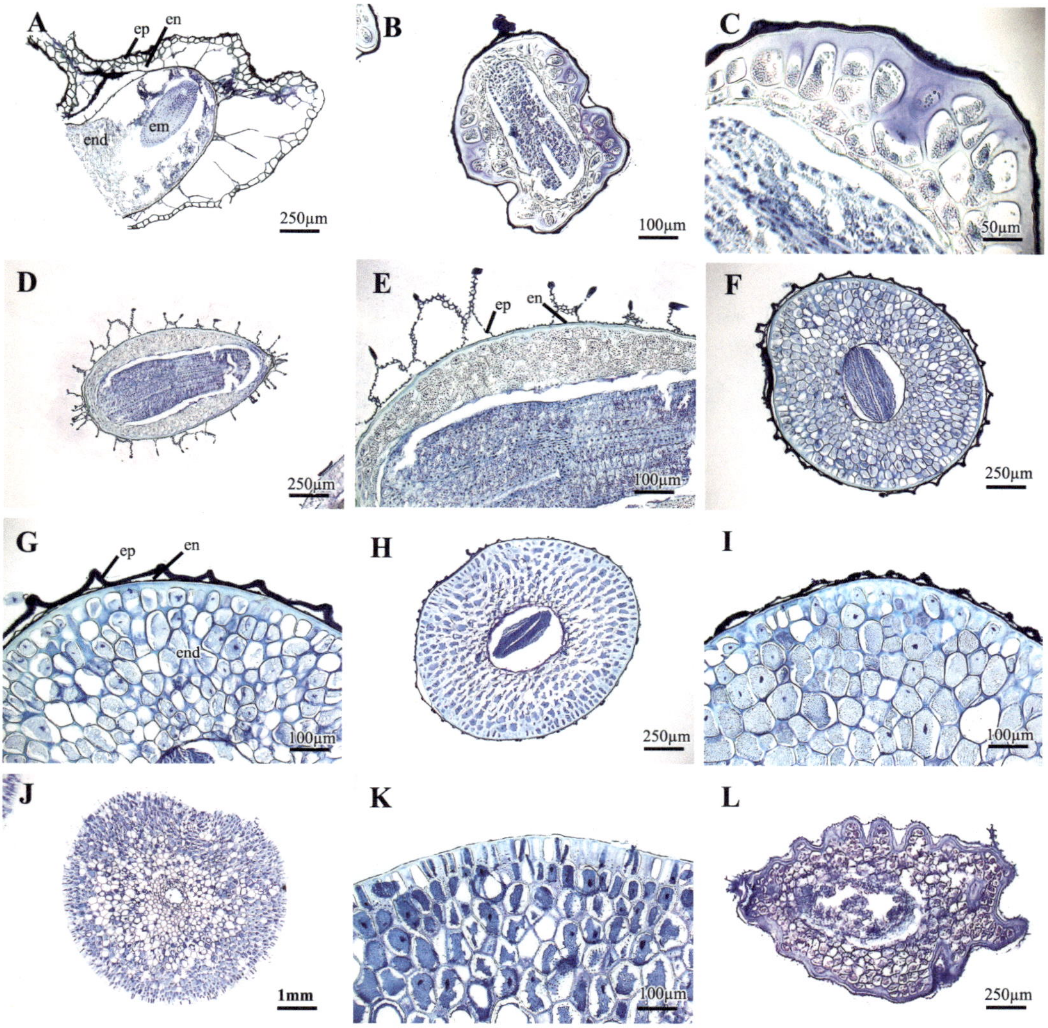


**Figure 10.** Seed anatomy of (A) *Linaria japonica*, (B, C) *Lindernia procumbens*, (D, E) *Phtheirospermum japonicum*, (F, G) *Pedicularis mandshurica*, (H, I) *P. resupinata*, (J, K) *Melampyrum roseum*, and (L) *Siphonostegia chinensis*. Abbreviations: em, embryo; en, endothelium; end, endosperm; ep, epidermis.

slightly concave, and finely granulate, whereas the anticlinal wall was slightly raised and finely granulate to folded.

### 3.5. *Paulownia* (Paulowniaceae)

An endemic species, *P. coreana*, was investigated. The seeds were small, fluffy winged, and pale yellow to white in color (**Figure 5K, L**). The testa surface was ribbed, and epidermal cells were polygonal-isodiametric in shape. The testa periclinal wall was flat to slightly concave, and its surface was smooth, whereas the anticlinal wall was raised, straight to sinuous, and unevenly thickened.

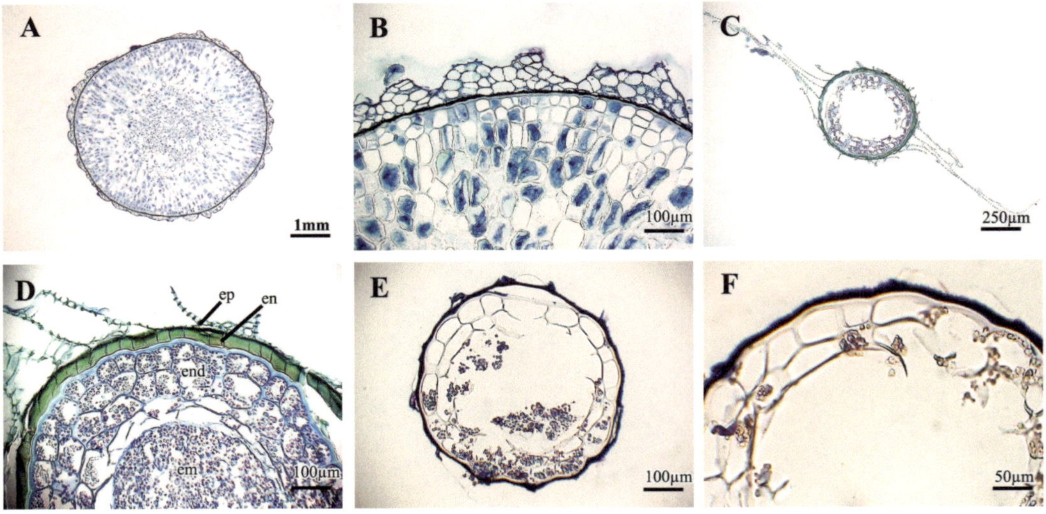


**Figure 11.** Seed anatomy of (A, B) *Lathraea japonica*, (C, D) *Paulownia coreana* and (E, F) *Mazus pumilus*. Abbreviations: em, embryo; en, endothelium; end, endosperm; ep, epidermis.

### 3.6. *Mazus* (Phrymaceae)

Only *Mazus pumilus* was included in this study, and seeds were small ovoid or ellipsoidal in shape (**Figure 8M–O**). The surface sculpture was colliculate with rectangular to polygonal epidermal cells. The periclinal wall was convex with fine folds, whereas the anticlinal wall was shallow with fine folds.

### 3.7. Seed anatomy

The seed coat in Scrophulariaceae s.s. was somehow distinct, thin, and fairly comparable in all studied species. The epidermis was represented by a layer of degenerated cells followed by one or two endothelium layers (**Figure 9A–C**). In most places, endothelium cells were degenerated.

In 18 taxa of Plantaginaceae, the seed coat was distinct and more well characterized than in *Scrophularia* except in three species of *Veronica* (*V. arvensis, V. didyma* var. *lilacina*, and *V. persica*), in which the seed coat was unclear and degenerated (**Figures 9D–O** and **10A**). The epidermis was well represented in all species except in the aforementioned species, and it was noticeable in *V. peregrina* and *Veronicastrum sibericum* (**Figure 9L, M**). In all species, the epidermis was followed by one or two endothelium layers which were represented by either distinct cells in layers or degenerated layers.

The seeds of the six *Melampyrum* taxa were easily distinguished from rest of the Orobanchaceae species. They had largest seeds among the studied genera and a very thin seed coat in transverse section (**Figures 10D–L** and **11A, B**). The epidermis was well characterized, and endothelium was present in eight taxa of the family, excluding all *Melampyrum* species.

http://dx.doi.org/10.5772/intechopen.70572

The seed coat was well represented in *Lindernia* spp., *P. coreana*, and *M. pumilus*, although the epidermal cells were more clearly noticeable in the former rather than the latter two (**Figures 10B, C** and **11C–F**). In *M. pumilus*, the epidermis was characterized by a degenerated layer of cells. The endothelium was present but poorly developed in all species.

### 3.8. Data analyses

The relationships among the taxa for the 13 seed characteristics were analyzed using correspondence analysis (CA) and cluster analyses (**Figures 12** and **13**). In CA, the first four axes explained 78.869% of the total variance of the analyzed data (**Table 3**). Axis 1 described 34.355% of the variance based on the values of primary surface sculpture (PSS), anticlinal wall (AW), periclinal wall (PW), periclinal wall ornamentation (PWO), seed coat anatomy (SCA), and endothecium (EN) (**Table 3**). Axis 2 explained 18.675% of the data variability, of which seed shape (SH), seed length (SL), seed length/width ratio (LWR), and hilum character (HC) were the significant variables for the ordination of the species. Correspondingly, axis 3 and axis 4 were explained by 13.041 and 12.213% of the data variability, respectively. CA biplots revealed a cluster of taxa corresponding with primary surface sculpture, periclinal wall ornamentation, and seed coat anatomy (**Figure 12**). When the distribution of seed surface characters on the CA biplots was observed, most of the taxa with a reticulate surface were grouped on the negative side of axis 2, whereas the taxa with a colliculate and other types of surface

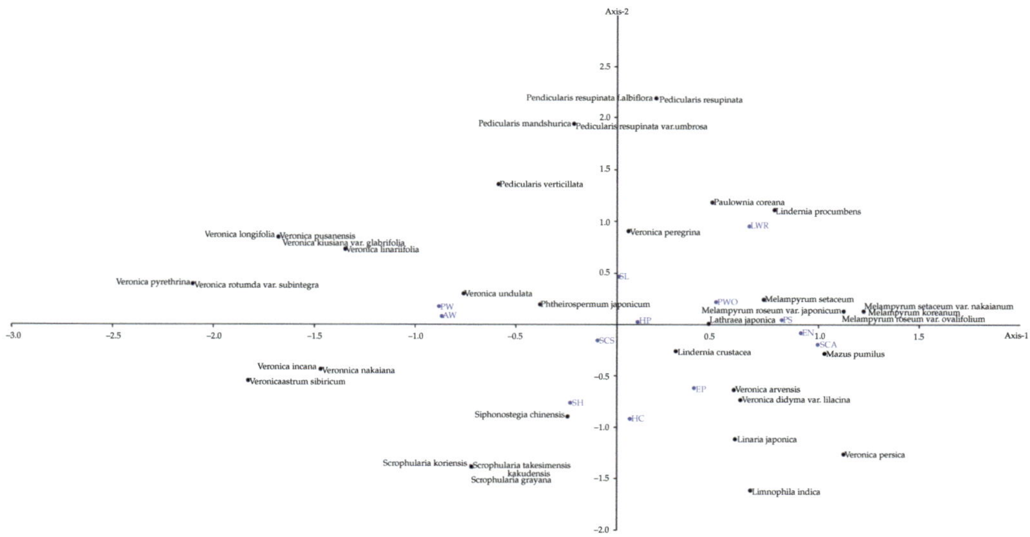


**Figure 12.** Correspondence analysis (CA) biplot of 13 seed characters sampled for 41 taxa of Scrophulariaceae s.l. Samples of different genera are represented by different symbols. AW, anticlinal wall; EN, endothecium; EP, epidermis; HC, hilum character; HP, hilum position; LWR, length/width ratio; PSS, primary surface sculpture; PW, periclinal wall; PWO, primary wall ornamentation; SCA, seed coat anatomy; ECS, epidermal cell shape; SH, seed shape; SL, seed length.

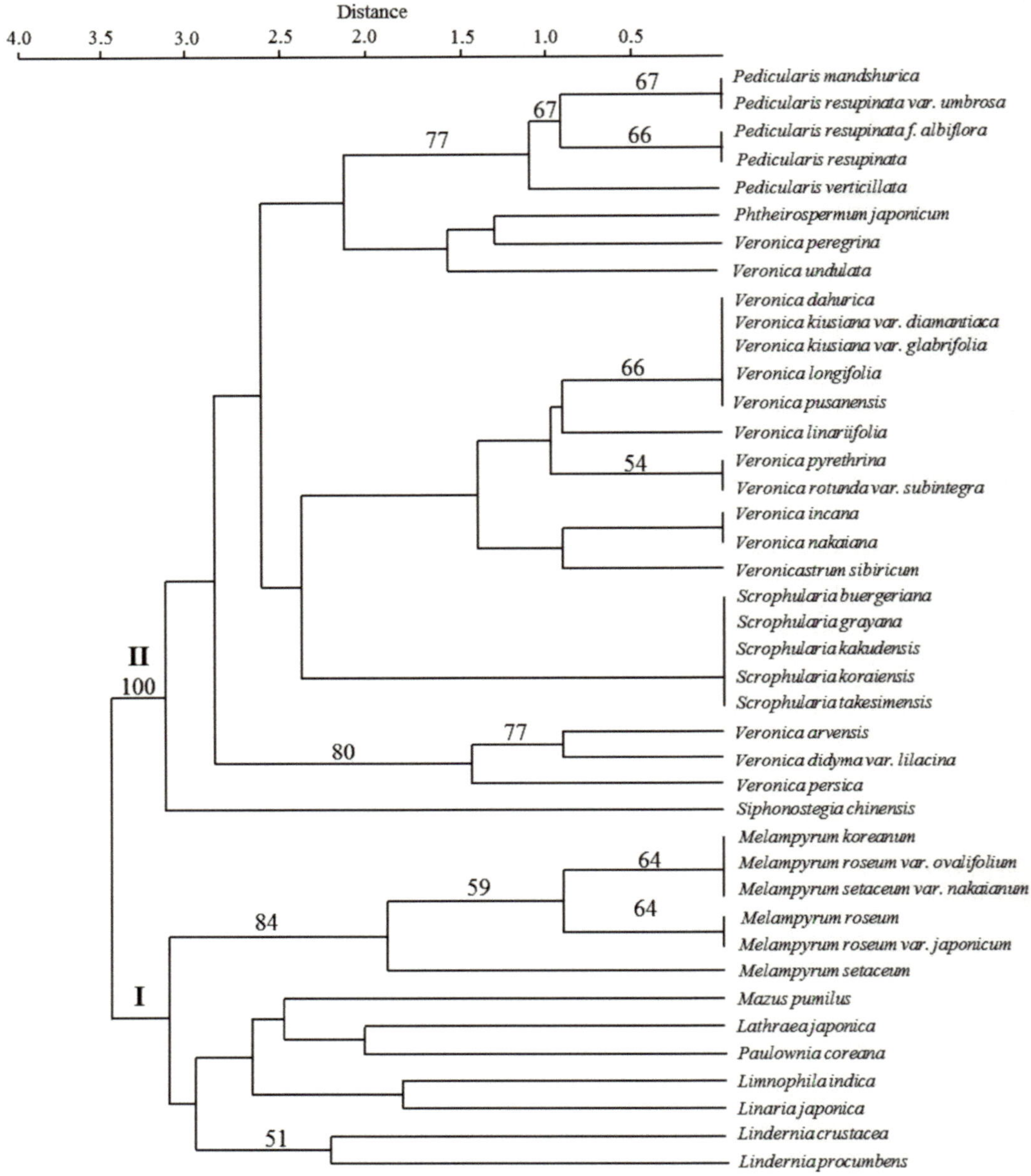


**Figure 13.** Seed morphological relationship among the taxa as displaced by UPGMA cluster diagram. Numbers above the branch represent bootstrap value.

were grouped on the opposite side. Likewise, all the taxa with striate periclinal walls were grouped on the negative side of axis 2, whereas taxa with papillate/granulate and smooth/folded walls, except *Veronica dahurica* and three *Pedicularis* species, were distributed on the positive side of axis 2. The cluster-based UPGMA tree revealed two main clusters [supported by a 100% bootstrap (BT) value]: cluster I included *Melampyrum, Mazus, Lathraea, Paulownia, Limnophila, Linaria,* and *Lindernia,* whereas cluster II comprised species of *Pedicularis, Veronica, Phtheirospermum, Veronicastrum, Scrophularia,* and *Siphonostegia* (**Figure 13**). In cluster I, there were two subclusters: the first one, which was supported by an 84% BT value, included *Melampyrum* species, and the second was formed by the remaining taxa of cluster I with bootstrap support of less than 50%. Correspondingly, in cluster II, there were three separate subclusters: *Pedicularis/Phtheirospermum/Veronica, Veronica/Veronicastrum/Scrophularia,* and three species of *Veronica*. *Siphonostegia chinensis* remained isolated and positioned at the bottom of the cluster. Interestingly, the subcluster formed by three *Veronica* species was supported by an 80% BT value, which was higher than for any other subclusters in cluster II which had bootstrap support <50%.

# 4. Discussion

## 4.1. Variations in seed morphology

This study demonstrated the high diversity of seed morphology in Scrophulariaceae s.l. in terms of seed shape, hilum character, primary ornamentation, epidermal cell characters, and seed coat anatomy. Variations are mainly found in seed primary sculpture, surface cell shape, and periclinal wall ornamentation. Seeds are minute, and most of the taxa are less or slightly larger than 1 millimeter in length except *Melampyrum* and *Pedicularis*. Within each taxon, seed size is variable; however, overall not much variation was found among the different species. That is why size is not very useful for the description of a particular taxon. Seed shape is also very heterogeneous, even with the same species. Mostly in studied species, the seed shape ranged from elliptical, broad elliptical, to ovoid. Elliptical seeds were cylindrical, navicular, flattened, or plano-convex. Seed of *Siphonostegia chinensis* is elliptical and beaked, that of *P. coreana* is winged, and seed of *V. didyma* var. *lilacina* is cupular. In general, however, in Scrophulariaceae s.l., seed shape related directly to its relative position in the fruit [30].

As far as surface sculpture and ornamentation are concerned, we found quite similar patterns in several species of same genera, particularly in *Veronica, Pedicularis, Scrophularia,* and *Melampyrum*. Our result agrees with previous studies regarding *Veronica* and *Pedicularis* [30, 32–34, 37, 38]. Out of 15 taxa of *Veronica* observed in this study, *V. arvensis, V. didyma* var. *lilacina,* and *V. persica* were reticulate-verrucate with a centrally located tubercle in epidermal cells, *V. undulata* had a reticulate-corrugate seed surface, and *V. peregrina* had a typical reticulate surface, whereas the rest of the species had a reticulate-striate surface with a cristate wall. In terms of the systematic significance of seed morphology in *Veronica,* Muñoz-Centeno et al. [32] described eight types of surface pattern in 123 species, and our samples represent four of them, although reticulate-striate is the most dominant pattern. Our results disagree with those of previous study [32] in terms of the surface pattern of *V. peregrina* as these previous authors mentioned a reticulate-corrugate surface pattern; in contrast, in our samples we observed a typical reticulate pattern. As regards seed anatomy data, *V. arvensis, V. didyma* var. *lilacina*, and *V. persica* had the most poorly differentiated seed coat, represented by a thin epidermal layer of almost degenerated cells, whereas *V. peregrina* and *V. undulata* had the most clearly defined epidermis and endothelium among the *Veronica* species. All the species with a reticulate-striate, cristate surface had well-characterized epidermis but indistinct endothelium.

Regarding *Pedicularis*, the results of our study agree with those of earlier results (for example, see [33]), that the regular-reticulate surface pattern is common among the studied taxa. Although gross primary sculpture looks like a reticulate pattern in all five *Pedicularis* taxa, there are substantial dissimilarities in secondary ornamentation and anticlinal wall formation. The anticlinal wall of *P. mandshuricais, P. verticillata*, and *P. resupinata* var. *umbrosa* was highly raised, whereas that of *P. resupinata* and *P. resupinata* f. *albiflora* was only slightly raised. The regular-reticulate, membranous-reticulate, cristate-reticulate, and the undulate primary ornamentations have been mentioned in previous studies [30, 33, 38–40], but the reticulate-tuberculate primary ornamentation found in *P. resupinata* f. *albiflora* is reported for the first time here. Nevertheless, the reticulate seed surface of *Pedicularis* is a common feature among genera of the families Orobanchaceae and Plantaginaceae [26, 30, 41–43]. The most usual and consistent feature observed among the five taxa of *Pedicularis* was a seed coat comprising a clearly defined epidermis and endothelium. On the other hand, most of the seed features were consistent within six *Melampyrum* taxa; variation was only observed in the gross surface sculpture with colliculate (*M. roseum* and *M. roseum* var. *japonicum*), rugose (*M. roseum* var. *ovalifolium* and *M. setaceum*), and rugose + colliculate (*M. koreanum* and *M. setaceum* var. *nakaianum*) surfaces.

In most cases, the seed coat comprised the epidermis and the endothelium. Nevertheless, in all *Melampyrum* and some *Veronica* species, the seed coat was very poorly represented and consisted only of a papery layer of epidermis. In this regard, our results agree with the findings of Juan et al. [30] who described the seed coat of Scrophulariaceae as being composed of the epidermis and endothelium; the latter is a useful character with which to distinguish the seeds of certain genera, particularly *Scrophularia* and *Verbascum*. Although they did not include any *Melampyrum* species, similar to our result, they indicated that some *Veronica* species consisted only of an epidermis, and no endothelium.

According to the CA analyses, the close affinities among the species of *Scrophularia* are well supported by their proximity to one another. Similarly, the proximity of *Melampyrum* species and of *Pedicularis* species is also apparent. Alternatively, in the CA plot, *Veronica* species are divided into two clusters. One is characterized by a concave periclinal wall, striate wall ornamentation, and a distinct seed coat, whereas the other, comprising only four species (*V. arvensis, V. persica, V. didyma* var. *lilacina*, and *V. peregrina*), has a convex periclinal wall and either papillate or smooth folded wall ornamentation. In addition, these four species differ in gross surface pattern as they have reticulate-verrucate or typical reticulate (*V. peregrina*) ornamentation instead of a reticulate-striate, cristate surface in the rest of the species (except *V. undulata*). Morphologically, *V. arvensis, V. peregrina*, and *V. persica* share an annual habit without rhizomes and flowers in terminal inflorescence; however, the latter species differs from the former two by having a longer pedicel than bract [44].

### 4.2. Systematic implications of seed characters

In the UPGMA tree based on the seed morphological and anatomical characters, two major clusters were obtained, of them cluster II was supported by a 100% BT value (**Figure 13**). Cluster I, which included *Lindernia, Linaria, Limnophila, Paulownia, Lathraea, Mazus,* and *Melampyrum,* was a highly heterogeneous group of plants in life form and nature, ranging from annual or perennial herbs to trees, erect or prostrate, and creeping or submerged amphibious forms. Within cluster I, two subclusters were distinguished: the first one, supported by an 84% BT value, contained only the genus *Melampyrum,* whereas the remaining genera formed a separate subcluster. Six *Melampyrum* species can be differentiated from each other by primary sculpture and the nature of the periclinal wall; and UPGMA tree indicated that *M. setaceum* remains isolated from the rest of the *Melampyrum* species (59% BT). Morphologically, *M. setaceum* differed from other five species by linear to linear-lanceolate leaves and lanceolate bracts, whereas the rest of the species had lanceolate leaves and ovate to ovate lanceolate bracts. In another subcluster of cluster I, two other subclusters contained *Lindernia* (51% BT) and the remaining five genera (*Linaria, Limnophila, Paulownia, Lathraea,* and *Mazus*), forming a very heterogeneous group containing members from four families. Out of the five genera, *Paulownia* is a deciduous or evergreen tree, and *Limnophila,* an annual or perennial herb, usually grows in marshy areas with erect, prostrate, or creeping stems.

Despite being clustered in the same clade, seeds of two *Lindernia* species are very different from each other in terms of shape, primary sculpture, and epidermal cell shape. The primary surface sculpture of *L. procumbens* is ribbed with rectangular/elongated surface cells, whereas that of L. crustacea is rugose and pitted with irregular cells. These two species also differ in their leaf morphologies as the leaves of *L. procumbence* are sessile, elliptical to oblong, and somewhat rhomboid with entire or weakly toothed margins, whereas the leaves of *L. crustacea* are shortly petiolate, triangular-ovate to broadly ovate, and shallowly crenate or serrate. After observing the seeds of 14 *Lindernia* species, [45] classified the genus with ribbed and unribbed seeds and also indicated that this character shows good correspondence with subdivision of the genus explained in [46].

In cluster II, the *Siphonostagia* was separated first from rest of the genera despite belonging to the same family, Plantaginaceae, with *Pedicularis*. *Pedicularis,* which is known to be hemiparasitic like *Melampyrum,* formed the topmost subcluster in the tree (77% BT) and remained quite far from *Melampyrum*. Apart from their parasitic nature, these two genera share some morphological features including leafy bracts, campanulate calyces, didynamous stamens, and capitate stigmas. Yet again, three *Veronica* species (*V. arvensis, V. didyma* var. *lilacina,* and V. persica) constituted a remarkable case making a separate subcluster with 80% BT, whereas the rest of *Veronica* species grouped either with *Scrophularia/Veronicastrum* or with *Pedicularis/Phtheirospermum*. Many of the *Veronica* species combined with *Veronicastrum* and *Scrophularia,* although BT support for this subcluster was <50%. In our results, *Scrophularia* shared some similar seed features with *Veronica* and *Veronicastrum;* however, from a morphological point of view, the position of the *Scrophularia* in this subcluster is very confusing. As *Veronica* and *Veronicastrum* are characterized by perennial rhizomatous herbs (although some *Veronica* are annual and without rhizosomes), four corolla lobes, two stamens, and capitate stigmas.

Instead, *Scrophularia* are characterized by perennial herbs without rhizomes, five corolla lobes, four didynamous stamens, and bifid stigmas.

As predicted, three *Veronica* species (*V. arvensis, V. didyma* var. *lilacina*, and *V. persica*) form an isolated group. Based on the infrageneric classification as done in [47], V. arvensis (with haploid chromosome, $n = 8$) belongs to subgenus *Chamaedrys* and V. didyma var. *lilacina* (as *V. polita*) and *V. persica*, both having base chromosome $n = 7$, belong to *Pocilla*. *V. linariifolia*, *V. longifolia*, *V. dahurica*, and *V. incana*, all have the chromosome number $n = 17$, and the reticulate-striate, cristate group belongs to *Pseudolysimachium*, whereas *V. peregrina* ($n = 9$) belongs to subgenus *Beccabunga*. The present results confirmed what was reported in previous studies on *Veronica* (for example, [32, 34]), that seed morphological data can be employed to evaluate the relationships of taxa within the genus and that seed characters can also provide additional support to infrageneric discrimination in *Veronica*.

The SEM investigations reveal that the reticulate surface pattern as a primary sculpture is a confusing case. The pattern is found in the members of Orobanchaceae, Plantaginaceae, and Scrophulariaceae, although quite variations are observed in secondary striation and anticlinal wall pattern. The current study likewise concurs that there is no even single seed features representing all investigated species. Nonetheless, few seed characters somehow nicely characterized the group of particular taxa, for instance, seeds of Orobanchaceae are either elliptical cylindrical or ovoid in shape and of Plantaginaceae are elliptical flattened or ovoid flattened. Correspondingly, reticulate-striate surface sculpture of *Veronica*, *Veronicastrum*, and *Scrophularia* is comparable, although later one is without cristae. The seed coat anatomy indicates that the epidermis and endothelium are informative characters. All the *Scrophularia* species have indistinct epidermis but distinct endothelium, whereas most of the *Veronica* has distinct epidermis and endothelium. On the other hand, all the *Melampyrum* are without both layers, but other Plantaginaceae have well-distinct epidermis and endothelium.

The concept of Scrophulariaceae s.l. has changed considerably since the application of molecular approaches in plant systematics. Studies have shown that the traditional Scrophulariaceae are an unusual assemblage of plants distributed throughout the phylogenetic tree of Lamiales [1, 2, 5–7]. Despite the limited number of taxa investigated, the present seed morphological analysis highlights the high heterogeneity existing within the studied taxa of Scrophulariaceae s.l., which may be observed at a higher taxonomic rank than genus. In particular, primary surface sculpture, anticlinal and periclinal walls of epidermal cell, epidermal cell shape, and seed coat layers are the important seed features observed in this study, and these features can be used to distinguish different groups of taxa in the family. The results of this study therefore suggest that seed coat micromorphology and anatomy have significant taxonomic importance.

## Acknowledgements

This study was supported financially by the project 'Studies on the Establishment of SeedBank base for the Asian network' at the Korea National Arboretum (KNA), Pocheon, Korea.

http://dx.doi.org/10.5772/intechopen.70572

# Appendix

| Plant name | Shape | Average size (mm) | Length/width ratio | Hilum | Primary surface sculpture |
|---|---|---|---|---|---|
| *Lindernia procumbens* | Elliptical, oblong | 0.35 × 0.16 | 2.11 | Basal, distinctly protuberant | Ribbed |
| *L. crustacea* | Ovoid, sub globose | 0.38 × 0.3 | 1.26 | Basal, distinctly protuberant | Rugose, pitted |
| *Mazus pumilus* | Ovoid, elliptical | 0.42 × 0.26 | 1.36 | Basal, distinctly protuberant | Colliculate |
| *Lathraea japonica* | Ovoid | 2.03 × 1.69 | 1.2 | Lateral, distinctly protuberant | Rugose |
| *Melampyrum koreanum* | Elliptical, cylindrical | 4.73 × 1.97 | 2.43 | Lateral, flat to indistinctly protuberant | Rugose/poorly colliculate |
| *M. roseum* | Elliptical, cylindrical | 4.96 × 1.98 | 2.51 | Lateral, flat to indistinctly protuberant | Colliculate |
| *M. roseum* var. *japonicum* | Elliptical, cylindrical | 3.79 × 1.65 | 2.32 | Lateral, flat to indistinctly protuberant | Colliculate |
| *M. roseum* var. *ovalifolium* | Elliptical, cylindrical | 4.35 × 1.74 | 2.51 | Lateral, flat to indistinctly protuberant | Rugose |
| *M. setaceum* | Elliptical, cylindrical | 4.15 × 1.55 | 2.7 | Lateral, flat to indistinctly protuberant | Rugose |
| *M. setaceum* var. *nakaianum* | Elliptical, cylindrical | 4.97 × 1.9 | 2.62 | Lateral, flat to indistinctly protuberant | Rugose/poorly colliculate |
| *Pedicularis mandshurica* | Elliptical, navicular | 2.99 × 1.27 | 2.38 | Basal, distinctly protuberant | Reticulate |
| *P. resupinata* f. *albiflora* | Elliptical, navicular | 2.83 × 1.27 | 2.25 | Basal, distinctly protuberant | Reticulate, tuberculate |
| *P. resupinata* | Elliptical | 2.69 × 1.32 | 2.05 | Basal, distinctly protuberant | Reticulate |
| *P. resupinata* var. *umbrosa* | Elliptical | 2.62 × 1.08 | 2.49 | Basal, distinctly protuberant | Reticulate |
| *P. verticillata* | Elliptical, navicular | 2.27 × 1.16 | 1.95 | Basal, distinctly protuberant | Reticulate |
| *Phtheirospermum japonicum* | Ovoid | 1.19 × 0.73 | 1.64 | Basal, distinctly protuberant | Reticulate |
| *Siphonostegia chinensis* | Elliptical, beaked | 0.92 × 0.46 | 2.05 | Basal, deep | Reticulate |
| *Paulownia coreana* | Elliptical, winged | 1.58 × 0.74 | 2.13 | Basal, distinctly protuberant | Ribbed |
| *Limnophila indica* | Ovoid, elliptical | 0.95 × 0.46 | 1.73 | Basal, distinctly protuberant | Rugose |
| *Veronica arvensis* | Ovoid, flattened | 0.96 × 0.64 | 1.51 | Basal, distinctly protuberant | Reticulate-verrucate/Rugose |
| *V. dahurica* | Broad elliptical, flattened | 1.02 × 0.89 | 1.16 | Lateral, distinctly protuberant | Reticulate-striate, cristate |
| *V. didyma* var. *lilacina* | Ovoid, cupular | 1.27 × 1 | 1.28 | Lateral, distinctly protuberant | Reticulate-verrucate |
| *V. incana* | Ovoid, flattened | 0.71 × 0.49 | 1.46 | Basal, distinctly protuberant | Reticulate-striate, cristate |
| *V. kiusiana* var. *diamantiaca* | Broad elliptical, flattened | 1.17 × 1.01 | 1.36 | Lateral, distinctly protuberant | Reticulate-striate, cristate |

| Plant name | Shape | Average size (mm) | Length/width ratio | Hilum | Primary surface sculpture |
|---|---|---|---|---|---|
| *V. kiusiana* var. *glabrifolia* | Broad elliptical, flattened | 1.17 × 1.01 | 1.16 | Lateral, distinctly protuberant | Reticulate-striate, cristate |
| *V. linariifolia* | Broad elliptical, flattened | 1.19 × 0.85 | 1.4 | Basal, distinctly protuberant | Reticulate-striate, cristate |
| *V. longifolia* | Broad elliptical, flattened | 1.19 × 0.99 | 1.54 | Lateral, distinctly protuberant | Reticulate-striate, cristate |
| *V. nakaiana* | Broad ovoid, flattened | 0.94 × 0.74 | 1.28 | Basal, distinctly protuberant | Reticulate-striate, cristate |
| *V. peregrina* | Elliptical, flattened | 0.71 × 0.41 | 1.73 | Basal, distinctly protuberant | Faintly reticulate |
| *V. persica* | Ovoid, flattened | 1.17 × 0.79 | 1.49 | Basal, distinctly protuberant | Reticulate-verrucate |
| *V. pusanensis* | Broad elliptical, flattened | 1.01 × 0.79 | 1.29 | Lateral, distinctly protuberant | Reticulate-striate, cristate |
| *V. pyrethrina* | Broad elliptical, flattened | 0.95 × 0.78 | 1.23 | Lateral, distinctly protuberant | Reticulate-striate, cristate |
| *V. rotunda* var. *subintegra* | Broad elliptical, flattened | 0.99 × 0.77 | 1.29 | Lateral, distinctly protuberant | Reticulate-striate, cristate |
| *V. undulata* | Broad ovoid, plano-convex | 0.62 × 0.40 | 1.56 | Lateral, distinctly protuberant | Reticulate-corrugate |
| *Veronicastrum sibiricum* | Ovoid, plano-convex | 0.73 × 0.46 | 1.61 | Lateral, distinctly protuberant | Reticulate-striate, cristate |
| *Linaria japonica* | Ovoid, flattened | 2.03 × 1.68 | 1.68 | Basal, deep | Colliculate |
| *Scrophularia buergeriana* | Ovoid | 10.3 × 0.71 | 1.47 | Basal, deep | Reticulate-striate |
| *S. grayana* | Ovoid | 1.08 × 0.76 | 1.45 | Basal, deep | Reticulate-striate |
| *S. kakudensis.* | Ovoid | 1.13 × 0.73 | 1.56 | Basal, deep | Reticulate-striate |
| *S. koraiensis* | Ovoid | 1.2 × 0.74 | 1.64 | Basal, deep | Reticulate-striate |
| *S. takesimensis* | Ovoid | 1.01 × 0.6 | 1.72 | Basal, deep | Reticulate-striate |

| Epidermal cell shape | Periclinal wall | Anticlinal wall | Seed coat | Epidermis | Endothelium |
|---|---|---|---|---|---|
| Rectangular, elongated | Concave, finely granulate | Slightly raised, straight, finely granulated | Distinct | Well represented | Absent |
| Irregular | Concave, finely granulate | Slightly raised, folded | Distinct | Well represented | Absent |
| Rectangular, Polygonal | Convex, finely folded | Shallow, finely folded | Distinct | Well represented | Present |
| Irregular | Concave, smooth to folded | Distinctly raised, straight to wavy, finely folded, unevenly thickened | Distinct | Well represented | Present |
| Irregular | Convex, smooth to finely folded | Shallow or slightly raised | Indistinct | Degenerated | Absent |
| Irregular | Convex, smooth to finely folded | Shallow | Indistinct | Degenerated | Absent |
| Irregular | Convex or pitted, smooth to finely folded | Shallow, smooth, unevenly thickened | Indistinct | Degenerated | Absent |

http://dx.doi.org/10.5772/intechopen.70572

| Epidermal cell shape | Periclinal wall | Anticlinal wall | Seed coat | Epidermis | Endothelium |
|---|---|---|---|---|---|
| Irregular | Convex or pitted, smooth to finely folded | Slightly raised, smooth, unevenly thickened | Indistinct | Degenerated | Absent |
| Irregular | Concave or pitted, smooth to finely folded | Slightly raised, smooth, unevenly thickened | Indistinct | Degenerated | Absent |
| Irregular | Convex, smooth to finely folded | Shallow | Indistinct | Degenerated | Absent |
| Rectangular, elongated | Flat, warty | Distinctly raised, straight, smooth to warty, unevenly thickened | Distinct | Well represented | Present |
| Rectangular, elongated | Flat with sessile tubercle | Slightly raised, straight | Distinct | Well represented | Present |
| Rectangular, elongated | Flat with sessile micro papillae | Slightly raised, straight, | Distinct | Well represented | Present |
| Rectangular, elongated | Flat with sessile micro papillae | Distinctly raised, straight, smooth or finely folded, unevenly thickened | Distinct | Well represented | Present |
| Rectangular, elongated | Flat, full of micro papillae | Distinctly raised, straight, smooth or finely folded, unevenly thickened | Distinct | Well represented | Present |
| Polygonal, isodiametric | Flat, smooth | Distinctly raised, straight to sinuous, unevenly thickened | Distinct | Well represented | Present |
| Polygonal, isodiametric | Flat, smooth to finely folded | Distinctly raised, straight to sinuous, unevenly thickened | Distinct | Degenerate | Present |
| Polygonal, isodiametric | Flat to slightly concave, smooth | Distinctly raised, straight to sinuous, unevenly thickened | Distinct | Well represented | Present |
| Irregular | Slightly convex, warty | Shallow | Distinct | Well represented | Present |
| Polygonal, elongated/isodiametric | Convex with central tubercle, smooth | Distinctly raised, straight to wavy, smooth | Indistinct | Degenerated | Present |
| Polygonal, elongated/isodiametric | Concave, striate | Distinctly raised, straight to wavy, smooth | Distinct | Well represented | Present |
| Polygonal, isodiametric | Convex with central tubercle, smooth | Distinctly raised, straight to wavy, smooth | Indistinct | Degenerated | Absent |
| Polygonal, elongated/isodiametric | Concave, striate | Distinctly raised, straight to wavy, folded | Distinct | Well represented | Present |
| Polygonal, elongated/isodiametric | Concave, striate | Distinctly raised, straight to wavy, folded | Distinct | Well represented | Present |
| Polygonal, elongated/isodiametric | Concave, striate | Distinctly raised, straight to wavy, folded | Distinct | Well represented | Present |
| Polygonal, elongated/isodiametric | Concave, striate | Distinctly raised, straight to wavy, folded | Distinct | Well represented | Present |

| Epidermal cell shape | Periclinal wall | Anticlinal wall | Seed coat | Epidermis | Endothelium |
|---|---|---|---|---|---|
| Polygonal, elongated/isodiametric | Concave, striate | Distinctly raised, straight to wavy, folded, unevenly thickened | Distinct | Well represented | Present |
| Polygonal, elongated | Concave, striate | Distinctly raised, straight to wavy, folded, unevenly thickened | Distinct | Well represented | Present |
| Polygonal, elongated/isodiametric | Flat, smooth/folded | Slightly raised, folded, unevenly thickened | Distinct | Well represented | Present |
| Polygonal, isodiametric | Convex with central tubercle, smooth | Slightly raised, folded, unevenly thickened | Indistinct | Degenerated | Absent |
| Polygonal, elongated/isodiametric | Concave, striate | Distinctly raised, straight to wavy, folded, unevenly thickened | Distinct | Well represented | Present |
| Polygonal, elongated | Concave, striate | Distinctly raised, straight to wavy, folded, unevenly thickened | Distinct | Well represented | Present |
| Polygonal, elongated | Concave, striate | Distinctly raised, straight to wavy, folded, unevenly thickened | Distinct | Well represented | Present |
| Polygonal, isodiametric | Concave, striate | Distinctly raised, folded, unevenly thickened | Distinct | Well represented | Present |
| Polygonal, elongated | Concave, striate | Distinctly raised, straight to wavy, folded, unevenly thickened | Distinct | Well represented | Present |
| Polygonal, isodiametric | Convex, finely folded | Shallow, finely folded, straight to wavy | Distinct | Well represented | Present |
| Polygonal, elongated | Concave, striate | Distinctly raised, straight to sinuous, unevenly thickened | Distinct | Degenerated | Present |
| Polygonal, elongated | Concave, striate | Distinctly raised, straight to sinuous, unevenly thickened | Distinct | Degenerated | Present |
| Polygonal, elongated | Concave, striate | Distinctly raised, straight to sinuous, unevenly thickened | Distinct | Degenerated | Present |
| Polygonal, elongated | Concave, striate | Distinctly raised, straight to sinuous, unevenly thickened | Distinct | Degenerated | Present |
| Polygonal, elongated | Concave, striate | Distinctly raised, straight to sinuous, unevenly thickened | Distinct | Degenerated | Present |

**Appendix 1.** Seed characters studied for Scrophulariaceae s.l.

| Plant name | 1 | 2 | 3 | 4 | 5 | 6 | 7 | 8 | 9 | 10 | 11 | 12 | 13 |
|---|---|---|---|---|---|---|---|---|---|---|---|---|---|
| *Lindernia procumbens* | 0 | 0 | 1 | 1 | 0 | 2 | 0 | 0 | 2 | 1 | 0 | 0 | 1 |
| *L. crustacea* | 1 | 0 | 0 | 1 | 0 | 2 | 2 | 0 | 2 | 1 | 0 | 0 | 1 |
| *Mazus pumilus* | 1 | 0 | 0 | 1 | 0 | 1 | 0, 1 | 0 | 0 | 2 | 0 | 0 | 0 |
| *Lathraea japonica* | 1 | 1 | 0 | 0 | 0 | 2 | 2 | 1 | 0 | 2 | 0 | 0 | 0 |
| *Melampyrum koreanum* | 0 | 2 | 1 | 1 | 1 | 2, 1 | 2 | 0 | 0 | 2 | 1 | 1 | 1 |
| *M. roseum* | 0 | 2 | 1 | 1 | 1 | 1 | 2 | 0 | 0 | 2 | 1 | 1 | 1 |
| *M. roseum* var. *japonicum* | 0 | 2 | 1 | 1 | 1 | 1 | 2 | 0 | 0 | 2 | 1 | 1 | 1 |
| *M. roseum* var. *ovalifolium* | 0 | 2 | 1 | 1 | 1 | 2 | 2 | 0 | 0 | 2 | 1 | 1 | 1 |
| *M. setaceum* | 0 | 2 | 1 | 1 | 1 | 2 | 2 | 0 | 2 | 2 | 1 | 1 | 1 |
| *M. setaceum* var. *nakaianum* | 0 | 2 | 1 | 1 | 1 | 2, 1 | 2 | 0 | 0 | 2 | 1 | 1 | 1 |
| *Pedicularis mandshurica* | 0 | 2 | 1 | 1 | 0 | 0 | 0 | 1 | 1 | 1 | 0 | 0 | 0 |
| *P. resupinata* f. *albiflora* | 0 | 2 | 1 | 1 | 0 | 0, 2 | 0 | 0 | 1 | 1 | 0 | 0 | 0 |
| *P. resupinata* | 0 | 2 | 1 | 1 | 0 | 0 | 0 | 0 | 1 | 1 | 0 | 0 | 0 |
| *P. resupinata* var. *umbrosa* | 0 | 2 | 1 | 1 | 0 | 0 | 0 | 1 | 1 | 1 | 0 | 0 | 0 |
| *P. verticillata* | 0 | 2 | 0 | 1 | 0 | 0 | 0 | 1 | 1 | 1 | 0 | 0 | 0 |
| *Phtheirospermum japonicum* | 1 | 1 | 0 | 1 | 0 | 0 | 1 | 1 | 1 | 2 | 0 | 0 | 0 |
| *Siphonostegia chinensis* | 0 | 0 | 0 | 1 | 2 | 0 | 1 | 1 | 1 | 2 | 0 | 0 | 0 |
| *Paulownia coreana* | 0 | 1 | 1 | 1 | 0 | 2 | 1 | 1 | 1 | 2 | 0 | 0 | 0 |
| *Limnophila indica* | 1 | 0 | 0 | 1 | 2 | 2 | 2 | 0 | 0 | 1 | 0 | 0 | 0 |
| *V. arvensis* | 1 | 1 | 0 | 1 | 0 | 0, 2 | 1 | 1 | 0 | 1 | 1 | 1 | 1 |
| *V. dahurica* | 0 | 1 | 0 | 0 | 0 | 0 | 1 | 1 | 2 | 0 | 0 | 0 | 0 |
| *V. didyma* var. *lilacina* | 1 | 1 | 0 | 0 | 0 | 0 | 1 | 1 | 0 | 1 | 1 | 1 | 1 |
| *V. incana* | 1 | 0 | 0 | 1 | 0 | 0 | 1 | 1 | 2 | 0 | 0 | 0 | 0 |
| *V. kiusiana* var. *diamantiaca* | 0 | 1 | 0 | 0 | 0 | 0 | 1 | 1 | 2 | 0 | 0 | 0 | 0 |
| *V. kiusiana* var. *glabrifolia* | 0 | 1 | 0 | 0 | 0 | 0 | 1 | 1 | 2 | 0 | 0 | 0 | 0 |
| *V. linariifolia* | 0 | 1 | 0 | 1 | 0 | 0 | 1 | 1 | 2 | 0 | 0 | 0 | 0 |
| *V. longifolia* | 0 | 1 | 0 | 0 | 0 | 0 | 1 | 1 | 2 | 0 | 0 | 0 | 0 |
| *V. nakaiana* | 1 | 0 | 0 | 1 | 0 | 0 | 1 | 1 | 2 | 0 | 0 | 0 | 0 |
| *V. peregrina* | 0 | 1 | 0 | 1 | 0 | 0 | 1 | 0 | 1 | 2 | 0 | 0 | 0 |
| *V. persica* | 1 | 0 | 0 | 1 | 0 | 0 | 1 | 0 | 0 | 1 | 1 | 1 | 1 |
| *V. pusanensis* | 0 | 1 | 0 | 0 | 0 | 0 | 1 | 1 | 2 | 0 | 0 | 0 | 0 |
| *V. pyrethrina* | 0 | 0 | 0 | 0 | 0 | 0 | 1 | 1 | 2 | 0 | 0 | 0 | 0 |
| *V. rotunda* var. *subintegra* | 0 | 0 | 0 | 0 | 0 | 0 | 1 | 1 | 2 | 0 | 0 | 0 | 0 |
| *V. undulata* | 1 | 1 | 0 | 0 | 0 | 0 | 1 | 1 | 2 | 2 | 0 | 0 | 0 |
| *Veronicastrum sibiricum* | 1 | 0 | 0 | 0 | 0 | 0 | 1 | 1 | 2 | 0 | 0 | 0 | 0 |
| *Linaria japonica* | 1 | 1 | 0 | 1 | 2 | 1 | 1 | 0 | 0 | 2 | 0 | 0 | 0 |

| Plant name | 1 | 2 | 3 | 4 | 5 | 6 | 7 | 8 | 9 | 10 | 11 | 12 | 13 |
|---|---|---|---|---|---|---|---|---|---|---|---|---|---|
| *Scrophularia buergeriana* | 1 | 1 | 0 | 1 | 2 | 0 | 1 | 1 | 2 | 0 | 0 | 1 | 0 |
| *S. grayana* | 1 | 1 | 0 | 1 | 2 | 0 | 1 | 1 | 2 | 0 | 0 | 1 | 0 |
| *S. kakudensis* | 1 | 1 | 0 | 1 | 2 | 0 | 1 | 1 | 2 | 0 | 0 | 1 | 0 |
| *S. koraiensis* | 1 | 1 | 0 | 1 | 2 | 0 | 1 | 1 | 2 | 0 | 0 | 1 | 0 |
| *S. takesimensis* | 1 | 1 | 0 | 1 | 2 | 0 | 1 | 1 | 2 | 0 | 0 | 1 | 0 |

**Appendix 2.** Data matrix of the character states used for analysis.

## Author details

Balkrishna Ghimire[1], Go Eun Choi[1], Hayan Lee[1], Kweon Heo[2] and Mi Jin Jeong[1]*

*Address all correspondence to: mjjeong121@korea.kr

1 Plant Conservation Division, Korea National Arboretum, Pocheon, Korea

2 Department of Applied Plant Sciences, Kangwon National University, Chuncheon, Korea

## References

[1] Tank DC, Beardsley PM, Kelchner SA, Olmstead RG. Review of the systematics of Scrophulariaceae s.l. and their current disposition. Australian Systematics Botany. 2006; **19**:89-307

[2] Olmstead RG, Reeves PA. Evidence for the polyphyly of the Scrophulariaceae based on chloroplast rbcL and ndhF sequences. Annals of Missouri Botanical Garden. 1995;**82**:176-193

[3] Bentham G. Scrophulariaceae. In: DeCandolle A, editor. Prodromus. Vol. 10. Reeve and Co: London, UK; 1846. p. 180-586

[4] Bentham G. Scrophulariaceae. In: Bentham G, Hooker JD, editors. Genera Plantarum. Vol. 2. Reeve and Co: London, UK; 1876. p. 913-980

[5] Olmstead RG, dePamphilis CW, Wolfe AD, Young ND, Elisons WJ, Reeves PA. Disintegration of the Scrophulariaceae. American Journal of Botany. 2001;**88**:348-361

[6] Olmstead RG. A Synoptical Classification of the Lamiales [Internet]. 2005 [Updated: Version 2.0]. Available from: http://depts.washington.edu/phylo/classifications.html [Accessed: 17 August 2009]

[7] Oxelman B, Kornhall P, Olmstead RG, Bremer B. Further disintegration of Scrophulariaceae. Taxon. 2005;**54**:411-425

[8] Cronquist A. An Integrated System of Classification of Flowering Plants. New York, USA: Columbia University Press; 1981

[9] dePamphilis CW, Young ND, Wolfe AD. Evolution of plastid gene rps2 in a lineage of hemiparasitic and holoparasitic plants: Many losses of photosynthesis and complex patterns of rate variation. Proceeding of National Academy Science USA. 1997;**94**:7367-7372

[10] Wolfe AD, dePamphilis CW. The effect of relaxed functional constraints on the photosynthetic gene rbcL in photosynthetic and nonphotosynthetic parasitic plants. Molecular Biology and Evolution. 1998;**15**:1243-1258

[11] Young ND, Steiner KE, dePamphilis CW. The evolution of parasitism in Scrophulariaceae/Orobanchaceae: Plastid gene sequences refute an evolutionary transition series. Annals of Missouri Botanical Garden. 1999;**86**:876-893

[12] Olmstead RG. Whatever happened to the Scrophulariaceae? Fremontia. 2002;**30**:13-22

[13] Fay MF, Bebbett JR, Dixon KW, Christenhusz MJM. Parasites, their relationships and the disintegration of Scrophulariaceae sensu lato. Curtis's Botanical Magazine. 2010;**26**:86-313

[14] APG IV. An update of the Angiosperm Phylogeny Group classification for the orders and families of flowering plants: APG IV. Botanical Journal of Linnean. Society. 2016;**181**:1-20

[15] Stevens PF. Angiosperm Phylogeny Website [Internet]. 2001. onwards [Updated: version 12]Available from: http://www.mobot.org/MOBOT/research/APweb/ [Accessed: July 2012]

[16] Sako Y, Regnier EE, Daoust T. Computer image analysis and classification of giant ragweed seeds. Weed Science. 2001;**49**:738-745

[17] Khalik KA, Van der Maesen LJG. Seed morphology of some tribes of Brassicaceae: Implications for taxonomy and species identification for the flora of Egypt. Blumea. 2002;**47**:363-383

[18] Constantinidis T, Psaras GK, Kamari G. Seed morphology in relation to infrageneric classification of *Consolida* (DC.) Gray (Ranunculaceae). Flora. 2001;**196**:81-100

[19] Mendum M, Lassnig P, Weber A, Christie F. Testa and seed appendage morphology in *Aeschynanthus* (Gesneriaceae): Phytogeographical patterns and taxonomic implications. Botanical Journal of Linnean Society. 2001;**135**:195-213

[20] Hufford L. A phylogenetic analysis of Hydrangeaceae based on morphological data. International Journal of Plant Science. 1997;**158**:652-672

[21] Arambarri A. A cladistic analysis of the new world species of Lotus L. (Fabaceae, Loteae). Cladistics. 2000;**16**:283-297

[22] Fukuhara T. Seed and funicle morphology of Fumariaceae-Fumarioideae: Systematic implications and evolutionary patterns. International Journal of Plant Science. 1999; **160**:151-180

[23] Simons MP, Hedin JP. Relationships and morphological character change among genera of Celastraceae *sensu lato* (including Hippocrateaceae). Annals of Missouri Botanical. Garden. 1999;**86**:723-757

[24] Thieret JW. The seeds of *Veronica* and allied genera. Lloydia. 1955;**18**:37-45

[25] Tiagi B. Development of the seed and fruit in *Rhinanthus major* and *R. serotinus*. American Journal of Botany. 1966;**53**:645-651

[26] Chuang TI, Heckard LR. Seed coat morphology in *Cordylanthus* (Scrophulariaceae) and its taxonomic significance. American Journal of Botany. 1972;**59**:258-265

[27] Juan R, Pastor J, Fernández I. Morphological and anatomical studies on fruits of Veronica from South-west Spain. Botanical Journal of Linnean Society. 1994;**115**:133-143

[28] Juan R, Férnandez I, Pastor J. Systematic consideration of micro-characters of fruits and seeds in the genus Verbascum (Scrophulariaceae). Annals of Botany. 1997;**80**:591-598

[29] Juan R, Férnandez I, Pastor J. Morphological and anatomical studies of *Linaria* species from south-west Spain: Seeds. Annals of Botany. 1999;**84**:11-19

[30] Juan R, Pastor J, Fernández I. SEM and light microscope observations on fruit and seeds in Scrophulariaceae from southwest Spain and their systematic significance. Annals of Botany. 2000;**86**:323-338

[31] Plaza L, Fernández I, Juan R, Pastor J, Pujadas A. Micromorphological studies on seeds of *Orobanche* species from the Iberian Peninsula and the Balearic Islands, and their systematic significance. Annals of Botany. 2004;**94**:167-178

[32] Muñoz-Centeno LM, Albach DC, Sanchez-Agudo JA, Martínez-Ortega MM. Systematic significance of seed morphology in *Veronica* (Plantaginaceae): A phylogenetic perspective. Annals of Botany. 2006;**98**:335-350

[33] Liu M-L, W-B Y, Li D-Z, Mill RR, Wang H. Seed morphological diversity of *Pedicularis* (Orobanchaceae) and its systematic significance. Plant Systematics and Evolution. 2013; **299**:1645-1657

[34] Hassan NM, Khalik KNA. Systematic significance of seed morphology in the genus *Veronica* (Plantaginaceae), with special reference to the Egyptian taxa. Journal of Systematics and Evolution. 2014;**52**:215-230

[35] Ahedor AR, Elisens W. Seed morphology and its taxonomic significance in the subtribe Gratiolinae (Plantaginaceae). Systematic Botany. 2015;**40**:845-852

[36] Hammer, Ø, Harper, DAT, Ryan, PD. PAST: Paleontological statistics software package for education and data analysis. Palaeontologia Electronica. 2001;**4**:9. DOI: http://palaeo-electronica.org/2001_1/past/issue1_01.htm

[37] Martínez-Ortega MM, Rico E. Seed morphology and its systematic significance in some Veronica species (Scrophulariaceae) mainly from the Western Mediterranean. Plant Systematics and Evolution. 2001;**228**:15-32

[38] Chen XL, Jing GH, Guo H. Ornamentation characteristics of seed coats in nineteen plants of *Pedicularis* from alpine meadow in east Qinghai-Xizang plateau and its ecological significance. Acta Prataculturae Sinica. 2007;**16**:60-68

[39] Pennell FW. Scrophulariaceae of the Western Himalayas. Lancaster, UK: Wickersham Printing Company; 1943. p. 163

[40] Berg RY. Development and dispersal of the seed of Pedicularis silvatica. Nytt Magasin for Botanikk. 1954;**2**:1-60

[41] Chuang TI, Heckard LR. Systematic significance of seed surface features in *Orthocarpus* (Scrophulariaceae-subtribe Castillejinae). American Journal of Botany. 1983;**70**:877-890

[42] Canne JM. Seed surface features in *Aureolaria, Brachystigma, Tomanthera,* and certain South American *Agalinis* (Scrophulariaceae). Systematic Botany. 1980;**4**:281-296

[43] Dong LN, Wang H, Wortley AH, Lu L, Li DZ. Phylogenetic relationships in the *Pterygiella complex* (Orobanchaceae) inferred from molecular and morphological evidence. Botanical Journal of Linnean Society. 2013;**171**:491-507

[44] Deyuan H, Fischer MA. *Veronica* Linnaeus. In: ZY W, Raven PH, editors. Flora of China (Scrophulariaceae through Gesneriaceae). Vol. 18. St. Louis: Missouri Botanical Garden Press; 1998. p. 65-80

[45] Sivarajan VV. Scanning electron microscopic studies on the seed coat surface patterns of South Indian species of *Lindernia* All. (Scrophulariaceae). Proceeding of Indian Academic Science (Plant Science). 1981;**90**:499-508

[46] Philcox D. Revision of the Malesian species of *Lindernia* All. (Scrophulariaceae). Kew Bulletin. 1968;**22**:1-72

[47] Albach DC, Martinez-Ortega MM, Fischer MA, Chase MW. A new classification of the tribe Veroniceae: Problems and a possible solution. Taxon. 2004;**53**:429-452

Chapter 12

# Seed Transmission of Tobamoviruses: Aspects of Global Disease Distribution

Aviv Dombrovsky and Elisheva Smith

Additional information is available at the end of the chapter

http://dx.doi.org/10.5772/intechopen.70244

**Abstract**

Global seed trade contributed to development and improvement of world agriculture. An adverse effect of global seed trade is reflected in disease outbreaks in new growing areas, countries, and continents. Among the seed-borne viruses, *Tobamovirus* species are currently considered a peril for crop production around the world. The unique tobamoviral particles confer stability to the RNA genome and preserve their infectivity for years. High titer of *Tobamovirus* species accumulates in reproductive organs leading to viral particles adsorbed to seed coat, which potentially establish a primary infectious source. *Tobamovirus*-contaminated seeds show very low virus transmission in grow-out experiments as detected by enzyme-linked immunosorbent assay (ELISA) and reverse transcription polymerase chain reaction (RT-PCR) analysis. Interestingly, in situ immunofluorescence analysis of *Cucumber green mottle mosaic virus* (CGMMV) reveals that the perisperm-endosperm envelope (PEE) is contaminated as well by the *Tobamovirus*. Indeed, chemical seed disinfection treatments that affect primarily the seed coat surface are efficient for several *Tobamovirus* species but apparently do not prevent seed transmission of CGMMV to occur. *Tobamovirus* infection of the seed internal layers, which rarely includes the embryo, may partially follow the direct invasion pathway of *Potyviruses* such as *Pea seed-borne mosaic virus* (PSbMV) to pea embryo.

**Keywords:** contaminated seeds, perisperm-endosperm envelope (PEE), seed coat, seed disinfection, seed testing, size exclusion limit (SEL), seedlings, symplast, phloem

## 1. Introduction

The contribution of global seed trade to modern agriculture is clearly emphasized in breeder's broad range of crop species that contribute to increased yield, long shelf life, pathogen resistance, and tolerance to extreme circumstances (e.g., salinity, drought, and arid climate conditions) around the world. However, spread of seed-borne viruses is an unfortunate side effect of this industry. Seed transmission of plant viruses has a great epidemiological significance causing disease outbreaks worldwide. Seed trade globalization has opened new pathways for the mobilization of crop produce between continents and countries. In the modern agriculture, new crop varieties are introduced into new growing areas and are cultivated in proximity to indigenous crops, which are exposed to the imported remote endemic diseases. Seed-borne plant viruses are a threat to world agriculture. Among them, species that belong to the genus *Tobamovirus* are considered a major peril to a range of cultivars especially to those belonging to the *Solanaceae* and *Cucurbitaceae* (cucurbit) families. These viruses have long been a threat to agriculture. For more than a century, tobacco, tomato, and pepper plants that belong to the *Solanaceae* family are infected by the *Tobacco mosaic virus* (TMV) [1, 2], *Tobacco mild green mosaic virus* (TMGMV) [3], *Tomato mosaic virus* (ToMV), [4] and *Pepper mild mottle virus* (PMMoV) [5]. Similarly, cucurbits are infected by the *Cucumber green mottle mosaic virus* (CGMMV) that was discovered in England in 1935 [6] and spread into neighboring countries in Europe and to the Middle East, Asia, and the Far East. In the last decade, CGMMV was detected in new countries and continents—Canada [7] and the USA [8] in North America and Australia [9]—recently reviewed in [10], achieving a global distribution and becoming a major threat to cucurbit industries worldwide. Among the main cucurbit-infecting tobamoviruses, CGMMV is the most economically important and currently considered a significant threat for the production of cucumber, melon, watermelon, gherkin, and pumpkin, which has been endangered by the globalized spread of the *Tobamovirus* (**Figure 1**). Additional *Tobamovirus* species that infect cucurbits need to be considered in seed transmission tests in countries that commercially produce seeds: the *Kyuri green mottle mosaic virus* (KGMMV) [11, 12], the *Zucchini green mottle mosaic virus* (ZGMMV) [13], and the *Cucumber fruit mottle mosaic virus* (CFMMV) [14].

The tobamoviruses are seed-borne, mechanically transmitted stable viruses. Infectious particles are primarily attached to the seed coat [4]. Indeed, viral inoculum is efficiently transmitted when it enters the embryo, and viruses attached to the seed coat may not survive germination when seed coat separates from the seedlings [15]. However, in nursery seedlings, the *Tobamovirus*-contaminated seed coat may affect the wounded roots, which occur upon transplantation [4]. Furthermore, low transmission rate to seedlings does occur when tobamoviruses contaminate the seeds. Seeds or seedlings are used in large scale continuously in regular farming. Therefore, even a low percentage of contaminated seeds can cause a multitude of infection foci [16]. Consequently, the primary infectious source can be spread rapidly by mechanical contacts, workers' hands, tools, greenhouse structure, and trellising ropes [17, 18] and the tractor path in open fields [19]. The infectivity of tobamoviruses is preserved in plant debris and in the contaminated soil and clay for months up to years [20].

The entry and the establishment of new *Tobamovirus* diseases inflict a major concern for growers around the world. Seed nurseries, especially those that produce grafted seedlings (a procedure that requires cutting and hand handling), and protected structures (e.g., greenhouses, net houses, walk-in tunnels), which are commonly used worldwide, mostly with monoculture

**Figure 1.** *Cucumber green mottle mosaic virus* (CGMMV)-infected cucurbit species. (**a, b, e, i**) Mild mottling and mosaic on infected leaves of cucurbits: (**a–b**) cucumber (*Cucumis sativus*), (**e**) melon (*Cucumis melo*), and (**i**) watermelon (*Citrullus lanatus*) plants. (**c–d**) Mosaic pattern developed on cucumber fruits. (**f–h**) Mosaic pattern and mottling developed on melon fruits. (**j–k**) Yellowing fruits and spongy flesh (double-headed arrows) accompanied by necrotic lesions on stems and peduncles (single-headed arrows) of watermelon fruits.

crops (without crop rotations between growing cycles), are highly susceptible to *Tobamovirus* infection. When a new *Tobamovirus* disease emerges in a new country, the first response is to try to eradicate the disease, a strategy that is unfortunately prone to failure in most of the scenarios. The major reasons for this failure may be attributed to multiple contaminated seed entry events that occur in parallel, in multiple growing areas/farms, and late detection and response. The second strategy upon the emergence of a new *Tobamovirus* disease is based on management of the disease, which combines several approaches to contain the disease and reduce the disease damage below the economic threshold. A proper management needs to cover all aspects of plant growth that are involved in disease spread: disinfection of contaminated soil from the previous growing cycle, trellising ropes and the greenhouse structure, worker hands and clothes, etc. This approach however, cannot ensure low disease damage and further emphasizes the importance of proper management of commercial seed production, accompanied by the appropriate diagnostic approaches in order to eliminate *Tobamovirus*-contaminated seed lots.

## 2. Resistance-breaking tobamoviruses

Plant viruses belonging to the genus *Tobamovirus* (currently comprised of 37 species) are rod-shaped particles encapsulating a single-stranded RNA (+ssRNA) genome of ~6.4 kb. The genome encodes four ORFs. ORF1 and ORF2, separated by a leaky stop codon, encode the 126 and 183 kDa replicase protein complex (recently reviewed in [21]). ORF3 encodes the 30 kDa movement protein (MP), and ORF4 encodes the 17.5 kDa coat protein (CP). The tobamoviruses endanger cultivars in the world. For more than a century, tobacco and tomato plants that belong to the *Solanaceae* family are infected by TMV [1, 2] and ToMV [4]. Resistance to these viruses was introduced to tomato plants by introgression [22]. However, the durability of resistance genes is compromised by the pathogen selection pressure that gradually breaks the plant defense system. In tomato plants, the durability of *Tm-1*, *Tm-2*, and $Tm\text{-}2^2$ [23] resistance genes has recently been jeopardized by the newly discovered tobamoviruses *Tomato mottle mosaic virus* (ToMMV), reported in Brazil [24], Mexico [25], and the USA [26], and the $Tm\text{-}2^2$ resistance-breaking *Tomato brown rugose fruit virus* (ToBRFV) reported in Jordan [27] and Israel [28] (**Figure 2**). Within the *Solanaceae* family, pepper crops were also affected in the last decades by *Tobamovirus* species, mostly by PMMoV that overcame the resistance conferred by the $L^{1\text{-}3}$ genes [29–31] and rarely the $L^4$ gene in Japan [32] and Israel [33]. In cucurbits, the production of cucumber, melon, and watermelon has been endangered by the globalized spread of CGMMV. Since its discovery by Ainsworth in 1935 [6], no commercial cultivars fully resistant to CGMMV are available although temperature-sensitive strains with specific resistance were reported.

**Figure 2.** *Tomato brown rugose fruit virus* (ToBRFV)-infected tomato (*Solanum lycopersicum*) plants. (**a, d**) Brown rugose symptoms developed on fruits. (**b, c**) yellow spots on fruits. (**a, e–g**) Mosaic pattern developed on leaves and narrowing accompanied by mottling leaves. (**g**) Necrotic symptoms on pedicle, calyces, and petioles.

## 3. Virus preservation in seeds

The viral particles of tobamoviruses are extremely stable (**Figure 3(a)**), and infectivity is preserved in seeds for up to several years. Most of the *Tobamovirus* species display low percentage of seed transmission, but even very low occurrence of seed transmission is enough to start a spread of the disease [16]. Seed transmission of viruses occurs primarily via infected embryos through paternal or maternal pathways. However, tobamoviruses mostly infect the seed coat (testa) and the endosperm [4, 34, 35]. Transmission of the virus occurs primarily by mechanical means through transplanting the seedlings and causing cuts in the roots that are then susceptible to infection by the contaminated seed coat. The rout of *Tobamovirus* transmission to the seeds is not quite clear. While the testa is of maternal origin, the endosperm is the outcome of fertilization. It is possible that tobamoviruses partially follow the rout of the symplasmic pathway suggested to occur in seed infection of *Pea seed-borne mosaic virus* (PSbMV) without the accomplishment of embryo invasion.

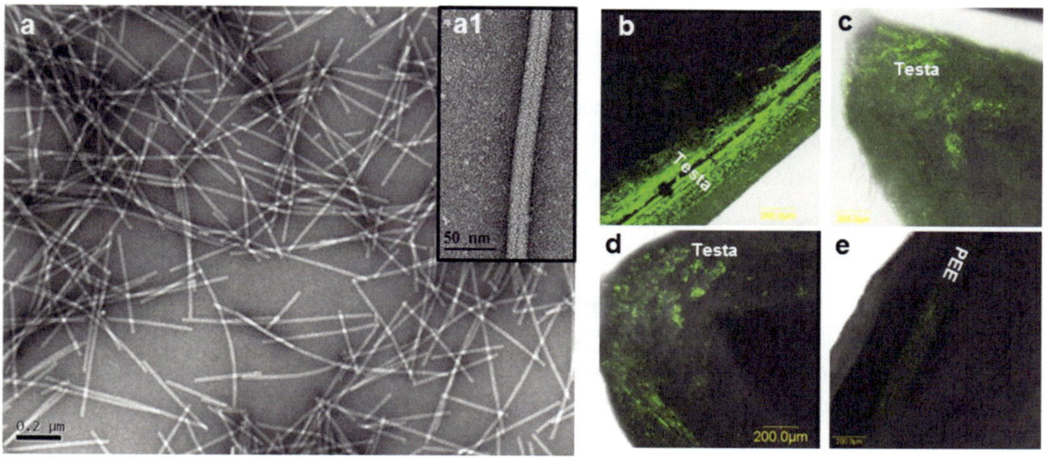


**Figure 3.** *Cucumber green mottle mosaic virus* (CGMMV) morphology and localization in cucurbit seeds. (**a**) Electron micrograph illustration of viral particles, (**a1**) high resolution of a single viral particle. (**b–e**) In situ immunofluorescence of CGMMV in infected melon (*Cucumis melo*) seeds using CGMMV-specific antibodies and secondary antibodies conjugated with Alexa Fluor 488. The fluorescent signal indicates the presence of CGMMV coat protein. Bars, 200 µm.

## 4. Symplasmic viral movement

The symplast is the conduit for virus movement from cell to cell and through the phloem vasculature. They follow source to sink transportation of photoassimilates albeit with a slower rate [36]. At early stages, seeds are sinks. Nutrients enter the seed through the vascular bundle in the funiculus that ends up in the chalazal vein, which consists of xylem strand surrounded by phloem elements [37]. The seed coat (testa) is of maternal origin and mediates the release of nutrients to the embryo. It is a development of the integument/s. Symplasmic transport that can be exploited by viruses occurs from the sieve elements of the chalazal vein into the parenchyma layer [37].

### 4.1. Cell-to-cell movement

Viruses exploit the plasmodesmata (PD) for their movement from cell to cell. PD are membrane-bound tunnels that interconnect the cytoplasm of adjacent cells. Flattened endoplasmic reticulum membrane or desmotubule, which is surrounded by actin and myosin molecules, traverses PD in the center and is continuous with the cortical endoplasmic reticulum (ER) [38]. The actin and myosin molecules connect the desmotubule to the plasma membrane [39]. There are simple PD and branched PD; the latter are more common in mature tissues. Conversion of simple PD to branched PD presumably involves formation of bridges between simple PD. Trafficking between cells occurs via the "cytoplasmic sleeve," which is the area of the cytoplasm between the desmotubule and the plasma membrane, and via the desmotubule that bridges between the ER of adjacent cells. The desmotubule exists in both appressed and dilated states. The neck region at both ends of the PD is frequently restricted. In experiments studying the transport of small dye molecules through the cytoplasmic sleeve, it was observed that the transport is not interrupted by cytochalasin B, an inhibitor of actin polymerization [40], indicating that this intercellular movement occurs by diffusion and the PD is rate limiting [38].

Actin and myosin localized in PD may serve as scaffold for active transport of proteins. Actin filaments are also closely associated with the ER [41, 42]. Directionality of the movement of ER-localized proteins may be attributed to myosin.

The PD has size exclusion limit (SEL) [43], which changes during development and in response to environmental stress [38]. The mobility of molecules through PD is determined by Stokes radius (Rs), which is the hydrodynamic radius that is influenced by both the molecular weight of a molecule and its side group [44]. The rate of diffusion is directly correlated to the radius of a molecule [38]. SEL is measured by injecting labeled size markers such as F-dextran [45]. Transient increases in SEL were observed, and it is commonly reduced during maturation [46, 47]. SEL differs between tissues in a plant and between plant species [38]. Actin cytoskeleton may participate in regulation of SEL. Depolymerization of actin cytoskeleton by cytochalasin widened the neck region of PD and increased SEL of tobacco PD from 1 kDa to over 20 kDa [48]. Actin filaments attached to the ER participate in controlling intracellular movement of TMV viral particles [38, 49]. The integrity of the cytoskeleton is apparently necessary for viral spread. Virus movement protein (MP) is necessary for viral spread from cell to cell [50]. Cytoskeletal components were suggested to be involved in the transport of MP to the PD site and in viral cell-to-cell movement via the PD [49, 51, 52]. However, study of TMV MP and CMV MP showed that these viral proteins have F-actin depolymerization activity that causes increase in PD SEL [53]. Since actin is required for MP targeting to PD [49], it was suggested that MP-induced F-actin degradation occurs at the orifice area [53].

The MP of many viruses and of TMV is an endoplasmic reticulum membrane protein [54, 55]. Exogenously expressed MP target the PD autonomously [56–58].

Microtubules participate in cellular distribution and long-distance movement of MP and viral replication complexes [59]. However, cytoskeleton involvement in active movement of viral particles, when the PD is dilated, predicts a fast spread of the particles between cells, in the order of seconds, which does not occur [51]. The model also predicts that in the presence of MP, cytoplasmic and ER membrane proteins will show enhanced cell-to-cell movement. This apparently did not occur when movement of cytoplasmic and ER membrane-fluorescent protein probes was studied [60], which may suggest the presence of passive diffusion of the viral complexes in the desmotubule [60].

Callose, a beta 1,3-glucan [61] deposition between the plasma membrane and the cell wall in the neck region of PD, participates in determining SEL [62, 63]. Class I beta-1,3-glucanase-deficient mutants have lower SEL and show decreased susceptibility to virus infection [64]. And targeted expression of class I beta-1,3-glucanase enhances TMV infection in tobacco plants [65].

The callose accumulation is often induced in the plants under stress conditions [64]. There is a controversy regarding the effect of TMV MP on increasing SEL [66]. It has been suggested that results showing the increase in SEL by MP depend on the technique of introducing the MP [67]. Apparently under conditions of unmodified SEL by MP, TMV-derived replicase supports the activity of the viral MP on induction of callose degradation [60] and increased the conductivity via the desmotubule. The involvement of RNA replicase in cell-to-cell movement of TMV has been observed [68]. RNA helicase domain in 126 kDa replicase protein is involved in execution of cell-to-cell movement independent of MP activity.

There are several other factors that are apparently involved in MP-induced dilation of PD and may shed a light on the mechanism of SEL modification by MP. Induction of callose degradation by the viral proteins may be mediated by plant cell proteins such as TGB12K interacting protein (TIP) that interacts with both the *Potato virus X* (PVX) movement protein TGB12K and beta-1,3-glucanase [69]. TMV MP has been shown to interact with pectin methyl transferase, which is essential for dilating PD [70, 71]. The carboxy terminus of TMV MP enhances cell-to-cell movement in *N. tobacco* cv. Xanthi plants [72]. Phosphorylation at the carboxy terminus is necessary for PD gating by MP [73]. MP dilation of PD may also be mediated by interacting with the *N. tabacum* noncell autonomous pathway protein 1 (NtNCAPP1) that its mutation interfered with TMV MP-induced SEL increase [74]. It has also been shown that mutation in Arabidopsis synaptotagmin, a $Ca^{2+}$-binding protein that is involved in endocytic recycling, interferes with TMV MP intercellular movement [75]. TMV MP also interacts with calreticulin, which resides in PD and sequester $Ca^{2+}$ [76]. This binding may indicate the involvement of local $Ca^{2+}$ levels in the control of TMV MP-induced PD dilation.

The above-illustrated role of *Tobamovirus* MP in dilating PD differs from that of MP of tubule-forming viruses that modifies PD by assembling tubules that mediate cell-to-cell viral transfer [77, 78]. And the tobamoviruses differ from the filamentous viruses that engage a homologue of Hsp70 and a complex of viral components for transport through PD [79].

### 4.2. Phloem movement

PD that mediate cell-to-cell movement partially mediate transport of viruses into the phloem vasculature for systemic infection. PD mediate symplasmic transport between epidermal or mesophyll cells through bundle sheath, vascular parenchyma to companion cells (CC). Between CC and the enucleated sieve elements (SE), specialized PD exist named pore plasmodesmal units (PPU), which consist of multiple channels on the CC side and a single channel facing the SE [80]. PPU have higher SEL than other PD in the phloem vasculature, but it is still not enough for viral particles to cross the boundary between CC and SE, and ribonucleoproteins cross this boundary [81]. Analyzing the form of the *Tobamovirus* CGMMV that is transported through the phloem exudate showed that the virus is in the form of virus particles [82]. Apparently particle formation is essential for long-distance movement of viruses [83]. Once in the SE, the viruses are transported by diffusion.

It has been suggested that different mechanisms are employed for virus loading and unloading from the phloem since viruses enter the phloem vasculature through all classes of veins, but virus exit is limited to major veins [84–87]. Accordingly, various mechanisms in the host plants, hindering or promoting viral long-distance movement, are differentially located in the loading and unloading sites of the phloem vasculature. The cadmium ion-induced glycine-rich protein is localized to the cell wall of CC and SE, and its blocking capacity of TMV and *Turnip vein clearing virus* (TVCV) long-distance movement is attributed to callose deposition. Mutation at the host VSM1 gene blocked the entry of TVCV into the phloem indicating a role of a host gene in promoting *Tobamovirus* loading into SE [88, 89].

On the other hand, at the phloem exit, protein degradation occurs and results in viral restriction to the phloem [90]. The 26S proteasome was found in pumpkin sap and Arabidopsis thaliana [91] and aminopeptidases and proteases found in pumpkin SE sap [91, 92]. Promotion of viral exit from the phloem vasculature occurs by the *Tobamovirus*-infected plant protein pectin methyl esterase (PME) [93].

The CP of many viruses, including tobamoviruses, is required for systemic movement of the viral particles in the host plants. CP ability to assemble the viral particles is important for the long-distance movement as well as the origin of assembly, indicating the importance of virion formation for transport through the phloem [94]. Analysis of CP mutants revealed that retaining the C-terminus is essential for high particle accumulation [95].

The RNA-dependent RNA polymerase activity by the 126 and/or 183 kDa proteins of TMV is necessary for the accumulation of viral particles in the phloem vasculature. Mutations occurred in the 126 kDa protein, and the 183 kDa protein, which is the read through of the amber termination of 126 ORF, prevented the accumulation of the virus in the phloem parenchyma and CCs, and systemic invasion of the virus was hindered [96]. Comparison between the systemic infection of TMV in tobacco plants and the nonsystemic infection of *Sunn-hemp mosaic virus* (SHMV) in these plants showed as well that replicase proteins are necessary for long-distance viral movement in the phloem vasculature [72, 97]. The 126/183 replicase is active also as a suppressor of silencing, which correlates with the ability of the virus to move systemically and to cause symptoms [98].

## 5. Symplasmic conductivity in seeds

Compartmentalization of symplasmic conductivity may differ between seeds of different plant varieties. The funiculus phloem reaches a phloem-unloading domain at the chalazal area of the outer integument, which either alone or together with the inner integument evolves to the seed coat, which differs between seed species as well. Using fluorescent probes, it has been shown that the entire outer integument at the end of the vascular bundle sheath of the funiculus is an extended symplasmic domain in *Arabidopsis thaliana* [99], and in *Crassulaceae* seeds [100], the inner integument and the embryo are separated symplasmic domains. In legumes, $^{14}C$-labeled photosynthates showed PD connections between SE of the chalazal vein and the parenchyma cells [101] that comprise most of the seed coat layers [37] and consist of chlorenchyma, ground parenchyma, and branched parenchyma. In peas (*Pisum sativum* L) phloem unloading occurs in the ground parenchyma [37]. The phloem mobile molecule pyranine was transported to the chlorenchyma but not to the branched parenchyma or to the epidermis [37]. Compartmentalization of the symplasmic domains raises the important question regarding transport of viruses in seeds. This question was primarily addressed in the study of seed transmission in pea cultivars susceptible to PSbMV infection.

Apparently, incidences of seed transmission of a specific virus vary between plant species and between cultivars. Comparison between cultivars allowed to delineate the sequence of events necessary for seed transmission. Infection of the vegetative tissues and of the maternal testa occurred irrespective of the capability of the virus to be transmitted via seeds. High incidence of seed transmission occurred in direct relation to virus invasion of an immature embryo.

PSbMV embryo invasion occurred at a post-fertilization stage. Pollen grains and maternal integuments were devoid of the virus. It has been suggested that the embryo, at early developmental stage, may have the symplasmic connections necessary for viral seed transmission. Immediately after fertilization zygotic divisions form a globular terminal cell and a suspensor. The integuments of the ovule develop into the testa. In the testa there are vascular strands that are continuous with the vascular bundle sheath of the funiculus. The suspensor consists of globular cell/cells that support the embryo and elongated basal cells that are in contact with the testa at the micropylar region. Using immunohistochemistry technique with antibodies raised against PSbMV particles, and in situ hybridization technique with an RNA probe specific for the viral RNA, it was shown that before fertilization PSbMV was localized in vascular tissues. Fertilization triggered invasion of the virus into the ovule along the vascular strand in both seed-transmitting and seed-un-transmitting cultivars. However, early developmental stages in seed-transmitting cultivars characteristically showed PSbMV release from vascular tissues to neighboring tissues infecting the micropyle area, concomitant to diminishing presence in the earlier infected tissues. This observation raises the question regarding the possible symplasmic connections between the maternal and filial tissues at the micropyle area. PSbMV cylindrical inclusions, which are commonly found at PD openings and were observed at the boundary between the testa and endosperm [102, 103], may only indicate the possible presence of such symplasmic connections. Crossing this boundary however does not ensure embryo invasion. It has been suggested that the suspensor is the conduit for viral transmission to the embryo [104], and pore-like structures that exist in the suspensor sheath wall at the boundary between the endosperm and the suspensor allow this transmission to occur [103].

High efficiency of PSbMV seed transmission occurred early in embryo development [105]. At this stage, the suspensor may establish continuity between virus present in the embryonic sac fluid, as was observed at late heart stage of embryo development, and the embryo [104, 105]. Viral contamination of the suspensor occurred in embryos at the globular to early heart stage [104]. The site of contact between the suspensor middle cell and the embryo became infected, further supporting the role of the suspensor in viral transmission to the embryo. The short time span the virus has before the suspensor degenerates and the fact that only passive accumulation of the virus and not its replication determines the amount of virus ready for transmission to the embryo [103, 104] are limiting conditions that hinder direct viral invasion of the embryo.

Direct viral invasion of the embryo differs from indirect embryo invasion in which the viruses infect gametes. Unlike the distribution of viruses that directly invade the embryo, such as PSbMV, viruses indirectly invading the embryo, e.g., *Pea early-browning virus* (PEBV), cause infection to the entire embryo [106, 107]. Apparently, indirect embryo invasion depends on the virus capability to invade the meristem [16]. In the plant meristem, there is a defense mechanism that protects the plant from viral RNA invasion to the germline [108]. Suppression of RNA silencing by the virus could allow meristem entry [108]. However, interfering with the plant systemic posttranscriptional gene silencing does not ensure meristematic invasion of the virus. Moreover, viral induction of DNA methylation associated with induction of posttranscriptional gene silencing does not correlate with viral invasion to meristematic cells [109]. Apparently, the signaling pathway for meristematic exclusion of viruses is not quite clear. Interestingly, the possibility that tobamoviruses could be transmitted indirectly to seeds through gamete infection was shown by mechanically applying CGMMV-infected pollen onto healthy flowering plants [110]. Under those specific experimental conditions, virus transmission rate via seeds was ~70%, which is not typical to seed transmission of tobamoviruses (**Table 1**).

| | | **Seed** | | | | |
|---|---|---|---|---|---|---|
| **Species/ acronym** | **Hosts** | **Contamination %** | | **Transmission %** | | **Reference** |
| *Cucumber green mottle mosaic virus* (CGMMV) | *Cucurbitaceae* | + | 84% | + | 2% | [141] |
| | | + | NT | + | NT | [148] |
| | | + | NT | + | 8% | [144] |
| | | + | 95% | + | <1% | [34] |
| | | + | 100% | + | 0.9% | [147] |
| *Kyuri green mottle mosaic virus* (KGMMV) | *Cucurbitaceae* | + | NT | + | NT | [148] |
| | | + | NT | NT | NT | [144] |
| *Odontoglossum ringspot virus* (ORSV) | *Orchidaceae, Solanaceae* | + | NT | + | NT | [148] |
| *Paprika mild mottle virus* (PaMMV) | *Solanaceae* | + | NT | NT | NT | [3] |
| *Pepper mild mottle virus* (PMMoV) | *Solanaceae* | + | NT | + | NT | [148] |
| | | + | NT | NT | NT | [3] |
| | | + | NT | + | (0.8–7.5%) | [123] |
| *Ribgrass mosaic virus* (RMV) | *Plantaginaceae* | + | NT | – | 0% | [145] |
| | | + | NT | + | NT | [148] |

http://dx.doi.org/10.5772/intechopen.70244

| Species/ acronym | Hosts | Seed Contamination % | | Transmission % | | Reference |
|---|---|---|---|---|---|---|
| *Sunn-hemp mosaic virus* (SHMV) | *Solanaceae* | + | NT | + | NT | [148] |
| | | + | NT | – | 0% | [130] |
| | *Fabaceae* | + | NT | + | 18% | [130] |
| *Tobacco mild green mosaic virus* (TMGMV) | *Solanaceae* | + | NT | NT | NT | [3] |
| *Tobacco mosaic virus* (TMV) | *Solanaceae* | + | NT | – | 0% | [4] |
| | | + | NT | + | 21.8% | [134] |
| | | + | 100% | – | 0% | [133] |
| | | + | NT | – | 0% | [142] |
| | | + | 50% | NT | NT | [149] |
| *Tomato mosaic virus* (ToMV) | *Solanaceae* | + | NT | + | NT | [35] |
| | | + | NT | – | 0% | [4] |
| | | + | NT | – | 0% | [143] |
| | | + | NT | + | NT | [148] |
| | | + | 38% | NT | NT | [149] |
| *Tropical soda apple mosaic virus* (TSAMV) | *Solanaceae* | + | >40% | + | 1–16% | [127] |
| *Turnip vein-clearing virus* (TVCV) | *Brassicaceae* | + | NT | + | 0% | [89] |
| *Zucchini green mottle mosaic virus* (ZGMMV) | *Cucurbitaceae* | + | 6% | + | 1.1% | [141] |
| | | + | NT | + | NT | [144] |

NT—not tested

**Table 1.** Selected seed-borne tobamoviruses.

# 6. Regulation of virus invasion

Virus invasion does not uniformly occur in plant cultivars. Invasion efficiency varies between virus strains and between various cultivars [15]. Introducing dominant resistance by introgression allows the production of cultivars resistant to virus infection. The R genes (nucleotide binding-site leucine-rich repeat, NB-LRR) so introduced confer the resistance [22]. The *N* gene in *Nicotiana glutinosa*, *Tm* gene in tomatoes, and *L* gene in pepper plants are the known R genes for *Tobamovirus* infection. However, virus strains that overcame these resistance genes have evolved. The R genes induce hypersensitive response (HR) that limits the virus movement to the inoculation area [22]. Metabolic accumulation of salicylic acid, reactive oxygen species, and $Ca^{2+}$ is characteristic to the HR response. Proteolysis through the ubiquitin pathway occurs, and induced caspase-like proteinases cause local cell death [22, 111]. In the case of *N* NB-LRR gene, translational arrest is involved [112]. The helicase domain of the replicase protein, MP, and CP is the avr proteins that elicit the HR of *N*, *Tm*, and *L* genes, respectively [22].

A new approach that introduces resistance may circumvent the resistance braking tobamoviruses. The clustered regularly interspaced short palindromic repeats (CRISPR-Cas) system is mutagenesis targeted by guide RNAs [113, 114]. The specific guide RNA has sequence homology to a desired DNA site, thus enabling the use of the CRISPR-Cas system to edit specific genome sites. Editing occurs by the introduction of double-stranded DNA breaks by the endonuclease Cas9 at a locus of interest while exploiting cellular repair mechanisms to cause high fidelity heritable genome sequence changes [113, 115]. Successful application of the Cas9 methodology was reported for the virus genera: *Geminivirus* [116], *Potyvirus*, and *Ipomovirus* [117].

The viral MP is a target for various regulatory events. During viral passage through PD, the MP is transiently synthesized, and it is degraded by 26S proteasome [118]. MP binding to viral RNA is associated with repression of viral RNA translation [119]. Phosphorylation of MP apparently has diverse regulatory consequences for MP. Phosphorylated MP no longer represses RNA translation [120], which may indicate temporal regulation of viral movement and synthesis. Kinases that phosphorylate the MP carboxy terminus stabilize the protein and promote its activity on PD dilation [73]. Interestingly the carboxy terminus of MP is susceptible to phosphorylation by the host cell wall-associated protein kinase that may be sequestering the MP to the cell wall [121]. Phosphorylation of ToMV MP at serine 37 apparently confers stabilization to the protein and is essential for its intracellular localization [122].

Suppression of posttranscriptional gene silencing (PTGS) is apparently an elementary mechanism for viral spread. Viruses are initiators of the endogenous PTGS, which degrades RNA. The viral RNA is processed into small interfering RNA (siRNA) by a drosophila Dicer-like enzyme, which is then incorporated into RNA-induced silencing complex (RISK) that degrades RNAs with sequence similarity to the siRNAs. A silencing signal, which may be the siRNAs, spreads between cells via PD and phloem [123]. This signal spread can be prevented by TMV [124]. Study of ToMV suppression of PTGS showed accumulation of siRNAs, suggested to indicate a block in PTGS downstream of siRNA production [125].

As mentioned above, indirect pathway to seed invasion is limited by meristematic exclusion. In addition, seeds apparently have an intrinsic inhibitory effect on virus infection [126].

## 7. Virus preservation in seeds

RNA viruses have high mutation rate [127] that contributes to rapid evolution dynamics which may ensure adaptation to new host plants or to stressful and fluctuating environments [128, 129]. Seed transmission may comprise a genetic bottleneck, which may reduce population size and induce the immergence of new virus strains [130]. Analysis of the effect of the vertical transmission bottleneck on *Zucchini yellow mosaic virus* (ZYMV), for example, showed that although high variability was observed in the 5′ untranslated region, the regions

http://dx.doi.org/10.5772/intechopen.70244

necessary for vector transmission were unchanged, indicating their contribution to virus spread [131]. Genetic bottlenecks may also cause reduction in virus virulence as suggested to occur in PSbMV seed transmission [132]. Interestingly, experiments on virus seed transmission carried out with CMV infecting Arabidopsis thaliana, under controlled conditions, showed that there is a reduction in virus virulence after several vertical passages [133]. Virus accumulation was reduced as well although seed transmission rate increased. These changes of the virus were concomitant to adaptation of the host plant to the evolving virus, showing an increase in vertical transmission. However, this host adaptation which favors a theory of coevolution of plants and viruses in vertical transmission exposes the cells to high virulence of non-evolved horizontally transmitted viruses [133].

## 8. Seed-transmitted tobamoviruses

Selected tobamoviruses reported to be transmitted via seeds are listed in **Table 1**. Very low transmission ratios or no transmission at all occurs in most *Tobamovirus* species although seeds are infected. Nevertheless, even a low transmission percentage may be critical for new growing areas. Seed transmission primarily occurs via transplantation procedure, commonly practiced by growers. Most of the tobamoviruses contaminate the seed coat. Study of PMMoV-contaminated pepper seeds using fluorescence probe revealed that the seed coat epidermis and parenchyma cells and the endothelium that surrounds the endosperm all are invaded by PMMoV, leaving the endosperm and the embryo clear of the virus [134, 135]. These cells are of maternal origin. This observation is unlike the reports on ToMV and CGMMV that invade the seed coat as well as the endosperm or the perisperm-endosperm envelope (PEE), respectively. Indeed the seed coat originates from maternal tissues, but the endosperm and its envelope are the result of fertilization process [136]. In most cases, washes of externally attached viruses are not sufficient to prevent seed transmission [4, 34].

CGMMV-contaminated seeds are morphologically modified, as observed by optical coherence tomography. Infected melon seeds show irregularities in the aleurone layer outside the endosperm, and cucumber seeds showed a narrower gap between the seed coat and the endosperm [137]. Interestingly, hair-like structures were observed on the infected seed surface of cucumber and muskmelon [138].

High seed transmission ratio of up to 16% was observed in transmission of *Tropical soda apple mosaic virus* (TSAMV) in tropical soda apple weeds (*Solanum viarum*) [139]. Importantly, the virus can cause leaf deformation and even plant death to the *Solanaceae* plants *Capsicum annuum* cv. Capistrano and *C. annuum* cv. Enterprise, respectively [139]. Weed preservation of tobamoviruses [140, 141] and weed seed transmission may comprise a hurdle difficult to handle by growers [10]. Interestingly, seeds treated with trisodium phosphate (TSP) did not transmit the virus, indicating that TSAMV infection occurred at the outer layers of the seed coat.

Most conspicuous are the results regarding seed transmission of *Sunn-hemp mosaic virus* (SHMV) in sunn-hemp and cowpea plants. While little viral transmission occurred through sunn-hemp seeds, up to 20% transmission ratio occurred when cowpea chlorotic spot isolate infected cowpea seeds. The virus was present in all seed parts including the embryo [142].

## 9. Seed disinfection treatments

Methods that are used in large-scale commercial seed production are mostly based on various chemical treatments—1–9% hydrochloric acid HCl, 1–5% calcium hypochlorite $Ca(OCl)_2$, 1–3% sodium hypochlorite NaOCl, tetramethylthiuram disulfide (TMTD) $(CH_3)_2NCSS_2CSN(CH_3)_2$, and the most commonly used in commercial seed production 10% trisodium phosphate (TSP) $Na_3PO_4$ mentioned above [5, 143]—which have been reported to provide satisfactory control of tobamoviruses in cucurbit [144] and solanaceous seeds (e.g., pepper [5, 145, 146] and tomato [147]). In addition to the chemical treatments, several heat treatment protocols at various temperature conditions ranging from 72 to 76°C for a minimum of 12 h up to 72 h [144] are also applied in large-scale production.

However, the recent global outbreaks of *Tobamovirus* diseases emphasize the incompetence of conventional seed disinfection treatments, which could be explained by the preservation of the virus in the inner tissues of the seeds [5]. Analyzing the efficacy of sequential treatments of 10% TSP followed by 72 h heat treatment at 72°C showed that CGMMV infectious particles are preserved in the seeds (**Figure 3(d–e)**) [34].

## 10. Diagnosis of tobamoviruses in seeds

A technique for the detection of plant viruses that relies on the serological method enzyme-linked immunosorbent assay (ELISA) [148–150] was adapted successfully for the detection of tobamoviruses in seeds. ELISA is considered to be a robust technique [151], and it enables the detection of viral CP subunits [152–159]. However, ELISA suffers from two main limitations: false-negative and false-positive results. The specificity and avidity of the antibodies used for the analysis can vary and may lead to false results. The International Seed Federation (ISF) http://www.worldseed.org/, the International Seed Testing Association (ISTA) https://www.seedtest.org/en/seed-health-methods-_content---1--1452.html, and International Seed Health Initiative for Vegetable Crops (ISHI-Veg) http://www.worldseed.org/our-work/phytosanitary-matters/seed-health/ishi-veg/#protocols adapted, developed, and validated protocols for pathogen diagnosis in seed stock materials (**Table 2**). For the tobamoviruses such as TMV, ToMV, and PMMoV or the *Potexvirus Pepino mosaic virus* (PepMV) that infect *Solanaceae*, proper testing requires 12 batches of 250 seeds each (3000 seeds in total), with detection threshold of 1:249 (1 infected seed/249 healthy seeds). For detection of cucurbit-infecting viruses such as the *Tobamovirus* CGMMV, the *Melon necrotic spot virus* (MNSV, *Carmovirus*), and *Squash mosaic virus* (SqMV, *Comovirus*), and for the detection of the PSbMV (*Potyvirus*) and PEBV (*Chlorovirus*) infecting pea seeds, 20 batches of 100 seeds each (2000 seeds in total) are required. For the latter virus species that belong to five genera (CGMMV, MNSV, SqMV, PSbMV, and PEBV), ELISA detection threshold of 1:99 (1 infected seed/99 healthy seeds) was reported and according to ISTA: "The theoretical detection limit is one seed in 100. To ensure a 95% probability that infestations of 0.15% or higher are detected, it is necessary to test 20 subsamples of 100 seeds." The efficacy of a steadfast seed diagnosis relies on several parameters: (i) random collection of seed samples from the homogenous seed lot and (ii) seed disinfection treatments that are calibrated and examined in parallel to germination assays. In commercial seed production,

| ***Tobamovirus* species (acronym)**<br>***primary hosts:** | **Sample size**<br>**Subsample size X number of subsamples** | **Assay** |
|---|---|---|
| *Tobacco mosaic virus* (TMV)<br>*Pepper/tomato | **3000**<br>250X12 | ****Bioassay:** local lesions; indicator plants<br>**Serological: ELISA**<br>**Molecular: RT-PCR/qRT-PCR** |
| *Tomato mosaic virus* (ToMV)<br>* Pepper/tomato | **3000**<br>250X12 | **ELISA**<br>****Bioassay:** local lesions; indicator plants<br>**RT-PCR/qRT-PCR** |
| *Pepper mild mottle virus* (PMMoV)<br>*Pepper/tomato | **3000**<br>250X12 | **ELISA**<br>****Bioassay:** local lesions; indicator plants<br>**Optional:** RT-PCR/qRT-PCR |
| *Cucumber green mottle mosaic virus* )<br>CGMMV(<br>*Cucurbit (cucumber, melon, watermelon) | **2000**<br>100X20 | **ELISA**<br>**Optional:** RT-PCR/qRT-PCR |
| **Other virus genera** | | |
| *Virus genus*<br>*Virus* species (acronym)<br>***Host:** | | |
| ***Potexvirus***<br>*Pepino mosaic virus* (PepMV)<br>* tomato | **3000**<br>250X12 | ELISA<br>****Bioassay:** indicator plant systemic infection in *Nicotiana benthamiana*<br>**Optional:** RT-PCR/qRT-PCR |
| ***Comovirus***<br>*Squash mosaic virus*<br>(SqMV)<br>*Cucurbit | **2000**<br>100X20 | **ELISA**<br>**Optional:** RT-PCR/qRT-PCR |
| ***Carmovirus***<br>*Melon necrotic spot virus* (MNSV)<br>*Cucurbit | **2000**<br>100X20 | **ELISA**<br>**Optional:** RT-PCR/qRT-PCR |
| ***Potyvirus***<br>*Pea seed-borne mosaic virus* (PSbMV)<br>*Pea | **2000**<br>100X20 | **ELISA** |
| ***Chlorovirus***<br>*Pea early-browning virus* (PEBV)<br>*Pea | **2000**<br>100X20 | **ELISA** |

***Primary host; **Bioassay**, e.g., *Nicotiana tabacum* "Xanthi NN"; *N. benthamiana*

**Table 2.** Recommended diagnostic ratios for seed-contamination tests.

applying conventional chemical disinfection treatments followed by extensive washes, which wash out the virus from the seed coat and decrease considerably the viral titer in the tested samples [145, 146], may end up in reducing the virus titer below the detection threshold. Seed analysis under these conditions will show false-negative results because viable viral particles still exist in the internal seed tissues, e.g., the perisperm-endosperm envelope in cucurbit seeds

[34] and the endosperm in *Solanaceae* seeds [4, 35]. The second limitation of the ELISA method, which is false-positive results, may occur in instances that show lack of correlation between the ELISA results and biological significance. The possibility of obtaining false-positive or false-negative ELISA results may lead to situations in which growers purchase treated seed lots that are contaminated. Since the ELISA assay detects only the CP subunit, it is not possible to draw a direct link between ELISA results and the status of the viral particles or the degree of infectivity de facto. It is a major obstacle even when more sensitive molecular-based methods such as reverse transcription polymerase chain reaction (RT-PCR) [3, 160, 161] or quantitative real-time polymerase chain reaction (qRT-PCR) [162–164] are applied, which detect amplified partial genome fragments. Therefore, especially when dealing with treated/disinfected commercial seed lots, the preferable scenario is to validate the ELISA-positive seed subsamples in biological assays on susceptible indicator plants in order to ensure the infectivity status of the tested seed lot before marketing or sowing.

## 11. Conclusions

The contribution of seed transmission to viral spread may be significant even under conditions of low seed-to-seedling transmission rate. The rapid spread of tobamoviruses by any mechanical contacts is reflected in the spread of the disease especially in plants that are grown trellised in protected structures. This mechanical virus spread may occur when handling seeds and transplanting seedlings. Furthermore, the modern monoculture agriculture contributes to virus buildup, preservation, and spread to new susceptible host plants which increase viral copy number leading to higher viral load in the growing area. Tobamoviruses contaminate both the seed coat and the PEE or the endosperm. The mechanism of *Tobamovirus* transmission in the seed is not quite clear, but it may follow the path suggested for the *Potyvirus* PSbMV in pea seeds. Seeds have regulatory mechanisms that may limit virus transmission, and dominant resistance genes block virus transmission. However, *Tm-2*$^2$ and L$^4$ resistance-breaking tobamoviruses have recently been spread, and new approaches for conferring resistance to cultivars against tobamoviruses are in demand. The CRISPR-Cas9 methodology offers exciting prospects and provides an alternative approach to conventional breeding for the acquisition of resistance to viruses. Hopefully, in the near future, it will be applied to *Tobamovirus* species in imported crops.

## Acknowledgements

We would like to thank Dr. Nadav Elad from the Electron Microscopy Unit, Departments of Chemical Research Support, Weizmann Institute of Science, Rehovot, Israel, for the great TEM analysis and to Victoria Reingold for the in situ immunofluorescence data of CGMMV. This work was supported by the Chief Scientist, Israel Ministry of Agriculture, and the Plant Production and Marketing Board for CGMMV initiative project number 132–1740 and *Tm-2*$^2$ Tobamo breaker in tomato plants, initiative project number 261–1159.

## Author details

Aviv Dombrovsky* and Elisheva Smith

*Address all correspondence to: aviv@volcani.agri.gov.il

Department of Plant Pathology, ARO, The Volcani Center, Rishon LeZion, Israel

## References

[1] Scholthof K-B. G. Tobacco Mosaic Virus: The Beginning of Plant Pathology. Online. 2008. APSnet Features. doi: 10.1094/APSnetFeatures-2008-0408

[2] Mayer A, Johnson J, Ivanovskiĭ DI, Beijerinck MW, Baur E. Über die Mosaikkrankheit des Tabaks (in German), Concerning the mosaic disease of tobacco. Die Landwirtschaftliche Versuchs-stationen. 1942;**32**:451-467

[3] Takeuchi S, Hamada H, Toyoda K, Suzuki K, Kiba A, Hikichi Y, et al. Discrimination between tobamoviruses and their pathotypes for L-gene-mediated resistance in green pepper (*Capsicum annuum* L.) by reverse transcription-polymerase chain reaction. Journal of General Plant Pathology. 2005;**71**(1):60-67

[4] Broadbent L. The epidemiology of tomato mosaic. XI. Seed transmission of TMV. Annals of Applied Biology. 1965;**56**:177-205

[5] Rast ATB, Stijger CCMM. Disinfection of pepper seed infected with different strains of capsicum mosaic-virus by trisodium phosphate and dry heat-treatment. Plant Pathology. 1987;**36**(4):583-588

[6] Ainsworth GC. Mosaic disease of cucumber. Annals of Applied Biology. 1935;**22**:55-67

[7] Ling KS, Li R, Zhang W. First report of *Cucumber green mottle mosaic virus* infecting greenhouse cucumber in Canada. Plant Disease. 2014;**98**(5):701

[8] Tian T, Posis K, Maroon-Lango CJ, Mavrodieva V, Haymes S, Pitman TL, et al. First report of *Cucumber green mottle mosaic virus* on melon in the United States. Plant Disease. 2014;**98**(8):1163

[9] Tesoriero LA, Chambers G, Srivastava M, Smith S, Conde B, Tran-Nguyen LTT. First report of *Cucumber green mottle mosaic virus* in Australia. Australasian Plant Disease Notes. 2015;**11**(1):1-3

[10] Dombrovsky, Aviv, Lucy TT Tran-Nguyen, and Roger AC Jones. "Cucumber green mottle mosaic virus: Rapidly Increasing Global Distribution, Etiology, Epidemiology, and Management." Annual Review of Phytopathology 2017;**55**(1):1-26

[11] Francki RIB, Hu J, Palukaitis P. Taxonomy of cucurbit-infecting tobamoviruses as determined by serological and molecular hybridization analyses. Intervirology. 1986;**26**:156-163

[12] Park G-S. Occurrence of two tobamovirus diseases in cucurbits and control measures in Korea. The Plant Pathology Journal. 2001;**17**(5):243-248

[13] Ryu KH, Min BE, Choi GS, Choi SH, Kwon SB, Noh GM, et al. Zucchini green mottle mosaic virus is a new tobamovirus; comparison of its coat protein gene with that of kyuri green mottle mosaic virus. Archives of Virology. 2000;**145**(11):2325-2333

[14] Antignus Y, Lachman O, Pearlsman A, Koren A. Containment of *Cucumber fruit mottle mosaic virus* (CFMMV) infection through roots by planting into a virus-free intermediating medium. Phytoparasitica. 2005;**33**(1):85-87

[15] Johansen E, Edwards MC, Hampton RO. Seed transmission of viruses - current perspectives. Annual Review of Phytopathology. 1994;**32**:363-386

[16] Maule AJ, Wang D. Seed transmission of plant viruses: A lesson in biological complexity. Trends in Microbiology. 1996;**4**(4):153-158

[17] Broadbent L, Fletcher JT. The epidemiology of tomato mosaic IV. Persistence of virus on clothing and glasshouse structures. Annals of Applied Biology. 1963;**52**:233-241

[18] Reingold V, Lachman O, Belausov E, Koren A, Mor N, Dombrovsky A. Epidemiological study of *Cucumber green mottle mosaic virus* in greenhouses enables reduction of disease damage in cucurbit production. Annals of Applied Biology. 2016;**168**(1):29-40

[19] Reingold V, Lachman O, Koren A, Dombrovsky A. First report of *Cucumber green mottle mosaic virus* (CGMMV) symptoms in watermelon used for the discrimination of non-marketable fruits in Israeli commercial fields. Plant Pathology. 2013;**28**:11

[20] Roberts A.G. 2014. Plant Viruses: Soil-borne. eLS. Encyclopedia of Life Sciences. John Wiley & Sons, Ltd. www.els.net; 2006

[21] Ishibashi K, Ishikawa M. Replication of tobamovirus RNA. Annual Review of Phytopathology. 2016;**54**(1):55-78

[22] de Ronde D, Butterbach P, Kormelink R. Dominant resistance against plant viruses. Frontiers in Plant Science. 2014;**5**:307

[23] Lanfermeijer FC, Warmink J, Hille J. The products of the broken Tm-2 and the durable Tm-$2^2$ resistance genes from tomato differ in four amino acids. Journal of Experimental Botany. 2005;**56**(421):2925-2933

[24] Moreira SR, Eiras M, Chaves AL, Galleti SR, Colariccio A. Caracterização de uma nova estirpe do Tomato mosaic virus isolada de tomateiro no Estado de São Paulo. Fitopatologia Brasileira, Brasília. 2003;**28**:602-607

[25] Li R, Gao S, Fei Z, Ling K-S. Complete genome sequence of a new tobamovirus naturally infecting tomatoes in Mexico. Genome Announcements. 2013;**1**(5):e00794-e00813

[26] Webster CG, Rosskopf EN, Lucas L, Mellinger HC, Adkins S. First report of tomato mottle mosaic virus infecting tomato in the United States. Plant Health Progress. 2015;**15**(2):1-2

[27] Salem N, Mansour A, Ciuffo M, Falk B, Turina M. A new tobamovirus infecting tomato crops in Jordan. Archives of Virology. 2016;**161**(2):503-506

[28] Luria N, Smith E, Reingold V, Bekelman I, Lapidot M, Levin I, et al. A new Israeli Tobamovirus isolate infects tomato plants harboring Tm-2$^2$ resistance genes. PloS One. 2017;**12**(1):e0170429

[29] Boukema IW. Allelism of genes controlling resistance to TMV in *Capsicum* L. Euphytica. 1980;**29**(2):433-439

[30] Garcia-Luque I, Ferrero M, Rodríguez J, Alonso E, De La Cruz A, Sanz A, et al. The nucleotide sequence of the coat protein genes and 3′ non-coding regions of two resistance- breaking tobamoviruses in pepper shows that they are different viruses. Archives of Virology. 1993;**131**(1-2):75-88

[31] Tsuda S, Kirita M, Watanabe Y. Characterization of a pepper mild mottle tobamovirus strain capable of overcoming the L3 gene-mediated resistance, distinct from the resistance- breaking Italian isolate. Molecular Plant-Microbe Interactions. 1998;**11**(4):327-331

[32] Genda Y, Kanda A, Hamada H, Sato K, Ohnishi J, Tsuda S. Two amino acid substitutions in the coat protein of pepper mild mottle virus are responsible for overcoming the L4 gene-mediated resistance in *Capsicum* spp. Phytopathology. 2007;**97**(7):787-793

[33] Antignus Y, Lachman O, Pearlsman M, Maslenin L, Rosner A. A new pathotype of *Pepper mild mottle virus* (PMMoV) overcomes the L 4 resistance genotype of pepper cultivars. Plant Disease. 2008;**92**(7):1033-1037

[34] Reingold V, Lachman O, Blaosov E, Dombrovsky A. Seed disinfection treatments do not sufficiently eliminate the infectivity of *Cucumber green mottle mosaic virus* (CGMMV) on cucurbit seeds. Plant Pathology. 2015;**64**(2):245-255

[35] Taylor RH, Grogan RG, Kimble KA. Transmission of tobacco mosaic virus in tomato seed. Phytopathology. 1961;**51**:837-842

[36] Maule AJ, Palukaitis P. Virus movement in infected plants. Critical Reviews in Plant Sciences. 1991;**9**(6):457-473

[37] Van Dongen JT, Ammerlaan AM, Wouterlood M, Van Aelst AC, Borstlap AC. Structure of the developing pea seed coat and the post-phloem transport pathway of nutrients. Annals of Botany. 2003;**91**(6):729-737

[38] Roberts A, Oparka K. Plasmodesmata and the control of symplastic transport. Plant, Cell & Environment. 2003;**26**(1):103-124

[39] Overall RL, Blackman LM. A model of the macromolecular structure of plasmodesmata. Trends in Plant Science. 1996;**1**(9):307-311

[40] Barclay GF, Peterson CA, Tyree MT. Transport of fluorescein in trichomes of *Lycopersicon esculentum*. Canadian Journal of Botany. 1982;**60**(4):397-402

[41] Quader H, Hofmann A, Schnepf E. Shape and movement of the endoplasmic reticulum in onion bulb epidermis cells: Possible involvement of actin. European Journal of Cell Biology. 1987;**44**(1):17-26

[42] Boevink P, Oparka K, Cruz SS, Martin B, Betteridge A, Hawes C. Stacks on tracks: The plant Golgi apparatus traffics on an actin/ER network. The Plant Journal. 1998;**15**(3):441-447

[43] Ding B, Itaya A, Woo Y-M. Plasmodesmata and cell-to-cell communication in plants. International Review of Cytology. 1999;**190**:251-316

[44] Terry B, Robards A. Hydrodynamic radius alone governs the mobility of molecules through plasmodesmata. Planta. 1987;**171**(2):145-157

[45] Angell SM, Davies C, Baulcombe DC. Cell-to-cell movement of potato virus X is associated with a change in the size-exclusion limit of plasmodesmata in trichome cells of *Nicotiana clevelandii*. Virology. 1996;**216**(1):197-201

[46] Rinne P, van der Schoot C. Symplasmic fields in the tunica of the shoot apical meristem coordinate morphogenetic events. Development. 1998;**125**(8):1477-1485

[47] Gisel A, Barella S, Hempel FD, Zambryski PC. Temporal and spatial regulation of symplastic trafficking during development in *Arabidopsis thaliana* apices. Development. 1999;**126**(9):1879-1889

[48] Ding B, Kwon MO, Warnberg L. Evidence that actin filaments are involved in controlling the permeability of plasmodesmata in tobacco mesophyll. The Plant Journal. 1996;**10**(1):157-164

[49] Wright KM, Wood NT, Roberts AG, Chapman S, Boevink P, MacKenzie KM, et al. Targeting of TMV movement protein to plasmodesmata requires the actin/ER network; evidence from FRAP. Traffic. 2007;**8**(1):21-31

[50] Lucas WJ. Plant viral movement proteins: Agents for cell-to-cell trafficking of viral genomes. Virology. 2006;**344**(1):169-184

[51] Kawakami S, Watanabe Y, Beachy RN. Tobacco mosaic virus infection spreads cell to cell as intact replication complexes. Proceedings of the National Academy of Sciences of the United States of America. 2004;**101**(16):6291-6296

[52] Liu J-Z, Blancaflor EB, Nelson RS. The tobacco mosaic virus 126-kilodalton protein, a constituent of the virus replication complex, alone or within the complex aligns with and traffics along microfilaments. Plant Physiology. 2005;**138**(4):1853-1865

[53] Su S, Liu Z, Chen C, Zhang Y, Wang X, Zhu L, et al. Cucumber mosaic virus movement protein severs actin filaments to increase the plasmodesmal size exclusion limit in tobacco. The Plant Cell. 2010;**22**(4):1373-1387

[54] Reichel C, Beachy RN. Tobacco mosaic virus infection induces severe morphological changes of the endoplasmic reticulum. Proceedings of the National Academy of Sciences. 1998;**95**(19):11169-11174

[55] Huang M, Zhang L. Association of the movement protein of alfalfa mosaic virus with the endoplasmic reticulum and its trafficking in epidermal cells of onion bulb scales. Molecular Plant-Microbe Interactions. 1999;**12**(8):680-690

[56] Atkins D, Hull R, Wells B, Roberts K, Moore P, Beachy R. The tobacco mosaic virus 30K movement protein in transgenic tobacco plants is localized to plasmodesmata. Journal of General Virology. 1991;**72**(1):209-211

[57] Kotlizky G, Katz A, van der Laak J, Boyko V, Lapidot M, Beachy RN, et al. A dysfunctional movement protein of Tobacco mosaic virus interferes with targeting of wild-type movement protein to microtubules. Molecular Plant-Microbe Interactions. 2001;**14**(7):895-904

[58] Schmitz J, Stussi-Garaud C, Tacke E, Prüfer D, Rohde W, Rohfritsch O. In situ localization of the putative movement protein (pr17) from potato leafroll luteovirus (PLRV) in infected and transgenic potato plants. Virology. 1997;**235**(2):311-322

[59] Heinlein M, Padgett HS, Gens JS, Pickard BG, Casper SJ, Epel BL, et al. Changing patterns of localization of the tobacco mosaic virus movement protein and replicase to the endoplasmic reticulum and microtubules during infection. The Plant Cell. 1998;**10**(7):1107-1120

[60] Guenoune-Gelbart D, Elbaum M, Sagi G, Levy A, Epel BL. Tobacco mosaic virus (TMV) replicase and movement protein function synergistically in facilitating TMV spread by lateral diffusion in the plasmodesmal desmotubule of *Nicotiana benthamiana*. Molecular Plant-Microbe Interactions. 2008;**21**(3):335-345

[61] Stone BA, Clarke AE (1992) Chemistry and biochemistry of (1-3)-β-Glucans. La Trobe University Press. Bundoora, Victoriane, Australia

[62] Delmer DP, Volokita M, Solomon M, Fritz U, Delphendahl W, Herth W. A monoclonal antibody recognizes a 65 kDa higher plant membrane polypeptide which undergoes cation- dependent association with callose synthase in vitro and co-localizes with sites of high callose deposition in vivo. Protoplasma. 1993;**176**(1):33-42

[63] Northcote D, Davey R, Lay J. Use of antisera to localize callose, xylan and arabinogalactan in the cell-plate, primary and secondary walls of plant cells. Planta. 1989;**178**(3):353-366

[64] Iglesias VA, Meins F. Movement of plant viruses is delayed in a β-1, 3-glucanase-deficient mutant showing a reduced plasmodesmatal size exclusion limit and enhanced callose deposition. The Plant Journal. 2000;**21**(2):157-166

[65] Bucher GL, Tarina C, Heinlein M, Di Serio F, Meins F, Iglesias VA. Local expression of enzymatically active class I β-1, 3-glucanase enhances symptoms of TMV infection in tobacco. The Plant Journal. 2001;**28**(3):361-369

[66] Wolf S, Deom CM, Beachy RN, Lucas WJ. Movement protein of tobacco mosaic virus modifies plasmodesmatal size exclusion limit. Science. 1989;**246**(4928):377-380

[67] Storms MM, Va C, De N, Schoot R, Prins M, Kormelink R, et al. A comparison of two methods of microinjection for assessing altered plasmodesmal gating in tissues expressing viral movement proteins. The Plant Journal. 1998;**13**(1):131-140

[68] Hirashima K, Watanabe Y. RNA helicase domain of tobamovirus replicase executes cell-to- cell movement possibly through collaboration with its nonconserved region. Journal of Virology. 2003;**77**(22):12357-12362

[69] Fridborg I, Grainger J, Page A, Coleman M, Findlay K, Angell S. TIP, a novel host factor linking callose degradation with the cell-to-cell movement of Potato virus X. Molecular Plant-Microbe Interactions. 2003;**16**(2):132-140

[70] Dorokhov YL, Mäkinen K, Frolova OY, Merits A, Saarinen J, Kalkkinen N, et al. A novel function for a ubiquitous plant enzyme pectin methylesterase: The host-cell receptor for the tobacco mosaic virus movement protein. FEBS Letters. 1999;**461**(3):223-228

[71] Chen MH, Sheng J, Hind G, Handa AK, Citovsky V. Interaction between the tobacco mosaic virus movement protein and host cell pectin methylesterases is required for viral cell-to-cell movement. The EMBO Journal. 2000;**19**(5):913-920

[72] Deom CM, He XZ, Beachy RN, Weissinger AK. Influence of heterologous tobamovirus movement protein and chimeric-movement protein genes on cell-to-cell and long-distance movement. Virology. 1994;**205**(1):198-209

[73] Waigmann E, Chen MH, Bachmaier R, Ghoshroy S, Citovsky V. Regulation of plasmodesmal transport by phosphorylation of tobacco mosaic virus cell-to-cell movement protein. The EMBO Journal. 2000;**19**(18):4875-4884

[74] Lee J-Y, Yoo B-C, Rojas MR, Gomez-Ospina N, Staehelin LA, Lucas WJ. Selective trafficking of non-cell-autonomous proteins mediated by NtNCAPP1. Science. 2003;**299**(5605):392-396

[75] Lewis JD, Lazarowitz SG. Arabidopsis synaptotagmin SYTA regulates endocytosis and virus movement protein cell-to-cell transport. Proceedings of the National Academy of Sciences. 2010;**107**(6):2491-2496

[76] Chen M-H, Tian G-W, Gafni Y, Citovsky V. Effects of calreticulin on viral cell-to-cell movement. Plant Physiology. 2005;**138**(4):1866-1876

[77] Wellink J, Van Lent J, Verver J, Sijen T, Goldbach R, Van Kammen A. The cowpea mosaic virus M RNA-encoded 48-kilodalton protein is responsible for induction of tubular structures in protoplasts. Journal of Virology. 1993;**67**(6):3660-3664

[78] Kasteel D, Perbal M-C, Boyer J-C, Wellink J, Goldbach R, Maule A, et al. The movement proteins of cowpea mosaic virus and cauliflower mosaic virus induce tubular structures in plant and insect cells. Journal of General Virology. 1996;**77**(11):2857-2864

[79] Dolja VV, Kreuze JF, Valkonen JP. Comparative and functional genomics of closteroviruses. Virus Research. 2006;**117**(1):38-51

[80] Oparka KJ, Turgeon R. Sieve elements and companion cells—traffic control centers of the phloem. The Plant Cell. 1999;**11**(4):739-750

[81] Santa Cruz S. Perspective: Phloem transport of viruses and macromolecules–what goes in must come out. Trends in Microbiology. 1999;**7**(6):237-241

[82] Simón-Buela L, García-Arenal F. Virus particles of cucumber green mottle mosaic tobamovirus move systemically in the phloem of infected cucumber plants. Molecular Plant-Microbe Interactions. 1999;**12**(2):112-118

[83] Spitsin S, Steplewski K, Fleysh N, Belanger H, Mikheeva T, Shivprasad S, et al. Expression of alfalfa mosaic virus coat protein in tobacco mosaic virus (TMV) deficient in the production of its native coat protein supports long-distance movement of a chimeric TMV. Proceedings of the National Academy of Sciences. 1999;**96**(5):2549-2553

[84] Roberts AG, Cruz SS, Roberts IM, Prior DA, Turgeon R, Oparka KJ. Phloem unloading in sink leaves of *Nicotiana benthamiana*: Comparison of a fluorescent solute with a fluorescent virus. The Plant Cell. 1997;**9**(8):1381-1396

[85] Cheng NH, Su CL, Carter SA, Nelson RS. Vascular invasion routes and systemic accumulation patterns of tobacco mosaic virus in *Nicotiana benthamiana*. The Plant Journal. 2000;**23**(3):349-362

[86] Oparka KJ, Cruz SS. The great escape: Phloem transport and unloading of macromolecules 1. Annual Review of Plant Biology. 2000;**51**(1):323-347

[87] Silva MS, Wellink J, Goldbach R, Van Lent J. Phloem loading and unloading of Cowpea mosaic virus in *Vigna unguiculata*. Journal of General Virology. 2002;**83**(6):1493-1504

[88] Lartey RT, Ghoshroy S, Citovsky V. Identification of an *Arabidopsis thaliana* mutation (vsm1) that restricts systemic movement of tobamoviruses. Molecular Plant-Microbe Interactions. 1998;**11**(7):706-709

[89] Lartey R, Ghoshroy S, Ho J, Citovsky V. Movement and subcellular localization of a tobamovirus in *Arabidopsis*. The Plant Journal. 1997;**12**(3):537-545

[90] Hipper C, Brault V, Ziegler-Graff V, Revers F. Viral and cellular factors involved in phloem transport of plant viruses. Frontiers in Plant Science. 2013;**4**:154

[91] Lin M-K, Lee Y-J, Lough TJ, Phinney BS, Lucas WJ. Analysis of the pumpkin phloem proteome provides insights into angiosperm sieve tube function. Molecular & Cellular Proteomics. 2009;**8**(2):343-356

[92] Batailler B, Lemaitre T, Vilaine F, Sanchez C, Renard D, Cayla T, et al. Soluble and filamentous proteins in Arabidopsis sieve elements. Plant, Cell & Environment. 2012;**35**(7):1258-1273

[93] Chen MH, Citovsky V. Systemic movement of a tobamovirus requires host cell pectin methylesterase. The Plant Journal. 2003;**35**(3):386-392

[94] Saito T, Yamanaka K, Okada Y. Long-distance movement and viral assembly of tobacco mosaic virus mutants. Virology. 1990;**176**(2):329-336

[95] Dawson WO, Bubrick P. and Grantham GL. Modifications of the tobacco mosaic virus coat protein gene affecting replication, movement, and symptomatology. Phytopathol. 1988;**78**:783-789

[96] Derrick PM, Carter SA, Nelson RS. Mutation of the tobacco mosaic tobamovirus 126-and 183-kDa proteins: Effects on phloem-dependent virus accumulation and synthesis of viral proteins. Molecular Plant-Microbe Interactions. 1997;**10**(5):589-596

[97] Deom C, Quan S, He X. Replicase proteins as determinants of phloem-dependent long-distance movement of tobamoviruses in tobacco. Protoplasma. 1997;**199**(1):1-8

[98] Ding XS, Liu J, Cheng N-H, Folimonov A, Hou Y-M, Bao Y, et al. The Tobacco mosaic virus 126-kDa protein associated with virus replication and movement suppresses RNA silencing. Molecular Plant-Microbe Interactions. 2004;**17**(6):583-592

[99] Stadler R, Lauterbach C, Sauer N. Cell-to-cell movement of green fluorescent protein reveals post-phloem transport in the outer integument and identifies symplastic domains in Arabidopsis seeds and embryos. Plant Physiology. 2005;**139**(2):701-712

[100] Wróbel-Marek, Justyna, et al. "Identification of symplasmic domains in the embryo and seed of Sedum acre L.(Crassulaceae)." Planta. 2017;**245**(3):491-505

[101] Patrick JW, Offler CE. Post-sieve element transport of photoassimilates in sink regions. Journal of Experimental Botany. 1996;**47**(Special Issue):1165-1177

[102] Roberts I, Wang D, Findlay K, Maule A. Ultrastructural and temporal observations of the potyvirus cylindrical inclusions (CIs) show that the CI protein acts transiently in aiding virus movement. Virology. 1998;**245**(1):173-181

[103] Roberts I, Wang D, Thomas C, Maule A. Pea seed-borne mosaic virus seed transmission exploits novel symplastic pathways to infect the pea embryo and is, in part, dependent upon chance. Protoplasma. 2003;**222**(1-2):31-43

[104] Wang D, Maule AJ. A model for seed transmission of a plant virus: Genetic and structural analyses of pea embryo invasion by pea seed-borne mosaic virus. The Plant Cell. 1994;**6**(6):777-787

[105] Wang D, Maule AJ. Early embryo invasion as a determinant in pea of the seed transmission of pea seed-borne mosaic virus. Journal of General Virology. 1992;**73**(7):1615-1620

[106] Wang D, Macfarlane SA, Maule AJ. Viral determinants of pea early browning virus seed transmission in pea. Virology. 1997;**234**(1):112-117

[107] Wang D, Maule AJ. Contrasting patterns in the spread of two seed-borne viruses in pea embryos. The Plant Journal. 1997;**11**(6):1333-1340

[108] Martín-Hernández AM, Baulcombe DC. Tobacco rattle virus 16-kilodalton protein encodes a suppressor of RNA silencing that allows transient viral entry in meristems. Journal of Virology. 2008;**82**(8):4064-4071

[109] Jones AL, Thomas CL, Maule AJ. De novo methylation and co-suppression induced by a cytoplasmically replicating plant RNA virus. The EMBO Journal. 1998;**17**(21):6385-6393

[110] Liu HW, Luo LX, Li JQ, Liu PF, Chen XY, Hao JJ. Pollen and seed transmission of *Cucumber green mottle mosaic virus* in cucumber. Plant Pathology. 2014;**63**(1):72-77

[111] Mandadi KK, Scholthof K-BG. Plant immune responses against viruses: How does a virus cause disease? The Plant Cell. 2013;**25**(5):1489-1505

[112] Bhattacharjee S, Zamora A, Azhar MT, Sacco MA, Lambert LH, Moffett P. Virus resistance induced by NB–LRR proteins involves argonaute4-dependent translational control. The Plant Journal. 2009;**58**(6):940-951

[113] Barrangou R, Fremaux C, Deveau H, Richards M, Boyaval P, Moineau S, et al. CRISPR provides acquired resistance against viruses in prokaryotes. Science. 2007; **315**(5819):1709-1712

[114] Brouns SJ, Jore MM, Lundgren M, Westra ER, Slijkhuis RJ, Snijders AP, et al. Small CRISPR RNAs guide antiviral defense in prokaryotes. Science. 2008;**321**(5891):960-964

[115] Mali P, Yang L, Esvelt KM, Aach J, Guell M, DiCarlo JE, et al. RNA-guided human genome engineering via Cas9. Science. 2013;**339**(6121):823-826

[116] Yin K, Han T, Liu G, Chen T, Wang Y, Yu AYL, et al. A geminivirus-based guide RNA delivery system for CRISPR/Cas9 mediated plant genome editing. Scientific Reports. 2015;**5**:14926

[117] Chandrasekaran J, Brumin M, Wolf D, Leibman D, Klap C, Pearlsman M, et al. Development of broad virus resistance in non-transgenic cucumber using CRISPR/Cas9 technology. Molecular Plant Pathology. 2016;**17**(7):1140-1153

[118] Reichel C, Beachy RN. Degradation of tobacco mosaic virus movement protein by the 26S proteasome. Journal of Virology. 2000;**74**(7):3330-3337

[119] Karpova O, Ivanov K, Rodionova N, Dorokhov YL, Atabekov J. Nontranslatability and dissimilar behavior in plants and protoplasts of viral RNA and movement protein complexes formed in vitro. Virology. 1997;**230**(1):11-21

[120] Karpova O, Rodionova N, Ivanov K, Kozlovsky S, Dorokhov YL, Atabekov J. Phosphorylation of tobacco mosaic virus movement protein abolishes its translation repressing ability. Virology. 1999;**261**(1):20-24

[121] Citovsky V, McLean BG, Zupan JR, Zambryski P. Phosphorylation of tobacco mosaic virus cell-to-cell movement protein by a developmentally regulated plant cell wall-associated protein kinase. Genes and Development. 1993;**7**(5):904-910

[122] Kawakami S, Padgett HS, Hosokawa D, Okada Y, Beachy RN, Watanabe Y. Phosphorylation and/or presence of serine 37 in the movement protein of tomato mosaic tobamovirus is essential for intracellular localization and stability in vivo. Journal of Virology. 1999;**73**(8):6831-6840

[123] Mlotshwa S, Voinnet O, Mette MF, Matzke M, Vaucheret H, Ding SW, et al. RNA silencing and the mobile silencing signal. The Plant Cell. 2002;**14**(suppl 1):S289-S301

[124] Voinnet O, Pinto YM, Baulcombe DC. Suppression of gene silencing: A general strategy used by diverse DNA and RNA viruses of plants. Proceedings of the National Academy of Sciences of the United States of America. 1999;**96**(24):14147-14152

[125] Kubota K, Tsuda S, Tamai A, Meshi T. Tomato mosaic virus replication protein suppresses virus-targeted posttranscriptional gene silencing. Journal of Virology. 2003; **77**(20):11016-11026

[126] Hajj B, Stevens W. The effects of legume seed extracts on plant virus infection. Cellular and Molecular Life Sciences. 1979;**35**(11):1460-1462

[127] Sanjuan R, Nebot MR, Chirico N, Mansky LM, Belshaw R. Viral mutation rates. Journal of Virology. 2010;**84**(19):9733-9748

[128] Hanssen IM, Mumford R, Blystad DR, Cortez I, Hasiow-Jaroszewska B, Hristova D, et al. Seed transmission of Pepino mosaic virus in tomato. European Journal of Plant Pathology. 2010;**126**(2):145-152

[129] Lauring AS, Frydman J, Andino R. The role of mutational robustness in RNA virus evolution. Nature Reviews Microbiology. 2013;**11**(5):327-336

[130] Ali, A. and Roossinck, M.J. Genetic bottlenecks. In: Plant Virus Evolution (Roossinck, M.J. ed.), Berlin: Springer. 2008. pp. 123-131

[131] Simmons H, Dunham J, Zinn K, Munkvold G, Holmes E, Stephenson A. Zucchini yellow mosaic virus (ZYMV, Potyvirus): Vertical transmission, seed infection and cryptic infections. Virus Research. 2013;**176**(1):259-264

[132] Fabre F, Moury B, Johansen EI, Simon V, Jacquemond M, Senoussi R. Narrow bottlenecks affect pea seedborne mosaic virus populations during vertical seed transmission but not during leaf colonization. PLoS Pathogens. 2014;**10**(1):e1003833

[133] Pagán I, Montes N, Milgroom MG, García-Arenal F. Vertical transmission selects for reduced virulence in a plant virus and for increased resistance in the host. PLoS Pathogens. 2014;**10**(7):e1004293

[134] Genda Y, Sato K, Nunomura O, Hirabayashi T, Ohnishi J, Tsuda S. Immunolocalization of Pepper mild mottle virus in *Capsicum annuum* seeds. Journal of General Plant Pathology. 2005;**71**(3):238-242

[135] Genda Y, Sato K, Nunomura O, Hirabayashi T, Tsuda S. Immunolocalization of Pepper mild mottle virus in developing seeds and seedlings of *Capsicum annuum*. Journal of General Plant Pathology. 2011;**77**(3):201-208

[136] Ramakrishna P, Amritphale D. The perisperm-endosperm envelope in *Cucumis*: Structure, proton diffusion and cell wall hydrolysing activity. Annals of Botany. 2005;**96**(5):769-778

[137] Lee SY, Lee C, Kim J, Jung HY. Application of optical coherence tomography to detect *Cucumber green mottle mosaic virus* (CGMMV) infected cucumber seed. Horticulture Environment and Biotechnology. 2012;**53**(5):428-433

[138] Lee SY, Win NKK, Cho DM, Lee SH, Jung HY. *Cucumber green mottle mosaic virus* (CGMMV) can induce hair-like tissues on genus *Cucumis* seeds. Scientia Horticulturae. 2012;**146**:76-80

[139] Adkins S, Kamenova I, Rosskopf EN, Lewandowski DJ. Identification and characterization of a novel tobamovirus from tropical soda apple in Florida. Plant Disease. 2007;**91**(3):287-293

[140] Boubourakas I, Hatziloukas E, Antignus Y, Katis N. Etiology of leaf chlorosis and deterioration of the fruit interior of watermelon plants. Journal of Phytopathology. 2004;**152**(10):580-588

[141] Shargil D, et al. "New weed hosts for Cucumber green mottle mosaic virus in wild Mediterranean vegetation." European Journal of Plant Pathology. 2017;**148**(2):473-480

[142] Van Regenmortel, Marc HV, and Heinz Fraenkel-Conrat, eds. The Plant Viruses: The Rod-shaped Plant Viruses. Springer Science & Business Media, 2013;**13**:249-264

[143] Herrera-Vásquez JA, Córdoba-Sellés MC, Cebrián MC, Alfaro-Fernandez A, Jordá C. Seed transmission of Melon necrotic spot virus and efficacy of seed-disinfection treatments. Plant Pathology. 2009;**58**:7

[144] Kim SM, Nam SH, Lee JM, Yim KO, Kim KH. Destruction of *Cucumber green mottle mosaic virus* by heat treatment and rapid detection of virus inactivation by RT-PCR. Molecules and Cells. 2003;**16**(3):338-342

[145] Demski J. Tobacco mosaic virus in seedborne in pimiento peppers. Plant Disease. 1981;**65**(9):723-724

[146] Tošić M, ŠUtić D, Pešić Z. Transmission of tobacco mosaic virus through pepper (*Capsicum annuum* L.) seed. Journal of Phytopathology. 1980;**97**(1):10-13

[147] Cordoba-Selle MDC, Garcia-Randez A, Alfaro-Fernandez A, Jorda-Gutierrez C. Seed transmission of *Pepino mosaic virus* and efficacy of tomato seed disinfection treatments. Plant Disease. 2007;**91**(10):1250-1254

[148] Clark MF, Adams MJ. Characteristics of the microplate method of enzyme-linked immunosorbent assay for the detection of plant viruses. Journal of General Virology. 1977;**34**:475-483

[149] Clark MF. Immunosorbent assays in plant pathology. Annual Review of Phytopathology. 1981;**19**:83-106

[150] Barbara DJ, Clark MF. A simple indirect ELISA using F(ab')2 fragments of immunoglobulin. Journal of General Virology. 1982;**58**:315-322

[151] Sutula CL, Gillett JM, Morrissey SM, Ramsdell DC. Interpreting elisa data and establishing the positive-negative threshold. Plant Disease. 1986;**70**(8):722-726

[152] Tuitert G. Assessment of the inoculum potential of polymyxa-betae and beet necrotic yellow vein virus (Bnyvv) in soil using the most probable number method. Netherlands Journal of Plant Pathology. 1990;**96**(6):331-341

[153] Choi GS. Occurrence of two tobamovirus diseases in cucurbits and control measures in Korea. The Plant Pathology Journal. 2001;**17**(5):243-248

[154] de Assis Filho F, Sherwood J. Evaluation of seed transmission of Turnip yellow mosaic virus and Tobacco mosaic virus in *Arabidopsis thaliana*. Phytopathology. 2000;**90**(11): 1233-1238

[155] McGovern R, Polston J, Mullahey J. Solanum viarum: Weed reservoir of plant viruses in Florida. International Journal of Pest Management. 1994;**40**(3):270-273

[156] Lecoq H, Desbiez C. Viruses of cucurbit crops in the Mediterranean region: An ever-changing picture. Advances in Virus Research. 2012;**84**:67-126

[157] Ford R, Beczner L, Hamilton R. Turnip, cucumber, and ribgrass mosaic viruses isolated from *Hesperis matronalis* in British Columbia. Plant Disease. 1988;**72**(2):101-106

[158] Webster CG, Wylie SJ, Jones MGK. Diagnosis of plant viral pathogens. Current Science. 2004;**86**(12):1604-1607

[159] Li JX, Liu SS, Gu QS. Transmission efficiency of *Cucumber green mottle mosaic virus* via seeds, soil, pruning and irrigation water. Journal of Phytopathology. 2015;**5**(5):300-309

[160] Maroon C, Zavriev S, editors. PCR-based tests for the detection of tobamoviruses and carlaviruses. X International Symposium on Virus Diseases of Ornamental Plants. 2000;**568**

[161] Kumar S, Udaya Shankar A, Nayaka S, Lund O, Prakash H. Detection of Tobacco mosaic virus and Tomato mosaic virus in pepper and tomato by multiplex RT–PCR. Letters in Applied Microbiology. 2011;**53**(3):359-363

[162] Schoen CD, Knorr D, Leone G. Detection of potato leafroll virus in dormant potato tubers by immunocapture and a fluorogenic 5' nuclease RT-PCR assay. Phytopathology. 1996;**86**(9):993-999

[163] Fang Y, Wu WH, Pepper JL, Larsen JL, Marras SAE, Nelson EA, et al. Comparison of real-time, quantitative PCR with molecular beacons to nested PCR and culture methods for detection of *Mycobacterium avium* subsp paratuberculosis in bovine fecal samples. Journal of Clinical Microbiology. 2002;**40**(1):287-291

[164] Henson JM, French R. The polymerase chain-reaction and plant-disease diagnosis. Annual Review of Phytopathology. 1993;**31**:81-109

Section 4

# Seed Applications

Chapter 13

# Biology of Seed Vigor in the Light of *-omics* Tools

Isaac Oludayo Daniel

Additional information is available at the end of the chapter

http://dx.doi.org/10.5772/intechopen.71258

**Abstract**

Seed vigor is a major agronomic trait measurable by seed longevity in storage, germination capacity, and seedling growth in the field. Seed vigor has potential to significantly elevate crop resilience to biotic and abiotic stresses. That is important for crop yields enhancement and other enterprises that involve seeds like plant breeding, research and education, germplasm conservation and the seed trade. With the availability of high precision *-omics* tools for biological research, lots of investigations are undertaken globally to answer the physiological questions underlying seed germination and invigoration. The increasing *-omics* datasets constitute important resources for the delivery of new seed vigor markers and advancing new seed vigor manipulation opportunities. There is need to regularly update the knowledge generated from these investigations for the scientific improvement of seed vigor. Thus, this chapter highlights the biological backgrounds involved in the development of seed vigor traits in the light of modern *-omics* tools. The chapter is sectioned into; 1. Attributes of seed vigor and the *–omics* sciences; 2. State of *-omics*-based knowledge on underlying mechanisms of seed vigor; 3. Future perspectives of *-omics* application to genetic engineering of seed vigor with an insight to the latest technique of genome editing, the CRISPR-Cas9 technology.

**Keywords:** seed vigor, seed aging, seed priming, molecular mechanisms, *-omics* application

## 1. Introduction

Seeds constitute the basic biological input for crop production. The most important potential seed attribute directly affecting crop productivity is the seed vigor, because a good crop stand establishment is required to deliver the genetic and yield potentials of the seed. Thus, seed vigor had been a target trait of economic and ecologic values in crop improvement projects since the green revolution era [1]. Scientific manipulations to innate seed vigor of crops can be a key to increase crop yields per unit area because it can improve crop resilience against climate change effects and biotic impediments to crop yields. More importantly, it will promote low-input agriculture by minimizing crop production inputs e.g. fertilizers, pesticides etc. that increases the environmental footprints of agriculture. However, seed vigor is a complex trait for genetic manipulation since it involves multiple physiological parameters and metabolic events of water uptake by the mature dry seeds to produce morphological events of radicle protrusion (germination) and seedling growth and development [2, 3]. Hence, exploring the fundamental biological processes underlying the trait at the various levels of these events is of intense scientific research interest.

In many recent seed biology literature, seed vigor studies engage modern biology tools using seed deterioration/invigoration experimental models. In the context of seed deterioration, seed vigor is often measured as seed longevity in storage which is relevant to seed conservation and seed bank operation efforts [4, 5]. The processes involved in the loss of seed vigor during seed deterioration elucidate the complex biological phenomenon of seed vigor. Thus, studies on the mechanism of seed vigor involve storing seeds under conditions that accelerate aging, e.g. the accelerated aging (AA) test [6] or the controlled deterioration (CD) tests to simulate the seed aging or seed vigor loss processes. In the context of seed invigoration, seed vigor has been widely studied towards increasing the performance of commercial seed lots. Seed priming is one of the most acknowledged technology-based seed invigoration treatments, which is mainly controlled imbibition of the seeds followed by dehydration back to their initial water content [7–9]. Priming treatments are hypothesized as kick-starting physiological and biochemical processes of seed germination, thus giving the treated seeds, a head-start which should increase resilience, reduce the time it takes seeds to sprout and elicit uniformity of seedling growth. All these can be achieved because primed seeds sort of "memorize" the metabolic signals they acquired from the priming process when stimulated to germinate later [10]. Many studies have proved the phenotypic advantages of priming in terms of early germination, seedling vigor [11] and stress tolerance [12]. The metabolic mechanisms underpinning priming are still been actively investigated, providing veritable data resources for scientific improvement of seed vigor. As will be discussed later in this chapter, many of the published reports suggest that cellular repair, detoxification and induction of protective proteins are the mechanisms underlying the seed vigor process [3, 13–16].

The advances in the science of biology brought a new phrase termed the *-omics* [17]. The *-omics* tools are essentially a hybrid of biological technologies encompassing liquid chromatography-mass spectrometry (LC-MS) and next-generation sequencing (NGS) [18]. They include all genomic and post-genomic approaches which in recent years have been further contributing to identifying genes and understanding their functions [19]. *-Omics* serves as an informal suffix to prefixes of specific biological fields of study such as genomics for genetics *-omics*, proteomics for protein *–omics*, etc. (**Figure 1**). In general, all *-omics* science aim at the collective characterization and quantification of pools of biological molecules that translate into the structure, function, and dynamics of living organisms [19].

With reference to seed vigor, several *-omics* dissections have been reported [20]. Here is a quick run of definitions of key *-omics* sciences that have been applied to seed vigor. *Genomics*, which is the study of the genome of an organism, have been severally used to dissect genomic regions that have significant genetic effects on seed vigor of many crops [21]. *Proteomics* is the study of the entire complement of proteins, modifications made to a particular set of proteins and their functions in the cell. Unlike the genome, which is fixed for most cells, the proteome is known to be dynamic, changing with internal or external (environmental) stimuli. The proteomics of seed vigor of many plant species have been widely reported [22]. Transcriptome involves all set of RNA molecules—mRNA, rRNA, tRNA, and other non-coding RNAs, produced in one or a population of cells, which is important in gene expression. *Transcriptomics* of seed vigor traits has been studied; Soeda et al. [12] used the microarrays technique to dissect the expression of seed vigor related genes during the priming of *Brassica oleracea* seed lots while Dinkova et al. [23] used the RNA-Seq, method to study translation initiation factors during maize seed germination. *Metabolomics* is the scientific study of chemical processes and metabolites (chemical fingerprinting). A related scientific branch to metabolomics that is well researched for seed vigor traits is *Metabonomics*, the quantitative measurement of the dynamic, multi-parametric metabolic responses to pathological and physiological stimuli or genetic modifications. Examples of metabolomic and metabonomic dissections of seed vigor are the investigations of metabolic pathways and signaling involved in seed vigor. Lipidome is the entire complement of cellular lipids, and *lipidomics* has been involved in studies of lipids and glycolipids and lipid metabolites during seed deterioration or invigoration [24]. *Glycomics* is the analyses of the glycome—collection of cellular glycans (sugars and carbohydrates) of an organism. A glycomics study that associated seed vigor with sugar metabolites involved has helped to identify carbohydrate and sugar biomarkers for seed vigor [25]. Most of the published studies utilized one or more genomic tools, which explains why so much seed vigor -*omics* references has been categorized under genomics (**Figure 1**).

With the advent of genomic and post-genomic technologies, high-throughput analyses of molecular profiles have been implemented at the protein, RNA and metabolite levels to dissect the biological processes involved in seed vigor development. The aim of this chapter is to review advances in seed vigor studies in the light of new high precision -*omics* tools. In this

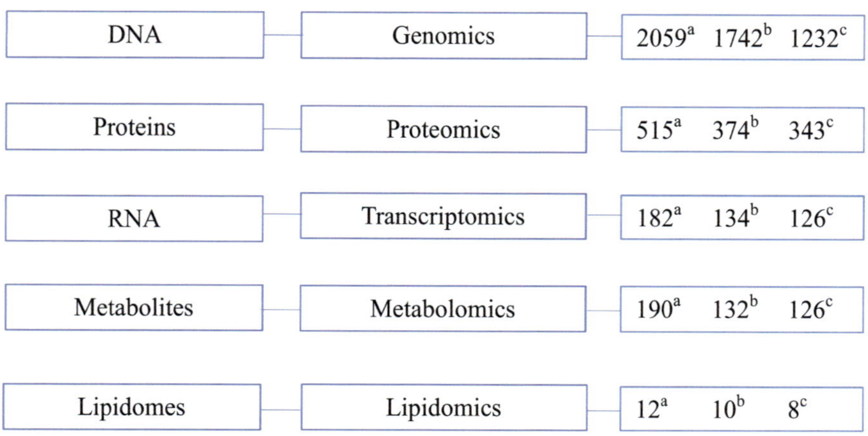


**Figure 1.** References of -*omics* sciences reported on seed vigor related traits in popular library databases as at July 2017. (a) Science Citation Index (expanded web of science), (b) ProQuest, and (c) PubMed.

chapter, I will be discussing some of the major advances made with specific *-omics* technologies towards dissecting the complex seed vigor traits in various crop plant species. I will also focus on the direction to which the current advances in seed vigor *-omics* might be pointing the seed industry in the near future. In discussing all the sections, I will pay attention to seed vigor reports from the experimental perspectives of seed deterioration (aging) and seed invigoration (priming).

## 2. *-Omics* technologies for seed aging/priming

### 2.1. Development of seed quality biomarkers

Rajjou et al. [20] argued that characterizing biomarkers of seed vigor as a component of breeding programs is an important strategy for producing seeds of the highest possible quality, particularly under the environmental stresses occasioned by climate change and increasing world population. Thus developing bio-markers for seed aging and vigor loss traits are the first notable achievements of various seed vigor *-omics* projects. **Table 1** summarizes key *-omics*-based biomarkers of seed aging and/or vigor that have gained significant attention in the seed research community.

Advances in *-omics* sciences have made the acquisition of seed aging signals feasible so that a number of sensitive and effective biomarkers has been developed to identify aging signals and evaluate aging status [26]. Fu et al. [27] summarized *-omics* based biomarkers for seed aging signals of 17 species and classified them into six categories namely: molecular, biochemical, physiological, metabolic, mitochondrial and morphological signals (**Table 1**). Catusse et al. [15] used comparative proteomics to identify 18 proteins during seed priming, germination and aging. Bentsink et al. [28] showed that the *DOG1* gene that controls seed dormancy in Arabidopsis is also a biomarker for seed longevity since the mutations within the *DOG1* gene, specifically were associated with a seed longevity phenotype. Prieto-Dapena et al. [29] found that transgenic Arabidopsis seeds that over-accumulate a heat stress transcription factor exhibit a heat shock protein (*HSP*) biomarker for enhanced longevity. Whereas Devaiah et al. [24] reported a high level of a membrane lipid-hydrolyzing phospholipase-D (*PLDα1*) as a biomarker for reduced seed longevity. Kranner et al. [30] proposed the concept of half-cell reduction potential ($E_{GSSG/2GSH}$) of glutathione (*GSH*) an antioxidant that scavenge ROS which increases to more oxidizing values during viability loss as biomarker signaling cascades that trigger cell death. Nagel et al. [5] engaged the *GSH* redox method as a biomarker for seed aging in barley. The activity of the protein L-isoaspartyl methyltransferase (*PIMT*), an enzyme repairing abnormal L-isoaspartyl residues in aging proteins of Arabidopsis is increasingly becoming a common biomarker in *-omics* both seed deterioration (aging) and seed invigoration (priming) studies in a number of crop plants [13, 14, 16, 22, 31] (**Table 1**).

Reports on seed priming-based *-omics* experiments have also been sources of seed vigor biomarkers. Chen et al. [32] reported protein profiling of spinach (*Spinacia oleracea* cv. Bloomsdale) seeds during priming with −0.6 MPa PEG at 15°C. The results showed two groups of proteins: Type I *i.e.* 37 and 35-kDa proteins which are major proteins in unprimed seeds gradually

http://dx.doi.org/10.5772/intechopen.71258

| Crop | Biomarkers | Description | Reference |
|---|---|---|---|
| Seed aging studies | | | |
| Arabidopsis | *DOG1* | Delay of germination | Bentsink et al. [28] |
| Arabidopsis | *HSFA9* | Heat stress transcription factor | Prieto-Dapena et al. [29] |
| Arabidopsis | *PLDα1* | Phospholipase D-alpha (membrane lipid enzyme) | Devaiah et al. [24] |
| Arabidopsis | *PIMT1* | l-Isoaspartyl methyltransferase (Repair) | Ogé et al. [22] |
| Maize | *eIF4E/ eIF(iso)4E* | Translation initiation factor | Dinkova et al. [23] |
| Chickpea | *PIMT2* | l -Isoaspartyl methyltransferase (Repair) | Verma et al. [13] |
| Barley | *GSH* | Glutathione di-sulfide (antioxidant) | Nagel et al. [5] |
| Seed priming studies | | | |
| Brassica | *AT5G06760* | Type 1 LEAs | Soeda et al. [12] |
| Spinach | LEAs | Stress tolerance | Chen et al. [32] |
| Maize/Spinach | *18S* | HGK | Chen et al. [33] |
| Rice | *PIMT1* | l-Isoaspartyl methyltransferase (Repair) | Wei et al. [16] |
| Rice | *PIMT2* | l-Isoaspartyl methyltransferase (Repair) | Petla et al. [14] |

**Table 1.** Biomarkers of that have gained attention in ***-omics'*** dissection of seed aging/vigor.

depleting as priming progresses, and Type II such as ~20 kDa doublets proteins having an accumulation patterns opposite to Type I during priming. The depletion and/or accumulation of Type I and Type II proteins during germination constitute biomarkers for seed vigor behaviors for the species. The type I proteins seed vigor biomarkers were seed maturation proteins, such as late embryogenesis abundant (LEA) or dormancy related proteins (e.g. short-chain dehydrogenase). The depletion of these proteins during germination and priming was also reported to mark seed vigor genes at transcription and translation levels in *B. oleracea* [12].

Chen et al. [33] identified a set of transcriptomic biomarkers for seed germination and vigor from quantitative real-time polymerase chain reaction (qRT-PCR) gene expression studies during germination of maize and spinach seeds. Since seed germination involves seeds transiting from dry and physiologically inactive state to hydrated and active state, the expression of house-keeping reference genes (HKG) may alter during the transition. From the study, the HKGs identified as valid reference genes and hence seed vigor biomarkers were *Actdf, UBQ, βtub, 18S, Act*, and *GAPDH*. The HKG *18S* notably maintained stability through the transition state and was stable for both maize and spinach.

As the biomarker capabilities of *PIMT* genes has been reported for seed aging studies, likewise some seed priming studies have reported PIMT as biomarkers for seed vigor [16, 34]. The potential importance of *PIMT* as one key candidate seed vigor biomarker will be discussed later in this chapter.

## 2.2. -*Omics* of regulatory mechanisms for seed aging/priming

Studies from the '70s and '80s identified the roles of regulatory hormones like ABA and GAs in seed germination control through mutations in Arabidopsis [35, 36]. With the advent of *-omics* methodologies, a significant in-depth understanding of regulatory mechanisms signaling seed deterioration and invigoration has been gained and applies to crops [37, 34]. Changes in specific sequences of highly polymorphic genetic markers in aging rice [38], tomatoes [39] and wheat seeds [40] provide hints of molecular hints on genetic influences behind seed aging.

Scanning through recent molecular studies on seed deterioration or invigoration, the mechanisms regulating seed vigor can be summarized into three systems: repair, protection, and detoxification systems [3, 23, 34, 41–44]. Research advances in manipulation of the cellular repair system for seed invigoration has been more pronounced, often intertwining with detoxification system research. The most apparently forward-looking cellular repair studies came out of the search for mechanisms underpinning the extra-ordinary longevity of sacred lotus (*Nelumbo nucifera*) seeds, which was found to be due to the repair activities of abnormal L-isoaspartyl residues accumulated in proteins during seed aging by *PIMT* in Arabidopsis [31]. PIMT combats protein mis-folding resulting from l-isoaspartyl formation by catalyzing the conversion of abnormal l-isoaspartyl residues to their normal l-aspartyl forms thus repairing an enzyme system which likely works with other anti-aging pathways to eliminate deleterious protein products and enable successful seedling establishment in the phenotypes [22, 31]. Studies of the role of *PIMT* in seed invigoration and longevity enhancement has extended to other crops, most of them reporting similar results. An immuno-localization study on rice concluded that the distinct *OsPIMT* isoform expression in embryo and aleurone layers of transgenic rice revealed its role in the restriction of deleterious isoAsp and age-induced ROS accumulation to improve seed vigor and longevity [14]. *PIMT* and *PIMT2* contains two genes (*At3g48330* and *At5g50240*) encoding protein-L-isoaspartate methyltransferase located on chromosome 5 and produces two proteins differing by three amino acids reported in Arabidopsis [22, 31], chickpea [13] and rice [16, 14]. The activities of two important *PIMT* coding genes (*At3g48330* and *At5g50240*) are gradually forming the bedrock for seed aging *-omics* and genetic engineering of seed vigor in many crop species [13, 22, 34].

Several *-omics* studies have generated information on the protective regulatory mechanism of seed aging/vigor. Protein and enzymes regulatory systems that are active in structural, membrane and genomic integrity were engaged in most of the studies that were attempting to dissect the protective system for seed invigoration [24, 29, 45]. Transcriptomics analysis of *de-novo* protein synthesis during priming-enhanced seed germination had shown the expression of aquaporins (AQPs) in abundance [46, 47]. AQPs are plasma membrane proteins known to regulate water transport, since they are mostly expressed in hydrated seeds. AQPs are either plasma membrane intrinsic proteins (−PIPs) or tonoplast intrinsic proteins (−TIPs) serving as water channels in membranes that control cell-to-cell water movement, plant cell expansion and organ development. The expression of four spinach (*S. oleracea*) AQP coding genes (SoPIP1;1, SoPIP1;2, SoPIP2;1, and SoδTIP) during osmopriming and germination under chilling drought and optimal conditions were investigated by [46]. The up-regulation of the four genes within 2–4 days of priming (phase II-imbibition) suggests that these proteins are essential for radicle protrusion and subsequent progress of seed germination vigor. The expression of vacuolar aquaporin genes increases a thousand times after the initiation of cell elongation in both orthodox and recalcitrant seeds [47]. During priming of *Beta vulgaris* L. (sugarbeet) seeds, Catusse et al. [15] used comparative proteomics to reveal 18 proteins exhibiting up-regulation during priming and down-regulation during aging and up-regulation again upon priming of the aged seeds. In the study, six translation initiation factors were found among the proteins exhibiting the highest levels of up-regulation upon priming the aged seeds, highlighting the roles of stored mRNAs and *de-novo* synthesized mRNAs in seed vigor protection regulatory mechanism. Proteomic analysis of sugarbeet seeds led to the identification of 758 proteins whose metabolic status in seed longevity protection can be inferred and reconstructed in further details [15]. Dinkova et al. [23] streamlined the translational control of seed germination in maize using the ratio of two cap binding proteins (*eIF(iso)4E* to *eIF4E*) in the corresponding eIF4F complex, *eIF(iso)4E* being more abundant in dry seeds and both cap-binding proteins being present at similar levels following 24-hour seed imbibition. Furthermore, Prieto-Dapena et al. [29] found that over-accumulation of heat stress transcription factor (HSPs) enhanced seed longevity in transgenic Arabidopsis seeds. Regente et al. [41] reported regulation of phospholipid accumulation in extracellular fluids of sunflower during priming and seed germination. Devaiah et al. [24] reported that the ablation of the gene for a membrane lipid-hydrolyzing phospholipase D (PLD$\alpha$1) in Arabidopsis enhanced seed germination and oil stability after storage or exposure of seeds to adverse conditions. The *PLD*$\alpha$*1*-deficient seeds exhibited a smaller loss of unsaturated fatty acids and lower accumulation of lipid peroxides than did wild-type seeds. However, *PLD*$\alpha$*1*-knockdown seeds were more tolerant of aging than were *PLD*$\alpha$*1*-knockout seeds. The results demonstrate the *PLD*$\alpha$*1* plays an important role in seed deterioration and aging in Arabidopsis. A high level of *PLD*$\alpha$*1* is detrimental to seed quality, and attenuation of *PLD*$\alpha$*1* expression has the potential to improve oil stability, seed quality and seed longevity.

Cellular detoxification mechanisms have been widely viewed as an important mechanism for seed invigoration. Several gene activities have been identified that controls these mechanisms [48, 49]. Detoxification genes/proteins that scavenge ROS are the most investigated system for mechanism for seed vigor enhancement. For example, Nagel et al. [5] linked seed aging to genetic backgrounds that regulate the production of ROS-scavenging antioxidants which are known to detoxify aging cells to enhance vigor in a similar fashion to cellular repair mechanisms. Antioxidants such as glutathione (GSH), tocochromanols and ascorbic acid scavenge ROS [42]. Decreases in the antioxidant capacity of GHS under continuous accelerated aging stress increase ROS, shifting the antioxidant redox state towards more oxidizing conditions. In agreement with this concept, the glutathione half-cell reduction potential (*EGSSG/2GSH*) increases to oxidizing values during viability loss, which is assumed to initiate further signaling cascades that trigger cell death [30]. The accumulation of oxidative damage in seeds was correlated with seed vigor loss [26]. At the molecular level, the process of carbonylation, in other words, increased protein oxidation often induces loss of functional properties of target seed proteins or enzymes thus increasing their susceptibility to proteolysis. Since the presence of ROS attacks proteins by oxidizing them, the important role of antioxidant systems through detoxification and protection of upstream mechanisms to maintain seed vigor is underscored.

### 2.3. Mapping the genes controlling seed aging/vigor

The use of molecular markers in modern plant breeding to increase selection efficiency through mapping genes to specific traits of interest was made possible by *-omics* precision tools. For many simply inherited traits of economic importance, fine-mapping and tagging with closely linked or gene-specific markers is straightforward simple. However, seedling vigor in crop plants is a complex quantitative trait under the control of large genotype and environment (GxE) effects. Hence, the advent of genomics tools for mapping and analyzing quantitative trait loci (QTL) is a major breakthrough for breeding seed vigor traits and gene identification for further experimentation. From many seed deterioration experiments, QTLs of seed longevity traits like $LD_{50}$ in Arabidopsis seeds [21], germination of aged wheat seeds [48, 49] and half-life ($P_{50}$) of aging barley seeds [50] have been mapped and linked to various genes. For the germination vigor of seeds, several seed priming experiments have found QTLs for germination of maize seeds [51] and QTLs for 30 vigor traits of rice seeds [52] to mention a few. These studies also provided useful information on chromosome regions and putative genes controlling various seed vigor traits in different crops.

QTL work on seed vigor began with the pioneering work of Clerkx et al. [21] on the model plant Arabidopsis, where QTL mapping was used to identify the loci controlling various aspects of seed longevity during storage and germination. Genotyping a recombinant inbred line population with 65 PCR-based markers and seed $LD_{50}$ of phenotypic marker erecta, they identified three QTLs affecting seed longevity after controlled deterioration on chromosomes 1, 3, and 4 for Arabidopsis. Nagel et al. [50] also reported large QTL effects associated with seed half-life ($P_{50}$) on chromosomes 5 and 7 in a doubled haploid mapping population of barley. Han et al. [51] found 65 QTLs in two maize populations mapped using single-nucleotide polymorphism (SNP) markers to four seed vigor traits under four germination treatment conditions. Integrating the QTLs into 18 meta-QTLs (mQTLs), 23 candidate genes associating with seed vigor phenotype coincides with 13 mQTLs controlling protein metabolism and the glycolytic pathway. They reported four seed vigor hotspots on chromosome regions for mQTL2, mQTL3-2, mQTL3-4, and mQTL5-2 with large QTL effects under various germination environments. There are a number of recent QTL studies on seed vigor of rice [52, 53]. Singh et al. [52] reported seed germination capacity of primed rice seeds derived from 253 $BC_3F_4$ lines of crosses between Swarna and Moroberekan, phenotyped for early vigor and genotyped with 194 SNP markers. They identified six seed vigor genomic regions on chromosomes 3, 4, 5, and 6 [52]. Two of the QTL regions namely chr3 (*id3001701-id300833*) and chr5 (*wd5002636-id5001470*) were identified and tagged QTL hotspots because they were expressed consistently in field and glasshouse conditions. In the chr3 hotspot, most of QTLs identified for early vigor-related traits were $qEV_{3.1}$, $qEUE_{3.1}$, $qSHL_{3.1}$, $qSL_{3.1}$, $qSFW_{3.1}$, $qTFW_{3.1}$, $qRDW_{3.1}$ associated with early vigor, early uniform emergence, shoot length, stem length, shoot fresh weight, total fresh weight and root dry weight respectively. The QTL hotspot on chr5 includes almost similar seed vigor traits as the first hotspot except total fresh weight and root dry weight but includes seed dry weight ($qSDW_{5.1}$) and total dry weight ($qTDW_{5.1}$).

From these QTL regions identified in the brief review above, putative candidate genes associated with many seed vigor traits in the hotspot QTL regions have been published for crops like wheat [35], maize [38] and rice [40]. Besides, Carrera et al. [54] used gene expression profiling -*omics* method to produce a list of candidate genes that signify seed germination which was used to produce TAGGIT, a spreadsheet based seed specific gene ontology that describes the seed germination signature. Other seed specific genomic resources for seed vigor are: PageMan/MapMan package which visualizes transcriptome changes in *Arabidopsis* [55] seeds during germination, and SeedNet which describes transcriptional interactions for seed vigor regulation [56].

## 3. Future perspectives for seed vigor improvement through -*Omics* results

This review has highlighted key advances provided by various -*omics* platforms for the dissection of the complex trait called seed vigor. With the current advances in -*omics* applications to seed vigor biomarkers, understanding of the regulatory mechanisms, gene mapping to traits, and genomic database resources for seed vigor, a unique platform for genetic manipulation of seed traits is emerging. We are moving towards a revolution of crop production that explores the complex traits of seed vigor for enhanced productivity in the face of environmental challenges, increasing human population and rising intensity of costs and land resources for food production.

The work of Xu et al. [31] on *PIMT* encoded genes (*PIMT1* and *PIMT2*), which display distinct expression patterns but similar biochemical properties of repairing IsoAsp accumulation in seed proteins is blazing a trail which researchers have validated for a number of crops [14, 16, 22]. For example, Rajjou et al. [20] confirmed that transgenic *Arabidopsis* seeds over-expressing *NnMT2a* and *NnMT3* displayed a remarkably improved resistance to accelerated aging treatment, indicating their significant roles in seed germination vigor. Wei et al. [16] worked on one of the two *PIMT* genes from rice (*Oryza sativa* L.) and found that over-expression of *OsPIMT1* in transgenic rice seeds reduced the accumulation of isoAsp-containing protein in embryos, and increased embryo viability. Petla et al. [14] also reported that transgenic rice constitutively over-expressing *OsPIMT1* and *OsPIMT2* exhibited improved seed vigor and longevity. These data indicated that engineering *OsPIMT*-related seed longevity improvement is a feasible option for producing enhanced vigor GMO seeds through target-gene methods. A way forward from understanding the clear role of *PIMT* in seed vigor improvement is the application of these findings for genetic engineering of *PIMT* towards improving seed vigor. Wu et al. [34] summarized current knowledge on *PIMT* gene modifications, specific genetic engineering methodologies and their outcomes for seed vigor improvement in three different crops (**Table 2**). Altering *PIMT* accumulation in seeds shows various effects of physiological significance in the various studies opening opportunities for genetically manipulating seed vigor. While *PIMT* offer opportunity for producing high vigor seeds through the repair mechanism, other candidate genes utilizing alternative strategies to producing seed vigor phenotypes are also waiting to be explored. Examples are the LEA and HSP proteins that use the protective gene mechanism, the detoxification mechanism that uses the ROS scavenging gene action [43] and the AQP water uptake mechanism to enhance seed vigor [46]. This leaves a wide research gap that are indeed opportunities to explore towards mapping and genetic engineering of these classes of proteins already identified as implicated in enhancing inherent seed vigor. The obvious research questions raised from this review are whether other single-gene manipulations methods can also produce such effects as *PIMT*. Other research concerns might be the investigation of the effects of enhanced expression of seed vigor genes/proteins on other seed traits like nutrient value, potential health risk as food and feed and ethical issues of GMO seeds for innate vigor.

Since 2013, newer *-omics* tools that allow genome-editing and gene targeting are poised to contain ethical concerns of GMOs because of its capacity for precise modulation of traits of interest with unprecedented control and efficiency. A set of techniques called clustered, regularly interspaced, short palindromic repeat (CRISPR) technology capable of making precise targeted changes in the genome of living cells appeared recently [57], and can be the next great opportunity for genetic manipulation of seed vigor. Coming out of this is the CRISPR-Cas9 which is the latest borderline technology based on a bacterial CRISPR-associated protein-9 nuclease (Cas9) from *Streptococcus pyogenes* [58]. This has already been successfully used to target important genes in many cell lines and organisms. The simplicity of this method lends it to wide applications in biology, currently, it is possible to introduce single point mutations (deletions or insertions) in a target DNA with a guide RNA (gRNA) [59]; and induce large genomic re-arrangements, such as inversions or translocations with a pair of gRNA-directed

| Species | Methodology | Outcome | Reference |
|---|---|---|---|
| *Arabidopsis thaliana* | T-DNA insertion line with increased *PIMT1* expression and transgenic lines with altered *PIMT1* expression | The physiological role of *AtPIMT1* in seed vigor and longevity has been established in *Arabidopsis*.<br>The higher PIMT1 amount in *pimt1–1* seeds correlates with lower isoAsp accumulation *in vivo* and increases both seed longevity and germination vigor, and *vice versa* | Ogé et al. [22] |
| *Cicer arietinum* | Seed-specific over-expression of *CaPIMT1* and *CaPIMT2* in *Arabidopsis* | The role of *CaPIMT2* in seed vigor and longevity has been elucidated<br>*CaPIMT2* enhances seed vigor and longevity by repairing abnormal isoAsp in the seed nuclear proteome | Verma et al. [13] |
| *O. sativa* | Overexpressing *OsPIMT1* lines and *OsPIMT1* RNAi lines | The role of *OsPIMT1* in seed vigor and longevity has been elucidated<br><br>Germination % after 21 days of CDT, overexpressing *OsPIMT1* transgenic seeds, increased 9–15%; *OsPIMT1* RNAi lines, rapid loss of germination. | Wei et al. [16] |
| *O. sativa* | Transgenic rice and *Arabidopsis* lines with altered expression of *OsPIMT1* and *OsPIMT2* | Transgenic rice and *Arabidopsis* lines with altered expression of *OsPIMT1* and *OsPIMT2*<br><br>Germination % after 4 days of CDT, control seeds, 8% (maximum); *OsPIMT1*, *OsPIMT2*, and *OsPIMT2* transformed seeds, 43–48%. | Petla et al. [14] |

**Table 2.** Seed vigor outcomes of different PIMT gene alteration methodologies from different reference sources in various crop species [34].

http://dx.doi.org/10.5772/intechopen.71258

Cas9 nucleases [60]. Proteins can also be targeted for transcriptional regulation using dCas9 version of the CRISPR-Cas9 system [57].

For seed vigor improvement, CRISPR-Cas9 can be used to manipulate gene functions that directly regulate DNA repair pathways like nucleotide and base excision repair, the non-homologous end joining and homologous recombination all of which play notable roles in seed vigor development [3]. Furthermore, the capacity of CRISPR-Cas9 to enable rapid genome-wide study of gene function by generating large gRNA libraries for genomic screening offer opportunity for large scale deployment of precision -*omics* technology for genetic engineering of seed vigor. Of particular relevance of the new technology to crop breeding is the possibility of removing targeted gene constructs by conventional breeding in subsequent generations of the modified plants, thus addressing concerns of GMO contaminations.

## 4. Conclusion

On a global scale, modern agriculture is currently pressurized to achieve food security with limited arable land due to the changing climate and increasing global population. For increases in crop yields with reduced inputs. This paper reviews the state of the art -*omics* results on seed vigor and offers insights towards up-scaling the laboratory results to field productivity. From the reviews, several biological studies that dissect seed vigor traits in crops are discussed narrowing down to few -*omics* approaches offering possibilities for genetically improving seed vigor by plant breeding. Availability of numerous candidate genes and/or proteins along with enormous seed-specific genomic libraries and high-precision -*omics* techniques like CRISPR-Cas9 constitute new resources for drastic improvement of the trait. One strategy mentioned in this review is genetic manipulation of a number of genes controlling cellular repair, protection, detoxification and enhanced membrane integrity in crops. In the near future, studies on reverse genetics coupled with high precision genetic engineering tools, will lead the way to breeding high vigor phenotypes of many crops on large-scale. The results are expected to produce exciting -*omics* contributions to the advancement of crop yields with less environmental damages when these techniques are up-scaled for agricultural applications. Application to important cereals such as wheat, rice, and maize may have a dramatic impact on global food security.

## Author details

Isaac Oludayo Daniel[1,2]*

*Address all correspondence to: drdayodaniel@yahoo.com

1 Department of Plant Breeding and Seed Technology, Federal University of Agriculture, Abeokuta, Nigeria

2 Department of Plant Science, University of Manitoba, Winnipeg, MB, Canada

## References

[1] Copeland LO, McDonald MB. Principles of Seed Science and Technology. 3rd ed. New York: Chapman and Hall; 1995 409 p

[2] Nonogaki H, Bassel GW, Bewley JD. Germination—Still a mystery. Plant Science. 2010; **179**:574-581

[3] Ventura L, Donà M, Macovei A, Carbonera D, Buttafava A, Mondoni A, Rossi G, Balestrazzi A. Understanding the molecular pathways associated with seed vigor. Plant Physiology and Biochemistry. 2012;**60**:196-206

[4] Daniel IO, Kruse M, Boerner A. Comparative longevity and viability modeling of *Solanum macrocarpon* seeds. Seed Science and Technology. 2011;**39**:680-685

[5] Nagel M, Kranner I, Neumann K, Rolletsche KH, Seal CE, Colville L, Fernández-marín B, Börner A. Genome-wide association mapping and biochemical markers reveal that seed ageing and longevity are intricately affected by genetic background and developmental and environmental conditions in barley. Plant, Cell & Environment. 2015;**38**:1011-1022. DOI: 10.1111/pce.12474

[6] Delouche JC, Baskin CC. Accelerated aging techniques for predicting the relative storability of seed lots. Seed Science and Technology. 1973;**1**:427-452

[7] Heydecker W, Higgins J, Gulliver RL. Accelerated germination by osmotic seed treatment. Nature. 1973;**246**:42-44

[8] Bradford KJ. Manipulation of seed water relations via osmotic priming to improve germination under stress conditions. Hortscience. 1986;**21**:1105-1112

[9] McDonald MB. Seed priming. In: Black M, Bewley JD, editors. Seed Technology and its Biological Basis, Sheffield Acad: Sheffield, UK; 2000. pp. 287-325

[10] Chen K, Arora R. Priming memory invokes seed stress-tolerance. Environmental and Experimental Botany. 2013;**94**:33-45

[11] Daniel IO, Adeniyan ON, Adetumbi JA, Olakojo SA, Okelana MA, Adekoya MA. Hydropriming improved germination and vigor of kenaf (*Hibiscus cannabinus* L) seeds. Journal of Food, Agriculture and Environment. 2012;**10**:760-763 www.world-food.net/scientificjournal.php

[12] Soeda Y, Konings MCJM, Vorst O, van Houwelingen, AMML, Stoopen GM, Maliepaard, CA. JK, Bino, RJ, Groot SPC, van der Geest AHM. Gene expression programs during *Brassica oleracea* seed maturation, osmopriming, and germination are indicators of progression of the germination process and the stress tolerance level. Plant Physiology. 2005;**137**:354-368

[13] Verma P, Kaur H, Petla BP, Rao V, Saxena SC, Majee M. Protein l-isoaspartyl methyltransferase2 gene is differentially expressed in chickpea and enhances seed vigor and longevity by reducing abnormal isoaspartyl accumulation predominantly in seed nuclear proteins. Plant Physiology. 2013;**16**:1141-1157

[14] Petla BP, Kamble NU, Kumar M, Verma P, Ghosh S, Singh A, Rao V, Salvi P, Kaur H, Saxena SC, Majee M. Rice protein l-isoaspartyl methyltransferase isoforms differentially accumulate during seed maturation to restrict deleterious isoAsp and reactive oxygen species accumulation and are implicated in seed vigor and longevity. New Phytologist. 2016;**211**:627-645. DOI: 10.1111/nph.13923

[15] Catusse J, Meinhard J, Job C, Strub JM, Fischer U, Pestsova E, Westhoff P, Van Dorsselaer A, Job D. Proteomics reveals potential biomarkers of seed vigor in sugarbeet. Proteomics. 2011;**11**:1569-1580

[16] Wei Y, Xu H, Diao L, Zhu Y, Xie H, Cai Q, Wu F, Wang Z, Zhang J, Xie H. Protein repair l-isoaspartyl methyltransferase 1 (*PIMT1*) in rice improves seed longevity by preserving embryo vigor and viability. Plant Molecular Biology. 2015;**89**:475. DOI: 10.1007/s11103-015-0383-1

[17] Ortiz R. Plant Breeding in the Omics Era. Cham Heidelberg New York Dordrecht London: Springer; 2015. 249 p. DOI: 10.1007/978-3-319-20532-8

[18] Haçarız O, Sayers GP. The omic approach to parasitic trematode research—A review of techniques and developments within the past 5 years. Parasitology Research. 2016; **115(7)**:2523. DOI: 10.1007/s00436-016-5079-1

[19] Moose SP, Mumm RH. Molecular plant breeding as the foundation for 21st century crop improvement. Plant Physiology. 2008;**147**:969-977

[20] Rajjou L, Duval M, Gallardo K, Catusse J, Bally J, Job C, Job D. Seed germination and vigor. Annual Review of Plant Biology. 2012;**63**:507-533

[21] Clerkx EJM, El-Lithy ME, Vierling E, Ruys GJ, Blankestijn-De VH, Groot SPC, Vreugdenhil D, Koornneef M. Analysis of natural allelic variation of Arabidopsis seed germination and seed longevity traits between the accessions Landsberg erecta and Shakdara, using a new recombinant inbred line population. Plant Physiology. 2005;**135**:432-443

[22] Ogè L, Bourdais G, Bove J, Collet B, Godin B. Protein repair L-isoaspartyl methyltransferase 1 is involved in both seed longevity and germination vigor in Arabidopsis. The Plant Cell. 2008;**20**:3022-3037

[23] Dinkova TD, Marquez-Velazquez NA, Aguilar R, Lazaro-Mixteco PE, de Jimenez ES. Tight translational control by the initiation factors eIF4E and eIF(iso)4E is required for maize seed germination. Seed Science Research. 2011;**21**: 85-93

[24] Devaiah SP, Hong Y, Roth M, Welti R, Wang X. Enhancing seed quality and viability by suppressing phospholipase D in Arabidopsis. The Plant Journal. 2007;**50**:950-957

[25] Bernal-Lugo I, Leopold AC. Changes in soluble carbohydrates during seed storage. Plant Physiology. 1992;**98**:1207-1210

[26] Rajjou L, Lovigny Y, Groot SPC, Belghazi M. Proteome-wide characterization of seed aging in Arabidopsis: A comparison between artificial and natural aging protocols. Plant Physiology. 2008;**148**:620-641

[27] Fu Y, Ahmed Z, Diederichsen A. Towards a better monitoring of seed ageing under *ex situ* seed conservation. Conservation Physiology. 2015;**3**:1-16. DOI: 10.1093/conphys/cov026

[28] Bentsink L, Jowett J, Hanhart CJ, Koornneef M. Cloning of *DOG1*, a quantitative trait locus controlling seed dormancy in Arabidopsis. Proceedings of the National Academy of Sciences of the United States of America. 2006;**103**:17042-17047

[29] Prieto-Dapena P, Castaño R, Almoguera C, Jordano J. Improved resistance to controlled deterioration in transgenic seeds. Plant Physiology. 2006;**142**:1102-1112

[30] Kranner I, Birtic S, Anderson KM, Pritchard HW. Glutathione half-cell reduction potential: A universal stress marker and modulator of programmed cell death? Free Radical Biology and Medicine. 2006;**40**:2155-2165

[31] Xu Q, Belcastro MP, Villa ST, Dinkins RD, Clarke SG, Downie AB. A second protein L-isoaspartyl methyltransferase gene in Arabidopsis produces two transcripts whose products are sequestered in the nucleus. Plant Physiology. 2004;**136**:2652-2664

[32] Chen K, Arora R, Arora U. Osmopriming of spinach (*Spinacia oleracea* L. cv. Bloomsdale) seeds and germination performance under temperature and water stress. Seed Science and Technology. 2010;**38**:36-48

[33] Chen K, Fessehaie A, Arora R. Selection of reference genes for normalizing gene expression during seed priming and germination using qPCR in *Zea mays* and *Spinacia oleracea*. Plant Molecular Biology Reporter. 2012;**30**:478-487

[34] Wu X, Ning F, Hu X, Wang W. Genetic modification for improving seed vigor is transitioning from model plants to crop plants. Frontiers in Plant Science. 2017;**8**:8. DOI: 10.3389/fpls.2017.00008

[35] Koornneef M. Gibberellin sensitive mutants in *Arabidopsis thaliana*. Arabidopsis Information Service. 1978;**15**:17-20

[36] Koornneef M, Jorna ML, Brinkhorst Van der Swan DLC, Karssen CM. The isolation of abscisic acid (ABA) deficient mutants by selection of induced revertants in non-germinating gibberellin sensitive lines of *Arabidopsis thaliana* (L.) Heynh. Theoretical and Applied Genetics. 1982;**61**:385-393

[37] Kermode AR. Regulatory mechanisms involved in the transition from seed development to germination. Critical Reviews in Plant Sciences. 1990;**9**:155-195

[38] Adeboye KA, Adabale OW, Adetumbi JA, Ayo-Vaughn MA, Daniel IO. SSR analysis of genetic changes during artificial ageing of rice seeds stored under gene bank management. Plant Breeding and Seed Science. 2015;**71**:37-45

[39] Daniel IO, Adabale OW, Adeboye KA, Aladele ES, Oduoye OT, Adetumbi JA, Börner A. Evaluation of genetic integrity of ageing tomato seeds by microsatellite markers. Nigerian Journal of Genetics. 2014;**28**:29-33

[40] Ahmed Z, Yang H, Fu Y. The associative changes in scutellum nuclear content and morphology with viability loss of naturally aged and accelerated aging wheat (*Triticum aestivum*) seeds. Frontiers in Plant Science. 2016;**7**:1474. DOI: 10.3389/fpls.2016.01474

[41] Regente M., Monzón, GC, de la Canal L. Phospholipids are present in extracellular fluids of imbibing sunflower seeds and are modulated by hormonal treatments. Journal of Experimental Botany. 2008;**59**:553-562. DOI: 10.1093/jxb/erm329.

[42] Colville L, Kranner I. Desiccation tolerant plants as model systems to study redox regulation of protein thiols. Plant Growth Regulation. 2010;**62**:241-255

[43] Mittler R, Vanderauwera S, Suzuki N, Miller G, Tognetti V, Vandepoele K, Gollery M, Shulaev V, Van Breusegem F. ROS signaling: The new wave? Trends in Plant Science. 2011;**16**(6):300-309

[44] Catusse J, Job C, Job D. Transcriptome- and proteome-wide analyses of seed germination. Comptes Rendus Biologies. 2008;**331**:815-822 http://france.elsevier.com/direct/CRASS3/

[45] Waterworth WM, Bray CM, West CE. The importance of safeguarding genome integrity in germination and seed longevity. Journal of Experimental Botany. 2015;**66**(12):3549-3558. DOI: 10.1093/jxb/erv080

[46] Chen K, Fessehaie A, Arora R. Aquaporin expression during seed osmopriming and post-priming germination in spinach. Biologia Plantarum. 2013;**57**:193-198

[47] Obroucheva NV, Sinkevich IA, Lityagina SV, Novikova GV. Water relations in germinating seeds. Russian Journal of Plant Physiology. 2017;**64**:625. DOI: 10.1134/S1021 44371703013X

[48] Landjeva S, Lohwasser U, Börner A. Genetic mapping within the wheat D genome reveals QTL for germination, seed vigour and longevity, and early seedling growth. Euphytica. 2010;**171**:129-143. DOI: 10.1007/s10681-009-0016-3

[49] Arif MAR, Nagel M, Neumann K, Kobiljski B, Lohwasser U. Genetic studies of seed longevity in hexaploid wheat using segregation and association mapping approaches. Euphytica. 2012;**186**:1-13

[50] Nagel M, Daniel IO, Gaebler M, Pasam RK, Rehman-Arif MA, Kilian B, Börner A. Seed longevity in a barley collection—Variation and gene identification. In: Börner AB, Kobijlski B, editors. European Cereals Genetics Co-Operative Newsletter. 2012. p. 151-158

[51] Han Z, Ku L, Zhang Z, Zhang J, Guo S, Liu H, Zhao R, Ren Z, Zhang L, Su H, Dong L, Chen Y. QTLs for seed vigor-related traits identified in maize seeds germinated under artificial aging conditions. PLoS One. 2014;**9**(3):e92535. DOI: 10.1371/journal.pone.0092535

[52] Singh UM, Yadav S, Dixit S, Ramayya PJ, Devi MN, Raman KA, Kumar AQTL. Hotspots for early vigor and related traits under dry direct-seeded system in rice (*Oryza sativa* L.). Frontiers in Plant Science. 2017;**8**:286. DOI: 10.3389/fpls.2017.00286. eCollection 201700286

[53] Mahender A, Anandan A, Pradhan SK. Early seedling vigour, an imperative trait for direct-seeded rice: An overview on physio-morphological parameters and molecular markers. Planta. 2015;**241**:1027-1050. DOI: 10.1007/s00425-015-2273-9

[54] Carrera E, Holman T, Medhurst A, Peer W, Schmuths H, Footitt S, Theodoulou FL, Holdsworth MJ. Gene expression profiling reveals defined functions of the ATP-binding cassette transporter COMATOSE late in phase II of germination. Plant Physiology. 2007;**143**(4):1669-1679 http://www.jstor.org/stable/40065381

[55] Joosen RVL, Ligterink W, Dekkers BJW, Hilhorst HWM. Visualization of molecular processes associated with seed dormancy and germination using MapMan. Seed Science Research. 2011;**21**:143-152

[56] Bassel GW, Lan H, Glaab E, Gibbs DJ, Gerjets T, Krasnogor N, Bonner AJ, Holdsworth MJ, Provart NJ. Genome-wide network model capturing seed germination reveals coordinated regulation of plant cellular phase transitions. Proceedings of the National Academy of Sciences of the United States of America. 2011;**108**:9709-9714

[57] Sander JD, Joung JK. CRISPR-Cas systems for editing, regulating and targeting genome. Nature Biotechnology. 2014;**32**:347-355. DOI: 10.1038/nbt.2842

[58] Cong L, Ran FA, Cox D, Lin S, Barretto R, Habib N, Hsu PD, Wu X, Jiang W, Marraffini LA, Zhang F. Multiplex genome engineering using CRISPR/Cas systems. Science. 2013;**339**(6121):819-823. DOI: 10.1126/science.1231143

[59] Nishimasu H, Ran FA, Hsu PD, Konermann S, Shehata SI, Dohmae N, Ishitani R, Zhang F, Nureki O. Crystal structure of Cas9 in complex with guide RNA and target DNA. Cell. 2014;**156**(5):935-949. DOI: 10.1016/j.cell.2014.02.001 Epub 2014 Feb 13

[60] Gratz SJ, Cummings AM, Nguyen JN, Hamm DC, Donohue LK, Harrison MM, Wildonger J, O'Connor-Giles KM. Genome engineering of drosophila with the CRISPR RNA-guided Cas9 nuclease. Genetics. 2013;**194**:1029-1035

Chapter 14

# The Potential of Garden Cress (*Lepidium sativum* L.) Seeds for Development of Functional Foods

Chandra Shekhar Singh and Vinod Kumar Paswan

Additional information is available at the end of the chapter

http://dx.doi.org/10.5772/intechopen.70355

**Abstract**

Garden cress (*Lepidium sativum* L.) belonging to Cruciferae family is widely grown in India, Europe, and United States. It has been used as an important medicinal plant since the Vedic era. Its seed, oil, and powder contain significant amount of protein, fat, minerals, fibers, and phytochemicals, which are incorporated in many functional beverages and foods. A number of clinical trials have been conducted on rats that also support the efficacy of garden cress seeds (GCSs). The seed of garden cress was used in the fortification of different food items but due to the lack of their physicochemical properties and medicinal value, the exploration of the potential of garden cress seed was limited. In the present review, we discuss the proximate chemical composition, physicochemical, medicinal properties, and the food product development with garden cress seed. The functional properties of garden cress seed stimulate us to review its different valuable properties and the fortified products developed by incorporating garden cress seeds.

**Keywords:** garden cress seed, galactogogue, natural antioxidants, functional foods, fortification

## 1. Introduction

Garden cress (*Lepidium sativum* L.) is a fast growing annual herb that is native to Egypt and west of Asia, and presently it is cultivated in all over the world. In local languages, garden cress (GC) is also known by Chandrasur, and it is considered as an important medicinal crop in India [1]. The plant is an erect, glabrous, annual, herbaceous growing up to the height of about 15–45 cm (**Figure 1**). It has small white flowers in long racemes and the pods are broadly or obovate, rotund, elliptic, emarginated, notched at apex, and winged. Garden cress

**Figure 1.** Garden cress plants and seeds.

can be sown and harvested several times throughout the year, the January, February, and November are the most suitable months of the year to sow in a Mediterranean climate [2]. Garden cress seed (GCS) possesses several of pharmacological properties like anti-anemic, antioxidant, galactogogues, etc. and has tremendous potential for the development of functional food by fortification with it [3, 4]. Generally, GC is consumed as cooking material and with salad. In recent years, efforts are made to develop human diets in such a way that it acts as medicinal foods in order to exploit several health benefits and to prevent increased diversity of diseases. Isothiocyanates are most important biochemical agents from the human health point of view as they are the major inducers of carcinogen-detoxifying enzymes. The most potent isothiocyanates are benzyl isothiocyanate (BITC), which is present in ample quantity in garden cress [5]. GCS have been used in traditional medicine since ancient times in India [6]. The GCS are galactogogue, bitter, thermogenic, depurative, rubefacient, aphrodisiac, ophthalmic, antiscorbutic, antihistaminic, diuretic, and act as tonic. Various diseases such as asthma, coughs with expectoration, diarrhea, dysentery, poultices for sprains, leprosy, skin disease, splenomegaly, dyspepsia, lumbago, leucorrhoea, scurvy, and seminal weakness can be treated using garden cress seed [7]. It is supplemented in the diet of lactating women to increase the milk secretion during postnatal period and also recommended for the treatment of diarrhea and dysentery [8, 9]. Seeds of GC are prescribed by Ayurvedic practitioners for the treatment of bronchial asthmatic patients. Garden cress seed oil (GCSO) has a balanced amount of polyunsaturated fatty acids (PUFA) (46.8%) and monounsaturated fatty acids (MUFA) (37.6%). It contains natural antioxidants like vitamin A, E, and eugenol, which help to protect cells from damage by free radicals [10]. It also protects oil from oxidation and causing rancidity. It was reported that GCS contain 22.5% protein, 27.5% fat, 30% dietary fiber, and 1193 mg/100 g potassium [11]. Hence, it can potentially be used as a functional food. The oil content of dried cress seed is 22.7% and the primary fatty acids found in cress oil are oleic (C18:1; 30.6%), linolenic (C18:3; 29.3%), palmitic (C16:0; 9.4%), linoleic (C18:2; 7.6%), erucic (C22:1; 3.0%), stearic (C18:0; 2.8%), and arachidic (C20:0; 2.3%) acids [4, 11]. GCSO contain high concentrations of $\gamma$-(1422 ppm) and $\alpha$ (356 ppm) tocopherols.

The fruit and vegetable juices are rich sources of vitamin and minerals, but these are limited in protein and fat content. For the compensation of these components, garden cress extract or powder can be added. As garden cress also acts as thickening agent, the combination of both juices and extract may lead to the formation of health promoting beverages having good textural, sensory attributes, and nutritional properties. A beverage was developed by combining lime juice and saccharin, honey, and garlic for the compensation of proteins and fat [12]. Similarly, Mohite et al. designed a health drink by combining GCS powder with skim milk powder for providing promising health benefits [13].

## 2. Chemical and nutritional composition of garden cress seeds

Proximate composition (%) of *L. sativum* seeds reported by Zia-Ul-Haq et al. indicates the presence of appreciable amounts of protein (24.2 ± 0.5), lipids (23.2 ± 0.2), carbohydrates (30.7 ± 1.2), fiber (11.9 ± 0.4), ash (7.1 ± 0.1), and moisture (2.9 ± 0.1) [14]. Proximate composition varies depending upon plant variety, agronomic practices, and stage of collection of seeds and climatic and geological condition of area from where seeds are collected. It is an important factor for the evaluation of nutritional status of fruits and seeds of plants and crops, and it dictates further studies on components, which seem more interesting [15]. Higher amounts of ash contents indicate that the GCS are good source of minerals. The low moisture content is an index of stability, quality, and increased shelf life of seeds [16]. Higher protein and lipid contents indicate that GCS have high food energy.

Qualitative and quantitative amino acid profile as presented in **Table 1** well introduces the nutritional quality of GCS protein [4, 14]. All essential amino acids are present in high amounts in garden cress, except tryptophan and S-containing amino acids, methionine and cysteine. Glutamic acid and aspartic acid are the major nonessential amino acids in the GCS. The total essential amino acid percentage (47.08%) suggests that this seed may contribute significantly to the supply of essential amino acids in the diet. Essential amino acid score is 28.53% with methionine being the most limiting amino acid. Aspartic and glutamic acids are present in significant amount in this oilseed. Glutamic acid is an important excitatory neurotransmitter, and it plays a vital role in the metabolism of sugars and fats [17]. The body uses methionine to derive the brain food and choline. It also aids in digestion, as well as serving as a fat burner. It can interact with other substances to detoxify harmful agents and is essential for the production of cysteine and taurine. It is also necessary for the production of niacin and is used by the body to make neurotransmitter and serotonin [18]. These play a very important role in human nutrition. Lysine helps in proper maintenance of nitrogen balance. L-Tryptophan acts as a sleep aid. The presence of tryptophan and cystine in GCS is also reported [14].

Mineral contents of seeds (**Table 1**) varied between species, but potassium constituted the major mineral in GCS 1236.51 mg/100 g, while zinc and manganese contents are low. GCS is a good source of calcium, phosphorus, and magnesium. GCS has the potential for providing essential nutrients for human and other animals, as the nutritional activity of any plant is usually related to the particular elements it contains [19]. With these minerals content, it can be utilized for the development of a number of supplementary food products.

Fatty acid composition (**Table 1**) reveals high content of linolenic acid (32.18%) and oleic acid (30.5%) in the garden cress seed oil (GCSO). Higher intake of oleic acid is associated with the decreased risk of coronary heart disease caused by high cholesterol level in blood [20]. The fatty acid composition of the GCSO is interesting from the nutritional point of view for their

| Amino acid profile (g/100 g protein) | |
|---|---|
| *Essential amino acids* | |
| Histidine | 3.87 ± 0.14 |
| Threonine | 2.66 ± 0.09 |
| Arginine | 4.51 ± 0.03 |
| Valine | 8.04 ± 0.03 |
| Methionine | 0.97 ± 0.02 |
| Phenyl alanine | 5.65 ± 0.03 |
| Isoleucine | 5.11 ± 0.03 |
| Leucine | 8.21 ± 0.01 |
| Lysine | 6.26 ± 0.39 |
| *Nonessential amino acids* | |
| Aspartic acid | 9.76 ± 0.03 |
| Glutamic acid | 19.33 ± 0.19 |
| Serine | 4.96 ± 0.09 |
| Glycine | 5.51 ± 0.07 |
| Alanine | 4.83 ± 0,02 |
| Tyrosine | 2.69 ± 0.09 |
| Proline | 5.84 ± 0.38 |
| **Mineral content (mg/100 g)** | |
| Calcium | 266.35 |
| Copper | 5.73 |
| Iron | 8.31 |
| Magnesium | 339.23 |
| Manganese | 2.00 |
| Phosphorus | 608.63 |
| Potassium | 1236.51 |
| Sodium | 19. 65 |
| Zinc | 6.99 |
| **Fatty acid profile (%)** | |
| Palmitic acid (16:0) | 10.30 ± 0.12 |
| Palmitoleic acid (16:1) | 0.70 ± 0.30 |
| Stearic acid (18:0) | 1.90 ± 0.19 |
| Oleic acid (18:1) | 30.50 ± 0.16 |
| Linoleic acid (18:2) | 8.60 ± 0.38 |

| **Amino acid profile (g/100 g protein)** | |
|---|---|
| Linolenic acid (18:3) | 32.18 ± 0.59 |
| Arachidic acid (20:0) | 2.10 ± 0.57 |
| Eicosaenoic acid (20:1) | 13.40 ± 0.66 |

Adopted from: Gokavi et al. [4] and Zia-Ul-Haq et al. [14].

**Table 1.** Amino acid profile, mineral content, and fatty acid profile of garden cress seed.

higher contents of unsaturated fatty acids; especially, it is rich in omega-3 fatty acid, which is beneficial for health. Palmitic acid is the most abundant saturated fatty acid, in the amounts of 10.3 ± 0.12 g/100 g in GCS. Similarly, palmitoleic acid is the least abundant unsaturated fatty acid, with values 0.70 ± 0.30 g/100 g in GCS.

## 3. Physicochemical properties of garden cress seed oil

Physicochemical parameters provide important information regarding storage, stability, and quality of the product. The physicochemical properties of garden cress seed oil (GCSO) extracted by the different methods are presented in **Table 2** [21]. Extracted oil by solvent extracted, supercritical $CO_2$, and cold expression were 21.54, 18.15, and 12.60%, respectively. The Soxhlet method yields the maximum oil content (21.54%). Oil yield of GCS is low when compared to the other oil seeds of Cruciferae family like mustard (25–40%), rapeseed (40–45%),

| Attributes | Cold-pressed oil | Soxhlet-extracted oil | Supercritical $CO_2$ extracted oil |
|---|---|---|---|
| Oil yield (% dry weight) | $12.6 \pm 0.87^{a}$ | $21.54 \pm 1.32^{c}$ | $18.15 \pm 1.20^{b}$ |
| Refractive index (nDt)[1] | 1.47 ± 0.001 | 1.47 ± 0.003 | 1.47 ± 0.002 |
| Specific gravity (g/ml)[2] | 0.91 ± 0.001 | 0.90 ± 0.001 | 0.91 ± 0.001 |
| Viscosity (η)[3] | $64.3 \pm 0.90^{a}$ | $55.5 \pm 0.37^{b}$ | $53.8 \pm 0.6^{b}$ |
| Peroxide value (mequiv peroxide/kg oil) | $0.70 \pm 0.13^{a}$ | $4.09 \pm 0.16^{c}$ | $2.63 \pm 0.81^{b}$ |
| Free fatty acid (% oleic) | $0.28 \pm 0.02^{a}$ | $0.39 \pm 0.04^{b}$ | $1.52 \pm 0.28^{c}$ |
| Saponification value (mg KOH/g) | $178.85 \pm 0.46^{a}$ | $182.23 \pm 0.73^{c}$ | $174 \pm 0.82^{b}$ |
| Unsaponifiable matter (g %) | $1.65 \pm 0.24^{a}$ | $1.39 \pm 0.10^{b}$ | $1.16 \pm 0.30^{c}$ |
| Iodine value (g of $I_2$ absorbed/100 g) | $122 \pm 0.70^{a}$ | $131 \pm 3.26^{b}$ | $123 \pm 1.68^{a}$ |

Each value is a mean ± SD of three determinations. Values within the same row with different alphabetical superscripts are significantly different at $P < 0.05$.

[1]nDt is the unit of refractive index (nD) for light with a wavelength equal to 589.3 nm at temperature, t = 24°C.

[2]The direct pycnometer determination at 33°C.

[3]Viscosity determined at 25°C MPa $s^{-1}$. Data compiled from: Diwakar et al. [21].

**Table 2.** Physicochemical properties of garden cress seeds extracted by various extraction methods.

and camelina or false flax (40–45%) [22]. The pungency of GCSO and mustard oil are similar, while the content is lower. The physicochemical properties of GCSO as summarized in **Table 2** establish the potential of garden cress for the development of novel products with several functional properties.

### 3.1. Physical properties of garden cress seed oil

Physical properties like color, odor, viscosity, specific gravity, and refractive index are important during the development of food products, because these properties may affect the different quality parameters of the developed products.

#### *3.1.1. Color*

Color of any oil determines the attraction of consumer and acceptability of the product. Garden cress oil color is dirty yellow due to the presence of chlorophyll and carotenoids pigments, which are unintentionally co-extracted during the oil extraction [23].

#### *3.1.2. Viscosity*

Viscosity of the GCSO ranges from 53.8 to 64.3 (**Table 2**). The cold-pressed GCO was more viscous than the oil extracted by the other two methods. Increasing extraction temperature up to a certain value increased viscosity, but at higher extraction temperatures viscosity decreased. The reduction of gum viscosity with temperature might be the result of irreversible change in molecular conformation [24]. It was concluded that high pH, low water, seed ratio, and mild extraction temperatures will give a high viscosity for *L. sativum* extract. It decides the flow behavior of the products and is considered much during the formulation of any liquid or semisolid products.

#### *3.1.3. Refractive index*

High refractive index value (1.47 ± 0.03) is the indication of substantial unsaturation and the presence of unusual components such as hydroxyl groups in GCSO [25]. It also provides useful information about the purity of oils. The refractive index of GCO is within the range of edible oils (**Table 2**); therefore, it can be a good fortifying agent for the product development.

#### *3.1.4. Specific gravity*

Specific gravity of garden cress seeds (0.91) resembles with the specific gravity value of milk. This suggests that drinks can easily be fortified with processed garden cress seeds powder and thus several health drinks can be formulated by incorporating GCS [13].

### 3.2. Chemical properties of garden cress seed oil

Chemical properties help in determining the stability of the GCSO and the developed blended products. It also helps in determining the shelf life of the food products.

#### *3.2.1. Free fatty acids and peroxide value*

The free fatty acids (FFA) and peroxide value (PV) of the cold-pressed GCSO is lowest compared to the oils extracted by solvent and supercritical $CO_2$ extraction (**Table 2**). The FFA content of supercritical $CO_2$ extracted GCSO is higher than the oil extracted by cold pressed and Soxhlet extraction. The acid value of oils depends upon the oil extraction methods apply. Lower the FFA, higher the stability of oil at room temperature (25 ± 2°C). The content of FFA of GCSO (**Table 2**) is in limit with the specifications of vegetable oils (1–7% of oleic acid). The high PV in Soxhlet extracted oil could be due to the exposure of the oil to high temperature (60–80°C) during extraction. The low PV of cold-pressed GCSO indicates that it is less prone to oxidative rancidity at room temperature.

#### *3.2.2. Iodine value*

The unsaturation of an oil or fat is measured by the iodine value (IV). IV depends on the unsaturated fatty acids present in the oil or fat. The IV of oil extracted by cold-pressed and supercritical fluid extracted of GCSO was relatively lower than solvent extracted (**Table 2**). The IV of the oil is affected due to the presence of many long chain unsaturated components like olefins, including carotenoids and squalenes [26]. The solvent-extracted oil contained a significantly higher amount of total carotenoids than the cold-pressed oil. Thus, the higher carotenoid content might be responsible for a high IV in solvent-extracted GCSO.

#### *3.2.3. Saponification value*

SAP value is the number of milligrams of potassium hydroxide required to saponify 1 g of fat or oil. It measures the average molecular weight (or chain length) of all the fatty acids present. The range of SAP value of GCSO is 174.00–182.23, indicating that the oil contained high molecular fatty acids (**Table 2**). The SAP value of GCSO is lower (178.36) than the palm oil (196–205), olive oil (188–196), sunflower oil (186–196), soybean oil (188–195), and safflower oil (186–198) [27]. Thus, the saponification value is appropriate to form or supplement in to other product.

#### *3.2.4. Unsaponifiable matter*

Unsaponifiable matter (USM) shows the pigments, chlorophyll, and other heterocyclic compounds present in the oil. The USM content in GCSO extracted by different method varied from 1.16 to 1.65 g/100 g (**Table 2**). The USM content was higher in GCSO than in sesame (1.2%), white melon (1.1%), corn (0.92%), cotton (0.52), palm (0.34%), peanut (0.33), palm kernel (0.22), and coco kernel (0.09) oil [28]. The presence of lignan (29.4%), crude fiber (16.5%), protein (24.3%), and minerals (5.4%) in GC seeds the USM contents was high [29].

### 3.3. Functional properties and other health benefits of garden cress seeds

#### *3.3.1. As antioxidants: free radical scavenging activity*

The antioxidant properties depend on the phenolic compounds present in garden cress seeds. The main phenolic compounds present in GCS extracts are tocopherols. Tocopherols act as biological scavengers of free radicals that inhibit oil oxidation. Tocopherols also help in preventing diseases, besides possessing an important nutritional function for human beings as a source of vitamin E [30, 31]. High amounts of tocopherols present in GCS can be responsible for the stabilization of fats and oils to prevent the oxidative deterioration and for its applications in dietary, pharmaceutical, or biomedical products [32]. Total tocopherol contents in GCSO is 139.73 ± 0.91 mg/100 g and δ tocopherol was the most abundant in the seed oil of GC. Vitamin E (tocopherol) is an important antioxidant, which protects vitamin A and essential fatty acids from oxidation and prevents breakdown of body tissues. Garden cress seeds possess maximum DPPH inhibition activity at concentrations of 100, 150, and 200 μg of methanolic extracts as reported during DPPH radical scavenging assay. These values are comparable with the standard free radical scavenger BHA at concentration 10, 50, and 100 μg [33]. Due to high free radical scavenging potential of GCS, its fortification to prepare balanced diet may help in incorporating and exploiting its rich nutritional as well as medicinal value to the developed food.

#### 3.3.2. As galactogogue and emmenagogue: for inducing milk secretion and menstruation

GCS can be used as a supplement for proper regulation of the menstrual cycle, because it has mild oestrogenic properties. It shows emmenagogue like herbal properties, which gave it an important place in Vedic era. Emmenagogues are herbs, which have the ability to provoke menstruation. They stimulate blood flow in the pelvic area and uterus and thus induce menstruation. GCS is used as emmenagogue in order to stimulate menstrual flow when menstruation is absent either due to pregnancy to cause an abortion or prevent pregnancy or for reasons other than pregnancy, such as hormonal disorders or conditions like oligomenorrhea. Similarly, consumption of GCS after birth of baby increases milk production and secretion in lactating mothers. Because of its high iron and protein content, it is often given post-partum as effective galactogogue to induce lactation in nursing mothers to meet the nutritional requirement of their children. Galactogogues promote lactation in humans and other animals. They exert their pharmacological effects through interactions with dopamine receptors, resulting in increased prolactin levels and thereby augmenting milk production [34].

#### 3.3.3. As gastrointestinal tract cleansing agent

Garden cress helps in cleansing gastro intestinal tract and stimulates appetites. The testa of these seeds contain mucilage which can be used during constipation as a laxative and a purgative. The paste of GCS and honey can be taken internally to treat amoebic dysentery. The irritation of the intestines in dysentery and diarrhea effectively reduces by the use of mucilage of the germinating seeds of GCS. Crushed GCS taken with hot water is beneficial to treat colic disease especially in infants. The plant is also used in treating bleeding piles.

#### 3.3.4. As haematic agent

The GCS is the rich source of iron, which is easily absorbed in intestine and helps to increase the hemoglobin level in blood. The bioavailability of iron content in GCS beneficial for anemia when taken daily. Vitamin C taking half an hour after consumption of these seeds enhances the iron absorption. L-ascorbic acid facilitates iron absorption by forming a chelate with ferric iron at acid pH converting them to ferrous state that remains soluble at the alkaline pH of the duodenum which gets easily absorbed [35, 36].

### 3.4. Other health benefits

Garden cress seeds are considered as a memory booster due to the presence of arachidic and linoleic acids. They help to increase the lean body mass because they are a good source of iron and protein. The absorbility of iron increases when GCS is soaked in lime water, which helps in strengthening of hair. The leaves are mildly stimulant and diuretic, useful in scorbutic (related to or resembling scurvy) diseases, and liver complaints. A paste of the seeds with water is effective against chapped lips and sunburn. As it is a good source of folic acid, it helps in synthesis of different nonessential amino acids. The well-documented antioxidant and phytochemical properties of this amazing plant make it a chemopreventive agent. The functional properties contained by GCS suggest that the regular consumption can greatly help to boost one's immunity and overcome to gamut of diseases. It acts as a general tonic and can also help to increase the libido naturally. As it is a good source of carotene, which is the precursor of vitamin A, it is good for the eyes. Therefore, it is advisable to add it raw to salads, sandwiches, and chutneys or to simply use it as a garnishing agent along with coriander leaves for any food item in order to utilize these health benefits.

### 3.5. Side effects of garden cress seed

It is an abortifacient (substance that induces abortion), if had in excess. Pregnant women should avoid taking garden cress in any form because it has the ability to induce uterine contractions and thereby trigger spontaneous abortion. It contains goitrogens that prevent iodine absorption in thyroids and hence can lead to hypothyroidism. Hence, it may not be suitable for patients suffering from hypothyroidism. If large quantities of garden cress are consumed, may cause digestive difficulties in some people. The oil extracted from GCS is edible and is used as a cooking medium; however, some people may experience symptoms of indigestion due to its use. To overcome these problems, people should discontinue using this oil or mix it with some other edible oil, so as to dilute it and reduce its adverse effects.

### 3.6. Fortification by garden cress for the development of new food products

Due to the contents of high nutritional and functional properties in garden cress, it can be used for the fortification with many drinks and foods. The fortified garden cress products are discussed as under:

#### *3.6.1. Development of garden cress fortified dahiwala bread*

Food product dahiwala bread was developed by Agarwal and Sharma using processed garden cress seeds [37]. It results in the development of the products with increased amount of protein, fat, calcium, iron, and phosphorous. Processing of the GCS helps in a significant decrease of antinutritional components like oxalate and total cyanogens, while phytic acid was reduced to a small extent. The developed food product was checked for its acceptability by semi-trained panels. Whole garden cress seed flour incorporated product was acceptable as standard. The developed food products using garden cress seeds could be beneficial for masses as nourishing as well as therapeutic agents due to the presence of various therapeutic properties like hypoglycemic, hypotensive, fracture healing, anticancerous, etc. [37]. The developed food product will benefit all age group individuals for nourishment and those at risk or suffering from anemia, fractures, diabetes mellitus, and the other chronic degenerative diseases to pursue prevention and management of these diseases.

*3.6.2. Development of omega-3 fatty acid–rich biscuits*

GCSO is rich source of $\alpha$-linolenic acid (ALA), thus it is highly prone for auto-oxidation. Microencapsulation of GCSO was difficult in spray drying method for preparation of whey protein concentrate with oil/protein ratio of 0.4. Microencapsulated GCO powder (MGCO) contained 25 g of GCSO/100 g with microencapsulation efficiency of 64.8% and particle size of 15.4 ± 9.1 microns. Omega-3 fatty acid–rich biscuits were prepared by adding MGCSO at 20 g/100 g or GCO at 5.0 g/100 g by replacing flour and fat or fat in biscuit formula [38]. In MGCSO- and GCSO-supplemented biscuits, the ALA content was found to be 1.02 g and 1.05 g/100 g, respectively. Metalized PET film (MPET) pouches are used for packaging and stored at three different storage conditions, viz., 90% RH/38°C for 3 months, 30–40% RH/38–40°C for 4 months, and 65%RH/27°C for 5 months. Prepared biscuits stored at 30–40% RH/38–40°C had 1 month shelf-life, whereas at 90% RH/38°C and 65% RH/27°C, they lasted after 4 and 5 months, respectively. The oxidation rate of ALA was high in GCSO-supplemented biscuits compared to MGCSO biscuits resulting the effect of GCO which prevent the oxidation of ALA. MGCSO-supplemented biscuit were acceptable by testing the sensory evaluation results of the panel.

*3.6.3. Improvement of dough rheology and quality parameters in rice-wheat bread*

Sahraiyan et al. have evaluated the effect of hydrocolloid of GCS and guar gum in improving dough rheology and quality parameters in composite rice-wheat bread [39]. It was reported that the rheological properties of GCS and guar gum seeds improved the quality of rice-wheat bread [40]. Four different composition (0, 0.3, 0.6, and 1%) of w/w Guar (G) and GCS (L) hydrocolloids were added to the flour. The levels of different combination were tested in order to obtain the following samples (the sub Index indicates the gum level): G0 L0, G0 L0.3, G0 L0.6, G0 L1, G0.3 L0, G0.3 L0.3, G0.3 L0.6, G0.3 L1, G0.6 L0, G0.6 L0.3, G0.6 L0.6, G0.6 L1, G1 L0, G1 L0.3, G1 L0.6, and G1 L1. The results of this research revealed that the effects of guar, GCS, and guar-*L. sativum* seed gum in order to substitute gluten in composite rice-wheat flour recipes. The properties of water absorption, dough development time, dough stability, and viscosity were increased by GCS. GCS gum highly affected the mixing tolerance index and gelatinization temperature as guar gum while these parameters decreased by addition of both hydrocolloids. The extensibility value dough increased with increasing hydrocolloid concentrations from 0.3 to 0.6%, and then decreased at the level of 1%. Crumb firmness decreased with increasing hydrocolloid concentration and increased with longer storage time, although the effect of GCS gum on crumb firmness reduction was more than guar gum. It was concluded that GCS gum can be a novel and useful gluten substitute for composite bread baking purposes. These properties may be useful in the preparation of several functional food preparations that overcome the milk secretion–related problems and anemic condition in women.

### 3.6.4. *Development of iron-rich biscuits*

Iron-rich biscuit was prepared by combining garden cress and rice flakes to prevent the anemia in adolescent girls from urban, rural, and tribal areas of Marathwada region of Maharashtra state [41]. The sensory properties of iron-rich biscuits were tested by sensory panel members. The high score sample were tested for proximate composition. The hemoglobin value of selected adolescent girls was found in the range of 8.7–10.96 mg/100 g. By comparing these values, they found the least value of hemoglobin content in tribal and low income group girls. The proximate composition of selected biscuits, which was highly accepted for all sensory attributes, is given in **Table 3**. It is evident from the study that the acceptability score ranged from 1.80 to 4.20. The mean values of different sensory characters reported that color scored maximum followed by taste and texture. Low scores were noted for flavor and overall appearance.

### 3.6.5. *Processed garden cress seeds fortified health drinks*

The edible whole seeds are known to have health promoting properties; hence, it was assumed that these seeds can serve as raw materials for functional foods contributing its peppery, tangy flavor, and aroma. Since it is rich in proteins, carbohydrates, and certain essential minerals like calcium, iron, and phosphorous along with crude dietary fiber (7.6%),

| Particulars | Nutrient content | |
|---|---|---|
| | Iron-rich biscuit [40] | Milk-based health drink [13] |
| Moisture (%) | 14.87 | 84.10 |
| Ash (%) | 6.11 | 0.85 |
| Fat (%) | 29.61 | 1.22 |
| Fiber (%) | 0.99 | – |
| Protein (%) | 2.80 | 3.44 |
| Calcium (mg/100 gm) | 17.63 | 127.20 |
| Phosphorus (mg/100 gm) | 15.48 | 106.20 |
| Iron (mg/100 gm) | 12.00 | 2.90 |
| Carbohydrate (%) | – | 10.30 |
| Energy (kcal/100 gm) | – | 65.63 |

**Table 3.** Chemical composition of garden cress seeds fortified iron-rich biscuit and milk-based health drink.

it can be used as health drink with milk as its base. By keeping this thing in mind, a health drink was developed with processed garden cress seeds that contain excess amount of minerals and nutrients [13]. The composition of health drink developed by adding 5% sugar (w/v) in skimmed milk with 1% fat and 3% of processed garden cress seeds powder had 8.75 overall consumer acceptability.

The health drink is considered as a type of fortified functional foods mainly needed for growing children, the aged, and the invalids and also certain convalescent patients. It enhances the nutritional properties of milk and also provides the entire essential factor that is needed by these people [42]. This type of finding help to develop many types of functional and fortified food products. The chemical composition of milk-based health drink prepared by Mohite et al. is presented in **Table 3** [13]. This type of drink helps to prevent nutrient deficiency and promotes lean muscle developments in persons doing regular exercise. Therefore, it is essential that the food prepared in such a way to meets the requirement of modern consumers. At the same time, it should appeal to the senses of the consumers by having pleasant organoleptic qualities.

#### *3.6.6. Vegetable oil blends with α-linolenic acid-rich garden cress oil*

Earlier finding of researcher concludes that the intake of high n-6 PUFAs in diet has change the physiological conditions like prothrombotic and proaggregatory, characterized by increase in blood viscosity, vasospasm, and vasoconstriction and decrease in bleeding time [40]. Atopic dermatitis, rheumatoid arthritis, asthma, ulcerative colitis, and cancer were caused by the deficiency of n-3 PUFA. However, sufficient intake of n-3 PUFAs alters membrane fluidity, downregulates inflammatory genes and lipid synthesis, and stimulates fatty acid degradation [43]. Humans cannot synthesize PUFAs like n-3 and n-6, which are essential fatty acids; therefore, they need to be supplemented through food. The eating of vegetable oils in diet (sunflower, corn oil, safflower oil, and soybean oil) rich in n-6 PUFA has shifted the n-6 to n-3 PUFA ratio to 50:1 instead of a recommended ratio of 10:1 or 2:1 [44, 45].

Umesha and Naidu developed a blended vegetable oil and studied its modulatory effects on lipid metabolism [46]. For the development of blended vegetable oil, different ratios of GCSO were blended with sunflower oil (SFO), rice bran oil (RBO), and sesame oil (SESO) to obtain n-6/n-3 PUFA ratio of 2.3–2.6 for the assessment of its modulatory effect on lipid metabolism. Wistar rats fed to Native and GCSO blended oils at 10% level in the diet for 60 days. Serum and liver lipids showed significant decrease in total cholesterol (TC), triglyceride (TG), and LDL-C levels in GCSO blended oil fed rats compared to native oil fed rats. In the GCSO blended oils fed rats, the ALA, eicosapentaenoic acid (EPA), and docosahexaenoic acid (DHA) contents were significantly increased, whereas linoleic acid (LA) and arachidonic acid (AA) levels decreased. The lipid profile of rats effectively modulated with blending of vegetable oils with GCSO by increasing the ALA and decreasing n-6 to n-3 PUFA ratio.

## 4. Conclusion

The content of biologically active compounds, as well as the antioxidant capacity of *L. sativum* has been investigated by several researchers and their findings indicated that seeds of garden cress plants are good source of amino acids, minerals, fatty acids and have the ability to act as *in vivo* as well as *in vitro* antioxidants due to their high content of phenolic compounds. The functional health benefits of GCS may be exploited by incorporating it in several food formulations and health drink preparations. Therefore, garden cress plant, seed as well as oil present us with wide scope for further investigations for their potential preventive effects toward chronic diseases and also as interesting ingredients for new functional food formulations.

## Author details

Chandra Shekhar Singh* and Vinod Kumar Paswan

*Address all correspondence to: chandrassingh007@gmail.com

Centre of Food Science and Technology, Banaras Hindu University, Varanasi, India

## References

[1] Tiwari PN, Kulmi GS. Performance of Chandrasur (*Lepidium sativum*) under different levels of nitrogen and phosphorus. Journal of Medicinal and Aromatic Plant Sciences. 2004;**26**:479-481

[2] Tuncay O, Esiyok D, Yagmur B, Bulent OB. Yield and quality of garden cress affected by different nitrogen sources and growing period. African Journal of Agricultural Research. 2011;**6**:608-617

[3] Vohora SB, Khan MSY. Pharmacological studies on *Lepidium sativum* Linn. Indian Journal of Physiology and Pharmacology. 1977;**21**:118-120

[4] Gokavi SS, Malleshi NG, Guo M. Chemical composition of garden cress (*Lepidium sativum*) seeds and its fractions and use of bran as a functional ingredient. Plant Foods for Human Nutrition. 2004;**59**(3):105-111

[5] Williams DJ, Critchley C, Pun S, Chaliha M, Timothy JO. Differing mechanisms of simple nitrile formation on glucosinolate degradation in *Lepidium sativum* and *Nasturtium officinale seeds*. Phytochemistry. 2009;**70**:1401-1409

[6] Mali RG, Mahajan SG, Mehta AA. *Lepidium sativum* (garden cress) a review of contemporary literature and medicinal properties. Oriental Pharmacy and Experimental Medicine. 2007;**7**(4):331-335

[7] Kirthikar KR, Basu BD. *Lepidium sativum* L. In: Indian Medicinal Plants in India, Lalith Mohan Basu, India; 1952. p. 174-175

[8] Sahsrabudde MB, The Wealth of India, Raw Materials. Publication and information Directorate, CSIR, De NN. Current Science. New Delhi. 1962;**12**:23-24

[9] Kirtikar KR, Basu BD, Indian medicinal plants, 1933, pp. 173-175, Vol. I, Dehradun: M/S Bishensingh Mahendra Palsingh

[10] Raghavendra RH, Akhilender NK. Eugenol and n-3 rich garden cress seed oil as modulators of platelet aggregation and eicosanoids in Wistar albino rats. The Open Nutraceuticals Journal. 2011;**4**:144-150

[11] Moser BR, Shah SN, Winkler-Moser JK, Vaughn SF, Evangelista RL. Composition and physical properties of cress (*Lepidium sativum* L.) and field pennycress (*Thlaspi arvense* L.) oils. Industrial Crops and Products. 2009;**30**:199-205

[12] Bhuiyan MHR, Shams-Ud-Din M, Islam MN. Development of functional beverage based on taste preference. Journal of Environmental Science & Natural Resources. 2012;**5**(1): 83-87 ISSN 1999-7361

[13] Mohite SY, Gharal DB, Ranveer RC, Sahoo AK, Ghosh JS. Development of health drink enriched with processed garden cress seeds. American Journal of Food Technology. 2012;**7**(9):571-576

[14] Zia-Ul-Haq M, Ahmad S, Calani L, Mazzeo T, Rio DD, Pellegrini N, Feo VD. Compositional study and antioxidant potential of Ipomoea Hederacea Jacq and *Lepidium sativum* L. seeds. Molecule. 2012;**17**:10306-10321

[15] Zia-Ul-Haq M, Cavar S, Qayum M, Imran I, DeFeo V. Compositional studies, antioxidant and antidiabetic activities of *Capparis decidua* (Forsk.) Edgew. International Journal of Molecular Sciences. 2011;**12**:8846-8861

[16] Marangoni A, Alli I. Composition of the seeds and pods of the tree legume *Prosopis Juliflora*. Journal of the Science of Food and Agriculture. 1988;**44**:99-110

[17] Garattini S. Glutamic acid, twenty years later. The Journal of Nutrition. 2000;**130**:901S-909S

[18] Reeds PJ. Dispensable and indispensable amino acids for humans. The Journal of Nutrition. 2000;**130**(7):1835S-1840S

[19] Sofowora A. Medicinal Plants and Traditional Medicine in Africa. Ibadan, Nigeria: Spectrum Books Ltd; 1993

[20] Corbett P. It is time for an oil change, opportunities for high oleic vegetables oils. Inform. 2003;**14**:480-481

[21] Diwakar BT, Dutta PK, Lokesh BR, Naidu KA. Physicochemical properties of garden cress (*Lepidium sativum L.*) seed oil. Journal of the American Oil Chemists' Society. 2010;**87**:539-548

[22] Budin JT, Breene WM, Putnam DH. Some compositional properties of Camelina (*Camelina sativa L. Crantz*) seeds and oils. Journal of the American Oil Chemists' Society. 1995;**72**:309-315

[23] Appelquist LA. Composition of seeds of Cruciferae oil crops. Journal of the American Oil Chemists' Society. 1971;**48**:851-859

[24] Esteves AM, Saenz C, Hurtado ML, Escobar B, Espinoza S, Suarez C. Extraction methods and some physical properties of mesquite (Prosopis Chilensis (Mol) Stuntz) seed gum. Journal of the Science of Food and Agriculture. 2004;**84**:1487-1492

[25] Pearson D. The Chemical Analysis of Foods. 7th ed. Edinburgh: Churchill Livingstone; 1981. p. 504-530

[26] Knothe G. Structure indices in FA chemistry, how relevant is the iodine value. Journal of the American Oil Chemists' Society. 2002;**79**:847-854

[27] White PJ. Fatty acids in oils and seeds (vegetable oils). In: Chow CK, editor. Fatty Acids in Food and their Health Implications. 3rd ed. Florida: CRC Press; 2007. p. 228-229

[28] Mathews S, Singhal RS, Kulakrni PR. Some physicochemical characteristics of *Lepidium sativum* (haliv) seeds. Nahrung. 1993;**1**:69-71

[29] Kapseu C, Parmentier M. Fatty acid composition of some vegetable oils from Cameroon. Sciences des Aliments. 1997;**17**:325-331

[30] Brigelius-Flohe R, Kelly FJ, Salonem JT, Neuzil J, Zingg JM, Azzi A. The European perspective on vitamin E: Current knowledge and future research. The American Journal of Clinical Nutrition. 2002;**76**:703-716

[31] Monahan FJ, Gray JI, Asghar A, Haug A, Shi B, Bukley DJ. Effect of dietary lipid and vitamin E supplementation on free radical production and lipid oxidation in porcine muscle microsomal fractions. Food Chemistry. 1993;**46**:1-6

[32] Sharma S, Agarwal N. Nourshing and healing powers of garden cress (*Lepidium sativum* Linn.) a review. Indian Journal of Natural Product Research. 2011;**2**:292-297

[33] Dandge PB, Kasabe PJ, Patil PN, Kamble DD. Nutritional, elemental analysis and antioxidant activity of garden cress seeds. International Journal of Pharmacy and Pharmaceutical Sciences. 2012;**4**(3):392-395

[34] Pattnaik AK, Effect of herbal additives on lactating ruminants with or without subclinical mastitis, Thesis submitted to Orissa university of Agriculture and Technology, Bhubaneswar (Orissa) India, 2003.

[35] Lynch SR, Cook JD. Interaction of vitamin C and iron. Annals of the New York Academy of Sciences. 1980;**365**:32-44

[36] Monsen ER. Iron nutrition and absorption: Dietary factors which impact iron bioavailability. Journal of the American Dietetic Association. 1988;**88**:786-790

[37] Agarwal N, Sharma S. Appraisal of garden cress (*Lepidium sativum* L.) and product development as an all pervasive and nutrition worthy food stuff. Annals of Food Science and Technology. 2013;**14**(1):2013

[38] Umesha SS, Manohar RS, Indiramma AR, Akshitha S, Akhilender Naidu K. Enrichment of biscuits with microencapsulated omega-3 fatty acid (alpha-linolenic acid) rich garden cress (*Lepidium sativum*) seed oil: Physical, sensory and storage quality characteristics of biscuits. LWT – Food Science and Technology. 2014;**30**:1-8

[39] Sahraiyan B, Naghipour F, Karimi M, Davoodi MG. Evaluation of *Lepidium sativum* seed and guar gum to improve dough rheology and quality parameters in composite riceewheat bread. Food Hydrocolloids. 2012;**30**:698-703

[40] Varsha SZ, Rohini D. Biofortification of biscuits with garden cress seeds for prevention of anaemia. Asian Journal of Home Science. 2007;**2**:1-5

[41] Camilla H, Anderson GS, Jacobsen S, Molgaard C, Henrik F, Sangild PT, Michaelson KF. The use of whey or SMP in fortified blended food for vulnerable group. The Journal of Nutrition. 2008;145-161

[42] FAO/WHO. Fats and Oil in Human Nutrition: Report of a Joint Expert Consultation, FAO food and nutrition paper. Vol. 57; 1994; 91 Rome

[43] Wien M, Raiaram S, Oda K, Sabate J. Decreasing the linoleic acid to alinoleic acid diet ratio increase eicosapentaeonic acid in erythrocyte in adults. Lipids. 2010;**45**:683-692

[44] Simopoulos AP. Essential fatty acid and human health and chronic diseases. The American Journal of Clinical Nutrition. 1999;**70**:S560-S569

[45] Gerd S, Josef E. The opposing effects of n-3 and n-6 fatty acids. Progress in Lipid Research. 2008;**47**:147-155

[46] Umesha SS, Naidu KA. Vegetable oil blends with $\alpha$-linolenic acid rich garden cress oil modulate lipid metabolism in experimental rats. Food Chemistry. 2012;**135**:2845-2851

Chapter 15

# Genetic Improvement of Oilseed Crops Using Modern Biotechnology

Diego Villanueva-Mejia and Javier Correa Alvarez

Additional information is available at the end of the chapter

http://dx.doi.org/10.5772/intechopen.70743

**Abstract**

In 2009, big challenges facing the agricultural sector in the twenty-first century were presented to the world. Human population growth, increased life expectancy, loss of biodiversity, climate change and accelerated land degradation are the main factors contributing to rethink agriculture system production. In that scenery, modern biotechnology has set a stage for the advancement of agricultural practices and it is clearly an important ally to apply a broad array of technologies and innovative systems where they are most needed, such as enhancing crop productivity, increasing yields, and ultimately ensuring food security. One of the biggest challenges is related to technify production systems, but with no doubt, developing genetic improvement toward getting an efficient and sustainable agriculture, generating new seed qualities (new traits), such as, among others, to upset fatty acids content in oilseed crops have been growing up significantly due to industry interest. In this study, a review about the main advances in genetic improvement of some oilseed crops, starting with omics to understand metabolic routes and to find out key genes in seed oil production, and also, getting in use of modern biotechnology to alter the production of fatty acids, and to face biotic challenges in oilseed crops is presented.

**Keywords:** seed oil, oilseed crops, fatty acids, genetic improvement, genetic engineering, modern biotechnology

## 1. Introduction

Over the last decades, the adoption of oilseed crops has been growing up significantly due to industry interest in the composition of their seed oils, which are made up of a wide range of fatty acids with six predominant types: 16 or 18 carbon palmitic, stearic, oleic, linoleic and linolenic acids, and 12 carbon lauric acid, as well as other unusual fatty acids produced by wild plant species include those with chain lengths between 8 and 24 carbons [1]. Due to their structure and composition, those oils are used as food/industrial feed [2] and as a range of product applications such as surfactants, soap, detergents, lubricants, solvents, paints, inks, chemical feedstocks and cosmetics [1]. In this study, a review about the main advances in genetic improvement of oilseed crops, starting with omics to understand metabolic routes and to find out key genes in seed oil production, and also, getting in use of modern biotechnology including genetic engineering and new breeding techniques (NBTs), a modern-breeding tool that has allowed the functional study of genes with potential application for breeding in agriculture, focusing on oilseed crop genetic improvement with high precision and less uncertainty (avoiding whole genomes crossing), and of course, in less time is presented; those scientific efforts where it was sought to upset fatty acids production or biotic tolerance will also be presented.

## 2. Oilseed crops

Seed oil is mainly obtained from plants recognized as oilseed crops, among them are: soybean (*Glycine max*), rapeseed/canola (*Brassica napus*), palm (*Elaeis guineensis*), mustard (*Sinapis hirta*), sunflower (*Helianthus annuus*), cottonseed (*Gossypium hirsutum*), flax (*Linum usitatissimum*), peanut (*Arachis hypogaea*), camelina (*Camelina sativa*), castor bean (*Ricinus communis*), jatropha (*Jatropha curcas*), tung tree (*Aleurites fordii*), jojoba (*Simmondsia chinensis*), sacha inchi (*Plukenetia volubilis*), niger seed (*Guizotia abyssinica*) and others [2]. Over the last decades, the adoption of these crops has been growing up significantly, reaching in 2014 more than 300 million hectares of oil crops worldwide cultivated [**Figure 1**] [3]. The main reason for this growth is due to seed oils are so attractive for industries, as mentioned in the previous paragraph, but also for the possibility to use their sub-products (metabolites) in biofuels development [4] and approaching their polyhydroxyalkanoates (PHAs) in response to petrol-based plastics waste's problems and harmful effects on the environment [5].

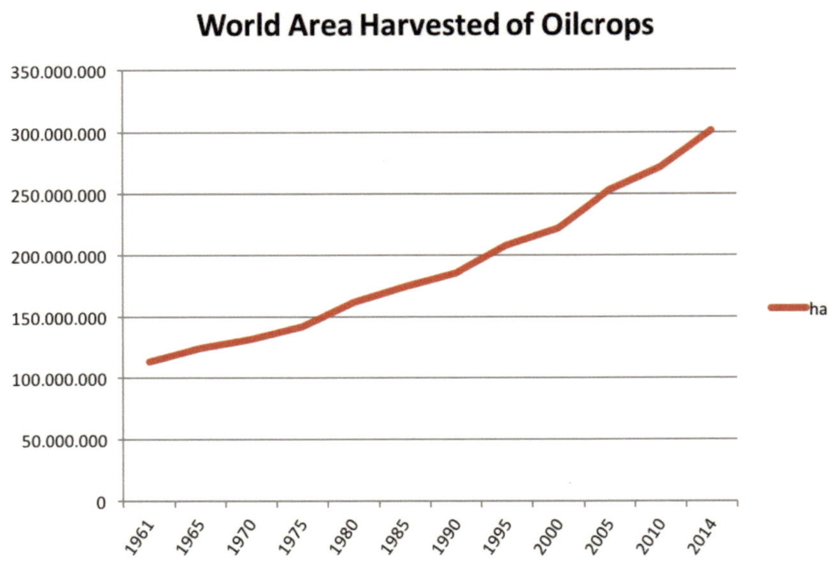


**Figure 1.** Total world area of oil crops harvested between 1961 and 2014 (FAOSTAT).

Seed oils applications depend on physical and chemical properties of their fatty acids composition [6]. Those oils are mainly composed of five major fatty acids, including the saturated palmitic (C16:0) and stearic (C18:0) acids, the monounsaturated oleic acid (C18:1), and the polyunsaturated LA (C18:2) and ALA (C18:3) [6]. A large variety of other less common but not less important fatty acids can be found in different species and used for various industrial applications. These fatty acids vary in the number of carbons in the chain (from 8 to 24), the number of double bonds, and the presence of epoxy, hydroxyl and other functional groups [7].

Despite the industry's significant interest in oil crops, it is reasonable to mention that agriculture has a challenging future. The Food and Agriculture Organization of the United Nations (FAO) in 2009 presented the big challenges facing the agricultural sector in the world for near future [8]. Human population growth, increased life expectancy, loss of biodiversity, climate change and accelerated land degradation are main factors contributing to rethink agriculture system production. Thus, there is a need to technify agricultural production systems, but without doubt, developing genetic improvement toward getting an efficient and sustainable agriculture, generating new seed qualities (new traits), such as, among others, high content of PUFAs in oilseed crops, it will be an aiming. Biotechnology will be fundamental to overcome these challenges. Genetic engineering techniques may play an important role by elevating the content of individual fatty acids or drastically changing the oil quality by the introduction of a new fatty acid,

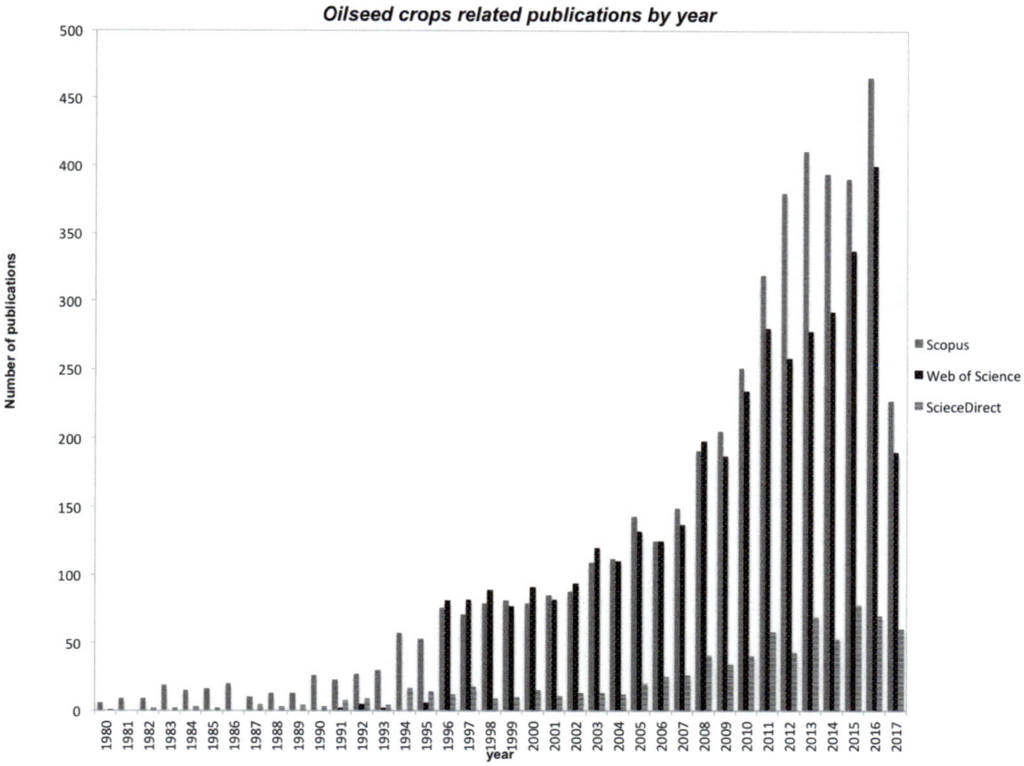


**Figure 2.** Growing in the number of publications with the words "oildseed crops" in the title, abstract or keywords of scientific articles. Records subtracted from literature databases: Scopus (bars in grey), Web of Science (bars in black) and ScienceDirect (bars with horizontal lines).

thus increasing raw materials available for oleochemistry. In this perspective, it has came growing research efforts of scientist around the world seeking to expand the knowledge barrier on oilseed crops. **Figure 2** shows continuous growth in the number of scientific publications in this field, records subtracted from literature databases: Scopus, Web of Science, and ScienceDirect.

## 3. Understanding metabolic routes in oilseed crops

Oilseed plants represent an important renewable source of fatty acids because they accumulate them in the form of triacylglycerol (TAG) as major storage components in seeds [9]. In plants, the reactions for de novo fatty acid synthesis begin in plastids [10] and then exported to the cytoplasm following two inter-related metabolic pathways: an acyl-CoA-dependent pathway and an acyl-CoA-independent pathway [11].

In the dependent pathway, commonly known as the Kennedy pathway, the priming and elongation of nascent acyl chains requires acetyl- and malonyl-CoA, respectively, as direct precursors up to eighteen carbons in length [12]. In this pathway, the glycerol-3-phosphate acyltransferase (G3PAT) is the first enzyme that catalyzes the transfer of a fatty acid to glycerol-3-phosphate (G3P) to form lysophosphatidic acid (LPA). Then, the LPA is acylated by the lysophosphatidic acid acyltransferase (LPAAT) to yield phosphatidic acid (PA). Next, PA is dephosphorylated by the phosphatidic acid phosphatase (PAP) to form diacylglycerol (DAG) and finally, a diacylglycerol acyltransferase (DGAT) catalyzes the acylation of DAG to the production of TAG [13]. In the acyl-CoA-independent pathway, an alternative enzyme is used for the final acylation reaction, termed phospholipid:diacylglycerol acyltransferase (PDAT). PDAT directly transfers an acyl group from phosphatidylcholine (PC) to DAG, producing TAG [14].

Desaturation steps for fatty acids are catalyzed by plastidial stearoyl-acyl carrier protein (ACP) desaturases. After termination, free fatty acids are activated to CoA esters, exported from the plastid, and assembled into glycerolipids at the endoplasmic reticulum (ER) [9]. In addition, further modifications (desaturation, hydroxylation, elongation, etc.) occur in the ER while acyl chains are esterified to glycerolipids or CoA [15]. The low polarity of TAG is believed to result in the accumulation of this lipid between bilayer leaflets leading to the budding of storage organelles termed oil bodies [9]. The accumulation of hydroxy fatty acids depends on many factors, including the performance of the desaturases and efficient channeling of hydroxy fatty acids into storage triacylglycerols [16]. Fatty acid dehydrogenase (FAD) catalyzes the desaturation reaction, leading to the formation of unsaturated FA. Interestingly, studies have revealed that some desaturase enzymes (such as the FAD2 and FAD3 genes) could be regulated at the transcriptional level or at the post-translation level in response to low-temperature induction in model plants [17, 18]. Other important enzyme is FAH12 which belongs to a large family of fatty acid modification enzymes that are related to the *Arabidopsis* oleate D12-desaturase (FAD2) protein, which is responsible for the synthesis of polyunsaturated fatty acids [19].

High-quality RNAseq data have allowed the identification and an accurate quantification of expression of transcription factors and key genes related with lipid metabolic pathways in soybean [20], *Jatropha curcas* [21], *Arabidopsis* [22], peanut [23] and castor bean [24]. In castor, comparison of expression between tissues allowed identification of candidate genes which may be important for triricinolein synthesis in seed in addition to the oleate-12 hydroxylase. Moreover, in purified endoplasmic reticulum from castor endosperm, the site of TAG synthesis, less than 10 genes were found being differentially expressed. Two of these genes, the DGAT2 and PDAT1A, were cloned in transgenic plants expressing the oleate-12 hydroxylase, increasing 18:1-OH incorporation into seed oils and also the expression of additional genes [24].

Gathering the RNAseq information and advances in plant transformation technology is possible now engineering plants for the production of oilseed fatty acids. In a remarkable research, Arabidopsis plant was modified introducing a fatty acid hydroxylase from castor plant, it leading to produce some ricinoleic acid and an unusual fatty acid in the seed [25]. Reactions in triacylglycerol biosynthesis have also been manipulated to increase seed oil content. Studies have suggested that the level of DGAT activity during seed development may have a substantial effect on the flow of carbon into seed oil. Thus, overexpression of cDNAs encoding either *Arabidopsis* DGAT1 or a variant of *B. napus* DGAT1 during seed development in *B. napus* resulted in increased seed oil content under both greenhouse and field conditions [26].

However, in the last decade, the scientists have realized that the manipulation of single genes only contribute with limited value to change the metabolic pathways. Nowadays, there are strategies focused on more complex approaches involving simultaneous overexpression or suppression of multiple genes to achieve optimal metabolic flux [27]. Understanding a metabolic network would facilitate the production of natural products and the synthesis of novel molecules in a predictable and useful manner [16]. For this reason, the metabolic engineering in oilseed plants has attracted industrial and academic researchers in the last decade.

## 4. Modern biotechnology for genetic improvement in oilseed crops

The Convention on Biological Diversity (CBD) has defined biotechnology as "any technological application that uses biological systems, living organisms, or derivatives thereof, to make or modify products or processes for specific use" [28]. In fact, biotechnology includes several agricultural as well as food manufacturing tools and techniques. However, when a biotechnology development uses new deoxyribonucleic acid (DNA) techniques, molecular biology, and reproductive technological applications ranging from gene transfer to DNA typing to cloning of plants and animals, it has been considerable modern biotechnology [29]. The potential of modern biotechnology is widely known, as it makes the use of recombinant DNA technology to generate modified microorganisms, plants and animals to make them more suitable for several potential applications: improved crops, production of new antibiotics and hormones, xenotransplantation, gene therapy, bioremediation, and genetic editing, one of the most recent techniques.

Genetic engineering crops based on recombinant DNA technology were first introduced for commercial production in 1990s. This technology uses the identification, isolation and manipulation followed by the introduction of desired gene(s) from one organism (for example, a plant or bacteria) to another, thus giving rise to a transgenic or genetically modified organism. This technique has been fast replacing plant breeding so as to incorporate characteristics that are impossible to achieve by breeding. Biotechnology has the potential to help overcome many of the short-comings of the species being promoted, especially where exogenous genes are needed because there are characters that are difficult to produce by traditional breeding, or where characters tissue-specific or temporal expression or suppression of endogenous genes would be valuable [30]. For oilseed crops, modern biotechnology should allow the production of plants with specific fatty acids content.

In the following paragraphs, main advances in plant genetic improvement using modern biotechnology, focused on oilseed crops, those scientific efforts in soybean (*Glycine max*), sunflower (*Helianthus annuus*), canola (*Brassica napus*), palm (*Elaeis guineensis*), castor bean (*Ricinus communis*), cotton (*Gossypium* spp.), peanut (*Arachis hypogaea*) and olive (*Olea europaea*) where it was sought to upset fatty acids production or biotic tolerance will be presented.

### 4.1. Soybean (*Glycine max* L.)

Soybean, *Glycine max* L. Merr., is a major crop that produces the best vegetable oil and protein for use in food and beverage production worldwide. Among legume species, soybean has the highest protein content (around 40%), while other species have a protein content between 20 and 30%. On the other hand, cereals have a protein content ranging from 8 to 15%. Other interesting point for oleochemistry is that soybean also contains about 20% oil [31]. Soybean oil is a complex mixture of five fatty acids: palmitic, stearic, oleic, linoleic, and linolenic acids. Nowadays, soybean oil is currently found in food products such as margarine, salad dressings and cooking oils, and industrial products such as plastics and biodiesel fuel. Lecithin, a natural emulsifier and lubricant extracted from soybean oil, is used in applications from pharmaceuticals to protective coatings [32].

Due to its importance as a crop, genetic transformation techniques have been used extensively to improve the crop's valuable traits. Herbicide-tolerant (Roundup Ready) Soybean (*Glycine max* L. Merrill) resistant to glyphosate (N-phosphonomethylglycine) was the first transgenic variety introduced for commercial production in 1995 [33]. In the contrary way to seek increase in fatty acids production, the goal was to give a competitive advantage to soybean favoring desirable plants (in this case soybean) and inhibiting undesirable plants by the application of glyphosate, the active ingredient of the non-selective herbicide Roundup. A glyphosate-tolerant soybean line was obtained through expression of the bacterial 5-enolpyruvylshikimate-3-phosphate synthase (EPSP synthase, EPSPS) enzyme from *Agrobacterium* sp. strain CP4, under the cauliflower mosaic virus 35S promoter (P-E35S), with the *Petunia hybrida* EPSPS chloroplast transit peptide (CTP) and a portion of the 3′ non-translated region of the nopaline synthase gene (NOS 3′) terminator. This soybean line was highly tolerant to glyphosate, showing no visual injury after application of up to 1.68 kg acid equivalent (a.e.) $ha^{-1}$ of glyphosate under field conditions.

In terms of genetic transformation methods, many reports describing soybean transformation by particle bombardment using meristems as the target tissue have been published. The Biolistics particle delivery system for soybean transformation was evaluated in two different regeneration systems: from shoot tips obtained from immature zygotic embryos of the cultivar Williams 82, and the second was somatic embryogenesis from a long-term proliferative suspension culture of the cultivar Fayette [34]. A method for high-frequency recovery of transgenic soybean by combining resistance to the herbicide imazapyr as a selectable marker, multiple shoot induction from embryonic axes of mature seeds and biolistics techniques was made by [35]. A targeting method to insert genes with biolistics to predefined soybean genome sites using the yeast FLP-FRT recombination system was made by [36]. A double-barreled gene *gus* device was used to improve transformation efficiency and then study soybean resistance (R) gene-mediated responses to effectors, induction and suppression of cell death by a wide variety of pathogen and host molecules [37]. The abscisic acid (ABA)-independent dehydration responsive element binding (DREB) gene family from *Arabidopsis thaliana* was inserted into soybean plants using biolistics to improve tolerance to abiotic stresses [38]. In the same way, [39] introduced an activated form of abscisic acid-responsive element binding protein (AREB1) into soybean plants to improve water deficit stress. Finally, the biotechnological potential of plastid genetic engineering was used to develop a reproducible method to generate plastid transformants in soybean. To sum up, transformation vectors were delivered to embryogenic cultures by the particle gun method and selection performed using the *aadA* antibiotic resistance gene, getting early homoplasmy and avoiding further selection cycles [40].

Genetic engineering approaches have been applied to enrich the content of soybean oil for a particular fatty acid or class of fatty acids. One of those examples was made by [41]. These authors developed transgenic soybean seeds by down-regulating the expression of *FAD2* genes that encode the enzyme that converts the monounsaturated oleic acid to the polyunsaturated linoleic acid. Those transgenic soybean seeds had oleic acid content of approximately 80% of the total oil, whereas conventional soybean oil contains oleic acid at levels of 25% of the total oil. With the same aim, [42] reported the creation of high oleic acid soybean varieties using targeted mutagenesis with transcription activator-like effector nucleases (TALENs) to bind and cleave specific DNA sequence targets in the *FAD2-1A* and *FAD2-1B* genes with high efficiency. These authors reported that mutant soybean plants produced nearly four times more oleic acid than the wild-type parents (80% vs. 20%, respectively). Furthermore, because they use a technique considered "genetic editing," the soybean lines lacked foreign DNA in their genomes and are thus not transgenic. Rather, they only have small deletions of coding sequence in the FAD2-1 gene targets.

On the other hand, regarding to biotic factors, [43] developed transgenic soybean to improve resistance against SMV. HC-Pro coding sequences were introduced within a RNAi inducing hairpin construct and *Agrobacterium*-mediated transformation system. Then, their response to viral infection was analyzed. The inhibition of HC-Pro expression enhanced viral resistance after viral infection, when compared to the resistance of virus-susceptible non-transgenic plants. RNAi induced by the hairpin construct of the SMV HC-Pro sequence effectively confers viral resistance. Among others, these results have proven the usefulness of RNAi-mediated resistance for crop improvement. Other cases of development of transgenic soybean have been reported by [44]. These authors show some scientific references where synthetic *Bacillus thuringiensis cry* genes were used to develop transgenic soybean and prevent agronomic loses caused by insects from Lepidoptera order such us *Anticarsia gemmatalis, Pseudoplusia includens* and *Helicoverpa zea*.

### 4.2. Sunflower (*Helianthus annuus* L.)

Sunflower is one of the most important oilseed crops cultivated on a global level. Its seeds have always been ground and pounded into flour for making bread, cracked and eaten as snacks, mixed with vegetables, and extracted for oil. The seeds are also a source of purple dye and have medicinal uses [45]. Sunflower seeds are composed by 20% protein and 50% fat. In this crop oil, up to 90% of its fatty acids are unsaturated, namely oleic (C18:1, 16–19%) and linoleic (C18:2, 68–72%) acids. The remaining 10% of its fatty acids are palmitic (C16:0, 6%), stearic (C18:0, 5%), and minor quantities of myristic (C14:0), myristoleic (C14:1), palmitoleic (C16:1), arachidic (C20:0), behenic (C22:0) [46].

Several scientific efforts have been made to develop genetic improvement methods in sunflower, using modern biotechnology.

Perhaps one of the earliest works in sunflower was developed by [47], which introduced plasmid into isolated sunflower protoplast. Another effort was made by [48] who used microprojectile bombardment of half-shoot apices followed by co-culture with *Agrobacterium tumefaciens*, to obtain transgenic shoots. However, [49] modified a step in which shaking of explants with glass beads replaced the microprojectile bombardment stage used by [48]. In an attempt to reduce or eliminate the *in vitro* regeneration component of a sunflower-transformation protocol, [50] infected 2-day-old seedlings, each with one cotyledon detached, with *A. tumefaciens* strain LBA4404 carrying a specific plasmid. On the other hand, to overcome the generation of chimeric plants, [51] used zygotic embryos, the latter being 4–6 mm in size and cut transversely below the cotyledons, and then, explants were cultured in the dark for 1 day before being bombarded with gold particles and co-cultured with *A. tumefaciens*. [52] reported an alternative procedure to wound cells of target sunflower explants that involved treatment of the explants with the cell wall-degrading enzymes Cellulase Onozuka R-10 (0.1% w/v) and pectinase Boerozyme M5 (0.05% w/v). After that, but before *Agrobacterium* inoculation, a sonication (50 MHz, 2, 4, 6 s) step of explants showed that transient expression of *gus* or *gfp* transgenes was increased.

One of the most important aspects of any transformation protocol is an efficient selection of transgenic plants. On the way to develop a procedure to minimize the number of transgenic escapes, [53] germinated sunflower seeds for 24 h on half-strength MS-based medium, before cutting the seeds to give two half embryos, each with one cotyledon. Once that, cotyledon explants were inoculated with *A. tumefaciens* carrying a vector with the *nptII* and the *gus* genes.

Leaving aside developments made around transformation methods, and focusing on advances toward genetic improvement with some functional characteristics, some efforts to improve oil production in sunflower have been made recently. Dağüstü et al. [54] introduced the *Erwinia uredovora* phytoene desaturase (*crtl*) and hydroxymethylglutaryl-CoA (*Hmgr-CoA*) genes into sunflower, which have potential to increase oil quality. On the other hand, [53] developed transgenic sunflower plants resistant to *Verticillium dahlia* and *Sclerotinia sclerotiorum* introducing antifungal genes, including *gln2* (a glucanase) from *Nicotiana tabacum*, a chitinase (*ch5B*) from *Phaseolus vulgaris*, an osmotin gene (*ap24*) from *N. tabacum*, and a gene coding for a ribosome inhibitor protein (rip). In the same way, [55] developed transgenic sunflower resistant to the herbicide phosphinothricin, herbicide resistance also being exploited to select the transgenic plants.

Some research interests have been around decreasing levels of palmitic and stearic acid of sunflower, due to their contribution on increasing the plasma cholesterol level in humans, associated with heart disease. Škorić et al. [45] induced mutations via seed treatment with γ-rays, X-rays, and mutagenic chemicals such as ethyl methanesulfonate (EMS) and dimethyl sulfate (DMS) to generate sunflower genotypes with high levels of C 18:2, C 18:1, C 18:0, C 16:1 and C 16:0.

### 4.3. Canola (*Brassica napus*)

Canola/rapeseed (*Brassica napus*) is considered one of the most important oil sources for edible or industrial uses, being the research to get better oil quality is important to improve rapeseed as a high-quality vegetable oil. Canola oil contains multiple fatty acids, such as palmitic acid, stearic acid, oleic acid, linoleic acid, α-linolenic acid, arachidic acid, erucic acid among others [56]. Due to its high nutritional value, Canola oil is included in human diets where it has been shown to reduce plasma cholesterol levels in comparison with diets containing higher levels of saturated fatty acids. It has demonstrated that consumption of canola oil also influences biological functions that affect various other biomarkers of disease risk [57].

A group of researchers developed transgenic canola seeds with significantly increasing of oil content [58]. Those authors showed that seed-specific overexpression of BnLEC1 and BnL1L genes (from canola), placed under the control of the truncated canola storage protein 2S-1 promoter, which is also known as the *napA* promoter, at an appropriate level substantially increases the seed oil content of the transgenic oilseed plant without detectable negative effects on other major agronomic traits.

In the same way to improve canola oil production, Qi et al. [59] isolated the RNA-binding motifs No2 (RRM2) of the flowering control locus A (FCA) protein (FCA-RRM2) from variety No. 1 "Nannongyou" of Canola, and then, it was introduced in cotyledon nodes using *Agrobacterium rhizogenes*, placed under a 35S-35S promoter (a variant of the cauliflower mosaic virus 35S promoter with higher transcriptional activity) to drive transgenic expression, into pBin438 vector with kanamycin resistance gene (for bacterial selection) and the hygromycin phosphotransferase gene (for plant transformation selection). These authors demonstrated that canola FCA-RRM2 increases plant size, organ size, cell size, plant productivity and oil content. According to the author of that research, these results provide a practical approach for the genetic improvement of this plant.

Aimed at not good perception of erucic acid (cis-13-docosenoic acid) in the canola oil triglycerides, because of presumptive effects on growth retardation and pathogenic changes to internal organs when fed at high concentrations to laboratory animals, a research was made to decrease erucic acid level in Canola plants. Shi et al. [60] reported the development of canola transgenic with change in fatty acids compositions, using *B. napus* cultivar "CY2" as the transgenic recipient of BnFAE1, a fragment involved in the synthesis of very long-chain fatty acids. These authors placed a BnFAE1 fragment driven by *napin A* promoters and then, they co-cultured hypocotyls with *Agrobacterium tumefaciens* EHA105 to introduce the genetic construct in canola cells. Due to CY2 that has high erucic acid (about 40%) and low oleic acid (about 20%) content, the researchers made seed-specific knockdown of BnFAE1, significantly changing the fatty acid composition. They demonstrated that the RNAi construct of BnFAE1 could effectively interfere with mRNA levels of BnFAE1 gene in F1 hybrid seeds derived from crosses between BnFAE1-Ri lines and high erucic acid cultivars. At the end of their research, they got canola transgenic lines with a dramatically decreased erucic acid (less than 3%).

### 4.4. Palm (*Elaeis guineensis*)

Cultivation of the oil palm (*Elaeis guineensis* Jacq.) has expanded tremendously in recent years such that it is considered second as a major source of the world supply of oils and fats [61]. Palm oil is one of the most price-competitive liquid cooking oils in many parts of the world due to its use in food products such as shortenings, margarines and spreads [62]. The benefits of palm oil on human health have been demonstrated scientifically by different authors. Qureshi et al. [63] indicated that palm oil lowers serum cholesterol levels to the same degree as sunflower oil, which is rich in polyunsaturated fatty acids. In the same way, Nesaretnam et al. and Kritchevsky et al. [64, 65] proposed anti-carcinogenic potential of palm oil presumably due to the presence of high levels of vitamin E and tocotrienols.

It is reported that currently palm oil accounts for about 20% of world oils and fats production. It was forecast that, with the increase in world population, the demand for palm oil would grow faster than the rise in supply, so that, the supply of palm oil would also need to be increased to meet the above demand. It was therefore crucial to increase the yield of palm oil, improve its oil quality, and produce novel products via genetic engineering, as it could be achieved faster this way than by conventional breeding ways.

According to [61], palm fruits have two storage tissues, mesocarp and kernel, that can be the target for accumulating genetically modified products. The substrates and intermediates implied in the production of storage oil or protein in these tissues may be channeled to alter the levels of existing products or to produce novel value-added products without deleterious effects on the plants. During oil palm fruit development, at level of period of oil accumulation as well as fatty acid composition, the mesocarp and kernel tissues show differences. A study reported by [66] showed the regulation of the gene expression during period of oil synthesis in both tissues. The expression profile of the mesocarp-specific gene in different oil palm tissues, as well as at different developmental stages of the mesocarp and at the cellular level indicated a strong correlation with that of a fatty acid biosynthetic gene, stearoyl-ACP desaturase. Using promoter-reporter constructs, assays and transformations, these authors demonstrated that this promoter is conducive to the development of genetic improvement research modern biotechnology, since specific genes can be located there generating high degree of expression.

Different efforts have been made to get biotechnological palm crops which can generate metabolites of interest. Genetic engineering in oil palm is relatively recent as it was initiated at Malaysian Palm Oil Board (MPOB) in the late 1980s [67]. Related to transformation protocols, particle bombardment and *Agrobacterium*-mediated transformation have been used to introduce genes into oil palm, and stable transformation has been achieved using both methods. Masani et al. [67] reports the development of a biolistic protocol for production of glufosinate-resistant transgenic oil palm, and considering its success, several hundreds of embryogenic calli have been bombarded with genetic constructs which contain genes involved in fatty acid biosynthesis to augment the accumulation of oleic acid, stearic acid, polyhydroxybutyrate (PHB) and polyhydroxyvalerate (PHBV). Likewise, [67] was able to develop some new transformation protocols based on DNA microinjection and PEG-mediated transfection demonstrating that protoplasts are appropriate as a target for oil palm genetic engineering. These authors successfully expressed a reporter gene encoding green fluorescent protein (GFP) allowing the rapid and efficient generation of non-chimeric transgenic callus without the use of standard selectable markers. In the same way, [68] reported that oil palm embryogenic calli were bombarded with a transformation vector, p35SCaMV-sgfpS65T, carrying a modified version of *gfp* gene driven by the 35S promoter. Those authors refer that upon bombardment, the expression of *gfp* in embryogenic calli was monitored visually while the regeneration of the embryogenic calli was ongoing.

Seeking to improve oleate oil palm production, some studies had some strategies where transformation vectors were constructed: (1) antisense palmitoyl-ACP thioesterase gene driven by CaMV35S promote, (2) antisense palmitoyl-ACP thioesterase and sense KAS II genes driven by mesocarp-specific promoter, (3) antisense palmitoyl-ACP thioesterase, sense KAS II and sense stearoyl-ACP desaturase genes driven by a mesocarp-specific promoter and, (4) antisense palmitoyl-ACP thioesterase, antisense oleoyl-CoA desaturase, sense KASII and sense stearoyl-ACP desaturase genes driven by a mesocarp-specific promoter, and then, those constructs were bombarded into oil palm embryogenic cultures. Molecular and biological assays were made and some plantlets were transferred to soil in the biosafety screenhouse [69]. It is important to refer that KAS II is one of the main enzymes contributing toward high palmitic acid.

Stearate oil palm is an another key oil with great industrial interest such as cocoa butter substitute and personal care products such as lotions, shaving cream, and rubbing oils. Biochemical studies have demonstrated that oil palm contains an active stearoyl-ACP desaturase, and therefore down-regulating the activity of stearoyl-ACP desaturase could reduce the conversion of stearoyl-ACP into oleoyl-ACP [62]. Three transformation vectors were constructed carrying an antisense stearoyl-ACP desaturase gene driven by ubiquitin, CaMV35S, and mesocarp-specific promoters [70]. These authors reported that those constructs were bombarded into oil palm embryogenic cultures and a few lines of resistant polyembryogenic cultures and then full plant regeneration were obtained. The transgenic plant's stearic acid content increased as a result of concomitant reduction of oleic acid levels.

Regarding to biotic factors, [71] introduced a synthetic *cryIA(b)* gene into oil palm while biobalistic. When *cryIA(b)* gene was expressed, it produced proteins which upon crystallization are highly toxic to Lepidoptera among which, *Metisa plana* is a major insect pest for oil palm. Those authors developed a rapid detection system for evaluating transgene expression among putative transformed tissues combining RT-PCR and Southern blotting. The procedure developed is reported very sensitive and rapid, and eliminates the long waiting period for transgenic plants to reach maturity.

### 4.5. Castor bean (*Ricinus communis* L.)

Castor or castor bean (*Ricinus communis* L.) is a major non-edible oilseed crop grown extensively in the tropical, subtropical, and temperate regions of the world for its highly valued oil.

Castor oil is rich (80–85%) in an unusual ricinoleic acid, hydroxyl fatty acid [72] which is used as an input (raw material) for the development of several different industrial products, among others, high quality lubricants for aircrafts, alternative to lubricants/additives in petroleum diesel, coatings, polishes, paints, textile dye, surfactants, plastics, resins, waxes, soaps, drugs, and cosmetics [73, 74].

Introduction of foreign genetic material defining specific agronomically important traits into castor through genetic transformation techniques have been attempted. Transgenic research in castor bean (*Ricinus communis* L.) has been undertaken for the development of insect resistant or ricin free genotypes.

Castor semilooper (*Achaea janata* L.) is a voracious feeder causing extensive defoliation and yield losses on castor bean in the semiarid tropics of India and other countries. Seeking to face this biotic challenge, [75] developed a stable genetic transformation system for introducing desired alien gene(s) into castor using embryo axes through *Agrobacterium*-mediated approach. They reported, the successful production of transgenic castor with *cry1Ab* gene affording resistance against castor semilooper. The synthetic delta endotoxin gene *cry1Ab* was under the control of CaMV35S promoter and NOS terminator, and then, cloned into the HindIII site of the vector pSB11bar containing herbicide-resistance gene bar. Focusing on main outcomes, [75] obtained transformed plants with levels of Cry1Ab protein concentration between 0.23 and 0.47 ng/mg in tissue. With respect to biological assays, these authors got insect mortality ranged from 88.91 to 97.25% in larvae fed on the leaves of primary transformants after 3 days of infestation. On the other hand, larvae fed on untransformed control leaves (which do not have Cry1Ab proteins) showed least (13.92%) insect mortality. Then, larval mortality was attributable to the varied levels of Cry1Ab protein expressed in different transformants.

In the same way, toward obtaining biotechnological strategies for the control of castor semilooper, [76] reported the use of *Agrobacterium*-mediated and biolistic bombardment methods for expression of the *cry1EC* gene in castor bean. Trying to face *Spodoptera litura*, [77] recently reported on the development of an *A. tumefaciens* mediated in planta transformation protocol for castor bean. These authors developed transgenic lines from 2-day-old seedlings infected with *Agrobacterium*, EHA105/pBinBt8 harboring *cry1AcF* gene and then, molecular and expression analysis confirmed the transgenic nature and identified high-expressing plants. Bioassay in greenhouse against *S. litura* corroborated strong resistance of transgenic castor bean.

On the other hand, two approaches were used to develop castor cultivars with reduced levels of toxin. [78] reported on "knocking out" the genes responsible for ricin production as well as genes responsible to produce ricinine and CB-1A. Then, conventional sexual hybridization was used to develop F6 lines of castor that have a 75–70% reduction in ricin and *Ricinus communis* agglutinin toxins. Those plants were combined with transgenic castor plants which had a great potential to reduce ricin content (for >99%). Subsequent selection in segregating generations resulted in a 99.9% reduction in protein toxins allowing development of castor cultivars with a very low level of this kind of toxin (ricin). For [78], these cultivars would allow to increase castor production in the U.S. while increasing the value of the high protein meal remaining after castor oil extraction. The denouement of this research is not clear, but recently, one of their researchers published a method to address the problem (concerns about the presence of the protein toxin ricin) with simple procedures to reduce or eliminate the toxin from the seed meal remaining after processing the seed for oil [79]. They used a protease treatment during extraction to provide an effective means to eliminate the ricin present after oil processing. In addition, heating intact seed via microwave irradiation was a way to demonstrate that they could inactivate considerable ricin activity.

### 4.6. Cotton (*Gossypium* spp.)

Cotton (*Gossypium* spp.) is one of most important fiber crops at world level. According to [80], cottonseed oil represents approximately 16% of the seed weight, and perhaps it is the most valuable product derived from cottonseed. Likewise, cottonseed oil is composed by 26% of saturated palmitic acid (C16:0), 15% of mono-unsaturated oleic acid (C18:1), and 58% of polyunsaturated linoleic acid (C18:2) [80]. In fact, this complementary product (cottonseed oil) has some advantages over soybean oil and rapeseed oil, like good quality and price, so that it is used in foods or as a raw material for biodiesel production [81]. However, some reports warn that cottonseed oil content around oversaturated, polyunsaturated, and monounsaturated fatty acids is unbalanced [80, 82].

Chapman et al. [82] reported the development of transgenic cotton plants with increased seed oleic acid levels. Using an *Agrobacterium*-mediated system transformation, these authors introduced a binary vector previously designed to suppress expression of the endogenous cottonseed enzyme fatty acid desaturase 2 (Fad2) by subcloning a mutant allele from a rapeseed fad2 gene. It is known that FAD2 enzyme, in the endoplasmic reticulum of plant cells, catalyzes conversion of oleic acid to linoleic acid so that, decreasing this enzyme activity would be an increase of oleic acid content in cottonseed oil. At the end of the research, these authors' increased seed oleic acid content ranged from 21 to 30% (by weight) of total fatty acid content in primary transformants and 47% of oleic acid content in their progeny, which represent an increasing of three times comparing with standard cottonseed oil.

Due to consumption of the saturated fatty acid, overall cholesterol levels increases, more specifically low-density lipoprotein (LDL) which is considered "bad cholesterol," and it is well known worldwide that its consumption increases risk of cardiovascular disease [83]; a group of researchers started a study to improve the quality of cottonseed oil. [84] used RNAi technology to regulate fatty acid metabolism of cottonseed inhibiting GhFAD2-1 and GhFATB gene expression levels, simultaneously. These genes encoding the microsomal oleate desaturase and palmitoyl-acyl carrier protein thioesterase, respectively, play significant roles in regulating the proportions of saturated and polyunsaturated fatty acids in cottonseed lipids. Using this technology, they decreased palmitic acid and linoleic acid content and increased oleic acid content, but unfortunately, they got an adverse effect on seed germination and seed vigor. In spite of achieving an adequate balance in the content of fatty acids, thinking in human consumption of cottonseeds oil, it is necessary to explore others effective regulating strategies to improve the quality of cottonseed oil.

On the other hand, recently, Wang et al. [81] reported a genome-wide analysis in several *Gossypium* species and possible ancestral diploids. In that study, authors analyzed a total of 40 Lysophosphatidic acid acyltransferase (LPAAT) genes and found that this gene is involved in increasing oil composition and content which was demonstrated in some experiments in transgenic yeast. This report shows an important way for further studies due to LPAAT genes that are involved in natural cottonseed oil content and variation which should open a possible strategy in development of genetically modified cotton crops with improvement of seed oil content and composition.

### 4.7. Peanuts (*Arachis hypogaea*)

Peanut (*Arachis hypogaea*) is grown worldwide as an oilseed crop. In many countries, peanut seeds do an important contribution to the people diet because they are a good source of proteins and lipids for human nutrition. [85] determined that peanut seeds which are rich in oil (about 50% of seed composition) and oleic acid (18:1), linoleic acid (18:2), palmitic acid (16:0), behenic acid (22:0), eicosenoic acid (20:1), stearic acid (18:0), arachidic (20:0) and lignoceric acid (24:0) are presented (sorted from highest to lowest content). However, the fatty acid composition of peanut oil varies depending on the seed maturity, genotype, growth location, climatic conditions, and they together [86].

Research about transgenic peanut crops has been undertaken for the development of fungi resistant. This crop is susceptible to many types of pathogens including those caused by fungi. Chenaulr et al. [87] reported the development of transgenic peanuts which were introduced two hydrolase genes, a glucanase from alfalfa (*Medicago sativa* L.), and a chitinase from rice (*Oryza sativa* L.) into somatic embryos using biolistic. Although the study focused on seedlings characterization (found up to 37% of hydrolase activity in transgenic lines), these authors assume that transgenic lines obtained could be promising due to high transgene expression what would exhibit some level of resistance to a broad range of fungal pathogens. Following with the same modified peanut lines, Chenault et al. [88] developed an assay under greenhouse conditions where these lines were tested for resistance to *Sclerotinia minor* by inoculation with a mycelial plug. There were lines up to 84% of resistance to the pathogen. On the other hand, the peanut lines considered more resistance kept going in race and were tested for *S. minor* resistance under field conditions [89]. In that report, three transgenic lines showed a significant resistance to the pathogen compared with the wild-type cultivar. Finally, Jonnala et al. [90] determined the oil composition of the best three transgenic lines obtained in the previous report. This author reported similar oil content of all transgenic peanut lines to that wild-type lines, indicating that genetic modification did not cause substantial unintentional changes in peanut chemical composition. In the same way, Ng et al. [91] examined chemical characteristics, volatile components, and olfactory characteristics of those three GM peanut lines (previously tested at field conditions) using gas chromatograph/mass spectrometer (GC/MS) equipped with an olfactory detector. These authors reported minimal variations in nutritional composition between GM peanuts and wild type, indicating that genetic modifications did not cause significant change in peanut.

### 4.8. Olive (*Olea europaea* L.)

Olive oil production and consumption are increasing in importance around the world. Spain is the largest producer with an average 1 million tons per year, followed by Italy and Greece with 560 and 350 thousand tons, respectively [92]. This crop contributes significantly not only to the global economy but also to food security in terms of its nutritional value. It is well known that olive and olive oil play an important role in prevention of coronary heart disease and certain cancers, due to their high levels of monosaturated fatty acids and phenolic compounds [93].

Olive is characterized by a long history of cultivation, as it was one of the first tree species to be domesticated, and by wide diversity. A very large number (over 1600) of cultivated varieties characterize this species. This diploid specie (2n = 46) has a small genome (2200 Mb) and it is predominantly allogamous in nature [94, 95]. During a long period of time, local and old cultivars have been evaluated by different genetics, morphological, and agronomics approaches. Recently, the huge genetic variability of this specie has been evaluated using molecular markers, including SSRs, particularly advantageous because olive is a clonally propagated, perennial, slow growing, highly heterozygous cultivated species with a very large uncharacterized genome [96].

As any other extensive crop, the olive has urgent challenges; they are summarized into six big areas such as: (i) olive growing; (ii) processing, byproduct, and environmental issues; (iii) virgin olive oil sensory quality; (iv) purity, authentication, and traceability; (v) health and nutrition; and (vi) consumers. Moreover, the olive varieties renewal have been hampered by the extreme longevity of olive trees, the long period of juvenility of their offspring, and the diffidence of the public to accept genotypes obtained with advanced biotechnological approaches. Modern biotechnological techniques are suitable for olive improvement because they both allow direct correction of main defects, combining with existing known superior cultivars, and can also support traditional breeding using the great genetic variability present in the species, to guide crossing of genotypes chosen among the olive populations of different sites [96]. Modern biotechnology along with traditional and *in vitro* technologies, can also provide new resistant cultivars to abiotic and biotic stresses, as water deficit and nematodes infection are becoming the major problems in field [97].

However, some aspects of the olive biotechnology remain challenging; for example, the olive propagation is still a laborious practice. As regards traditional propagation, rooting of cuttings and grafting stem segments onto rootstocks are possible. The regeneration of whole plants from ovules, on the other hand, is used only occasionally. Micropropagation of olive is not easy mainly due to explant oxidation, difficulties in explant disinfection, and labor-oriented establishment of *in vitro* shoot cultures [96].

## 5. Conclusion

As it was seen throughout the review, last three decades, for these oilseed crops: soybean (*Glycine max*), sunflower (*Helianthus annuus*), canola (*Brassica napus*), palm (*Elaeis guineensis*), castor bean (*Ricinus communis*), cotton (*Gossypium* spp.), peanut (*Arachis hypogaea*) and olive (*Olea europaea*), it has been evidenced a strong development supported by modern biotechnology, and there should be no doubt that, carefully undertaken, genetic engineering represents a very safe, fast and, low-cost method to enrich important oilseed crops for essential nutritional contents. Ongoing and future research will have to face big challenges in agriculture.

## Acknowledgements

This book chapter was supported by "Component No.1 - Biotechnology", project code 4600 000480 financed by Colombian general system of royalties (SGR, Spanish acronym), in partnership with Secretaría de Agricultura y Desarrollo Rural of Departamento de Antioquia, Universidad Pontificia Bolivariana, Universidad Nacional de Colombia and Universidad EAFIT.

## Author details

Diego Villanueva-Mejia* and Javier Correa Alvarez

*Address all correspondence to: dvillanu@eafit.edu.co

Department of Biological Sciences, School of Sciences, Universidad EAFIT, Medellin, Antioquia, Colombia

## References

[1] Carlsson AS. Plant oils as feedstock alternatives to petroleum—A short survey of potential oil crop platforms. Biochimie. 2009;**91**:665-670

[2] Maheshwari P, Kovalchuk I. Genetic engineering of oilseed crops. Biocatalysis and Agricultural Biotechnology. 2014;**3**:31-37

[3] FAO. FAOSTAT. http://www.fao.org/faostat/en/#home [Accessed March 26, 2017].

[4] Durrett TP, Benning C, Ohlrogge J. Plant triacylglycerols as feedstocks for the production of biofuels. The Plant Journal. 2008;**54**:593-607

[5] Ciesielski S, Mozejko J, Pisutpaisal N, et al. Plant oils as promising substrates for polyhydroxyalkanoates production. Journal of Cleaner Production. 2015;**106**:408-421

[6] Murphy DJ. Biogenesis, function, and biotechnology of plant storage lipids. Progress in Lipid Research. 1994;**33**:71-85

[7] Metzger JO, Bornscheuer U. Lipids as renewable resources: Current state of chemical and biotechnological conversion and diversification. Applied Microbiology and Biotechnology. 2006;**71**:13-22

[8] FAO. La agricultura mundial en la perspectiva del año 2050. Como Alimentar Al Mundo En 2050.

[9] Thelen JJ, Ohlrogge JB. Metabolic Engineering of Fatty Acid Biosynthesis in Plants. Metabolic Engineering. 2002;**4**:12-21

[10] Ohlrogge JB, Kuhn DN, Stumpf PK. Subcellular localization of acyl carrier protein in leaf protoplasts of Spinacia oleracea. Proceedings of the National Academy of Sciences of the United States of America. 1979;**76**:1194-1198

[11] Kroon JTM, Wei W, Simon WJ, et al. Identification and functional expression of a type 2 acyl-CoA:diacylglycerol acyltransferase (DGAT2) in developing castor bean seeds which has high homology to the major triglyceride biosynthetic enzyme of fungi and animals. Phytochemistry. 2006;**67**:2541-2549

[12] Kennedy EP. Biosynthesis of complex lipids. Federation Proceedings. 1961;**20**:934-940

[13] Cagliari A, Margis-Pinheiro M, Loss G, et al. Identification and expression analysis of castor bean (*Ricinus communis*) genes encoding enzymes from the triacylglycerol biosynthesis pathway. Plant Science. 2010;**179**:499-509

[14] Dahlqvist a SU, Lenman M, et al. Phospholipid:diacylglycerol acyltransferase: an enzyme that catalyzes the acyl-CoA-independent formation of triacylglycerol in yeast and plants. Proceedings of the National Academy of Sciences of the United States of America. 2000;**97**:6487-6492

[15] Lacey DJ, Beaudoin F, Dempsey CE, et al. The accumulation of triacylglycerols within the endoplasmic reticulum of developing seeds of *Helianthus annuus*. The Plant Journal. 1999;**17**:397-405

[16] Lu C, Fulda M, Wallis JG, et al. A high-throughput screen for genes from castor that boost hydroxy fatty acid accumulation in seed oils of transgenic Arabidopsis. The Plant Journal. 2006;**45**:847-856

[17] O'Quin JB, Bourassa L, Zhang D, et al. Temperature-sensitive post-translational regulation of plant omega-3 fatty-acid desaturases is mediated by the endoplasmic reticulum-associated degradation pathway. The Journal of Biological Chemistry. 2010;**285**:21781-21796

[18] Horiguchi G, Fuse T, Kawakami N, et al. Temperature-dependent translational regulation of the ER omega-3 fatty acid desaturase gene in wheat root tips. The Plant Journal. 2000;**24**:805-813

[19] Okuley J, Lightnffiger J, Feldmann K, et al. Arabidopsis FAD2 gene encodes the enzyme that is essential for polyunsaturated lipid synthesis. The Plant Cell. 1994;**6**:147-158

[20] Severin AJ, Woody JL, Bolon Y-T, et al. RNA-Seq Atlas of *Glycine max*: A guide to the soybean transcriptome. BMC Plant Biology. 2010;**10**:160

[21] Costa GGL, Cardoso KC, Del Bem LEV, et al. Transcriptome analysis of the oil-rich seed of the bioenergy crop *Jatropha curcas* L. BMC Genomics. 2010;**11**:462

[22] Beisson F. Arabidopsis genes involved in acyl lipid metabolism. A 2003 census of the candidates, a study of the distribution of expressed sequence tags in organs, and a web-based database. Plant Physiology. 2003;**132**:681-697

[23] Gupta K, Kayam G, Faigenboim-Doron A, et al. Gene expression profiling during seed-filling process in peanut with emphasis on oil biosynthesis networks. Plant Science. 2016;**248**:116-127

[24] Brown AP, Kroon JTM, Swarbreck D, et al. Tissue-specific whole transcriptome sequencing in castor, directed at understanding triacylglycerol lipid biosynthetic pathways. PLoS One. 2012;**7**. [Epub ahead of print]. DOI: 10.1371/journal.pone.0030100

[25] Burgal J, Shockey J, Lu C, et al. Metabolic engineering of hydroxy fatty acid production in plants: RcDGAT2 drives dramatic increases in ricinoleate levels in seed oil. Plant Biotechnology Journal. 2008;**6**:819-831

[26] Weselake RJ, Shah S, Tang M, et al. Metabolic control analysis is helpful for informed genetic manipulation of oilseed rape (*Brassica napus*) to increase seed oil content. Journal of Experimental Botany. 2008;**59**:3543-3549

[27] Capell T, Christou P. Progress in plant metabolic engineering. Current Opinion in Biotechnology. 2004;**15**:148-154

[28] United Nations. Convention on Biological Diversity. The Secretariat of the Convention on Biological Diversity. Rio de Janeiro, 1992. p.30

[29] McCullum C, Benbrook C, Knowles L, et al. Application of modern biotechnology to food and agriculture: Food systems perspective. Journal of Nutrition Education and Behavior. 2003;**35**:319-332

[30] Gressel J. Transgenics are imperative for biofuel crops. Plant Science. 2008;**174**:246-263

[31] Liu K. Soybeans: Chemistry, Technology, and Utilization. Singapore: Springer Science & Business Media; 1997

[32] American Soybean Association. 2017 SoyStats: A Reference Guide to Important Soybean Facts and Figures. 2017. http://soystats.com/wp-content/uploads/17ASA-006_Soy-Stats-2017_1F-web.pdf.

[33] Padgette SR, Kolacz KH, Delannay X, et al. Development, identification, and characterization of a glyphosate-tolerant soybean line. Crop Science. 1995;**35**:1451

[34] Sato S, Newell C, Kolacz K, et al. Stable transformation via particle bombardment in two different soybean regeneration systems. Plant Cell Reports. 1993;**12**:408-413

[35] Rech EL, Vianna GR, Aragão FJL. High-efficiency transformation by biolistics of soybean, common bean and cotton transgenic plants. Nature Protocols. 2008;**3**:410-418

[36] Li Z, Xing A, Moon BP, et al. Site-specific integration of transgenes in soybean via recombinase-mediated DNA cassette exchange. Plant Physiology. 2009;151. [Epub ahead of print]. DOI: 10.1104/pp.109.137612

[37] Kale SD, Tyler BM. Assaying effector function in planta using double-barreled particle bombardment. Methods in Molecular Biology. 2011;**712**:153-172

[38] Engels C, Fuganti-Pagliarini R, Marin SRR, et al. Introduction of the rd29A:AtDREB2A ca gene into soybean (*Glycine max* L. Merril) and its molecular characterization in leaves and roots during dehydration. Genetics and Molecular Biology. 2013;**36**: 556-565

[39] Leite JP, Barbosa EGG, Marin SRR, et al. Overexpression of the activated form of the AtAREB1 gene (AtAREB1ΔQT) improves soybean responses to water deficit. Genetics and Molecular Research. 2014;**13**:6272-6286

[40] Dubald M, Tissot G, Pelissier B. Plastid transformation in soybean. Methods in Molecular Biology. 2014;**1132**:345-354

[41] Kinney AJ. Genetic Engineering of Oilseeds for Desired Traits. In: Genetic Engineering. Boston, MA: Springer US. p. 149-166

[42] Haun W, Coffman A, Clasen BM, et al. Improved soybean oil quality by targeted mutagenesis of the fatty acid desaturase 2 gene family. Plant Biotechnology Journal. 2014;**12**:934-940

[43] Kim HJ, Kim MJ, Pak JH, et al. RNAi-mediated Soybean mosaic virus (SMV) resistance of a Korean Soybean cultivar. Plant Biotechnology Report. 2016;**10**:257-267

[44] Homrich MS, Wiebke-Strohm B, Weber RLM, et al. Soybean genetic transformation: A valuable tool for the functional study of genes and the production of agronomically improved plants. Genetics and Molecular Biology. 2012;**35**:998-1010

[45] Škorić D, Jocić S, Sakač Z, et al. Genetic possibilities for altering sunflower oil quality to obtain novel oils. Canadian Journal of Physiology and Pharmacology. 2008;**86**:215-221 [This article is one of a selection of papers published in the special issue Bridging the Gap: Where Progress in Cardiovascular and Neurophysiologic Research Meet]

[46] Davey MR, Jan M. Sunflower (*Helianthus annuus* L.): Genetic improvement using conventional and in vitro technologies. Journal of Crop Improvement. 2010;**24**:349-391

[47] Moyne A-L, Tagu D, Thor V, et al. Transformed calli obtained by direct gene transfer into sunflower protoplasts. Plant Cell Reports. 1989;**8**:97-100

[48] Knittel N, Gruber V, Hahne G, et al. Transformation of sunflower (*Helianthus annuus* L.): A reliable protocol. Plant Cell Reports. 1994;**14-14**:81-86

[49] Grayburn WS, Vick B. Transformation of sunflower (*Helianthus annuus* L.) following wounding with glass beads. Plant Cell Reports. 1995;**14**:285-289

[50] RAO K, VK ROHINI. Agrobacterium-mediated transformation of sunflower (*Helianthus annuus* L.): A simple protocol. Annals of Botany. 1999;**83**:347-354

[51] Müller A, Iser M, Hess D. Stable transformation of sunflower (*Helianthus annuus* L.) using a non-meristematic regeneration protocol and green fluorescent protein as a vital marker. Transgenic Research. 2001;**10**:435-444

[52] Weber S, Friedt W, Landes N, et al. Improved Agrobacterium-mediated transformation of sunflower (*Helianthus annuus* L.): Assessment of macerating enzymes and sonication. Plant Cell Reports. 2003;**21**:475-482

[53] Radonic LM, Zimmermann JM, Zavallo D, et al. Introduction of antifungal genes in sunflower via Agrobacterium. Electronic Journal of Biotechnology. 2008;**11**:8-9

[54] Dağüstü N, Fraser P, Enf?ssi E, et al. Screening for high callus induction and agrobacterium-mediated transformation of sunflower (*Helianthus annuus* L.). Biotechnology and Biotechnological Equipment. 2008;**22**:933-937

[55] Neskorodov YB, Rakitin AL, Kamionskaya AM, et al. Developing phosphinothricin-resistant transgenic sunflower (*Helianthus annuus* L.) plants. Plant Cell, Tissue and Organ Culture. 2010;**100**:65-71

[56] Wang Z, Qiao Y, Zhang J, et al. Genome wide identification of microRNAs involved in fatty acid and lipid metabolism of *Brassica napus* by small RNA and degradome sequencing. Gene. 2017;**619**:61-70

[57] Lin L, Allemekinders H, Dansby A, et al. Evidence of health benefits of canola oil. Nutrition Reviews. 2013;**71**:370-385

[58] Tan H, Yang X, Zhang F, et al. Enhanced seed oil production in canola by conditional expression of *Brassica napus* LEAFY COTYLEDON1 and LEC1-LIKE in developing seeds. Plant Physiology. 2011;**156**:1577-1588

[59] Qi W, Zhang F, Sun F, et al. Over-expression of a conserved RNA-binding motif (RRM) domain (csRRM2) improves components of *Brassica napus* yield by regulating cell size. Plant Breeding. 2012;**131**:614-619

[60] Shi J, Lang C, Wu X, et al. RNAi knockdown of fatty acid elongase1 alters fatty acid composition in *Brassica napus*. Biochemical and Biophysical Research Communications. 2015;**466**:518-522

[61] Wahid MB, Nor S, Abdullah A, et al. Oil palm—Achievements and potential. Plant Production Science. 2005;**8**:288-297

[62] Parveez GKA. Novel products from transgenic oil palm. AgBiotechNet. 2003;**5**:1-8

[63] Qureshi AA, Qureshi N, Wright JJ, et al. Lowering of serum cholesterol in hypercholesterolemic humans by tocotrienols (palmvitee). The American Journal of Clinical Nutrition. 1991;**53**:1021S-1026S

[64] Nesaretnam K, Guthrie N, Chambers AF, et al. Effect of tocotrienols on the growth of a human breast cancer cell line in culture. Lipids. 1995;**30**:1139-1143

[65] Kritchevsky D, Weber MM, Klurfeld DM. Influence of different fats (soybean oil, palm olein or hydrogenated soybean oil) on chemically-induced mammary tumors in rats. Nutrition Research. 1992;**12**:S175-S179

[66] Akmar ASN, Zubaidah R, Manaf MA, Nurniwalis W. Targeting Gene Expression in Mesocarp and Kernel of the Oil Palm Fruits. In: Murata N, Yamada M, Nishida I,

Okuyama H, Sekiya J, Hajime W. (eds) Advanced Research on Plant Lipids. 2003. Springer, Dordrecht. p. 411-414

[67] Masani MYA, Noll GA, Parveez GKA, et al. Efficient transformation of oil palm protoplasts by PEG-mediated transfection and DNA microinjection. PLoS One. 2014;**9**:1-11

[68] Majid NA, Parveez GKA. Regeneration of transgenic oil palm carrying GFP gene used as a visual selectable marker. Journal of Oil Palm Research. 2016;**28**:415-430

[69] Parveez GKA, Rasid OA, Hashim AT, et al. Tissue culture and genetic engineering of oil palm. In: Lai O-M, Tan C-P, Akoh CC, editors. Palm Oil: Production, Processing, Characterization, and Uses. Elsevier Science, Champaign, 2015. p. 87-135

[70] Parveez GKA, Omar R, Masani AMY, et al. Transgenic oil palm: Where are we? In: Murata N, Yamada M, Nishida I, editors. Advanced Research on Plant Lipids. Dordrecht: Springer Netherlands. p. 415-418

[71] Lee M-P, Yeun L-H, Abdullah R. Expression of *Bacillus thuringiensis* insecticidal protein gene in transgenic oil palm. Electronic Journal of Biotechnology. 2006;**9**:117-126

[72] Sujatha M, Reddy TP, Mahasi MJ. Role of biotechnological interventions in the improvement of castor (*Ricinus communis* L.) and *Jatropha curcas* L. Biotechnology Advances. 2008;**26**:424-435

[73] Singh AS, Kumari S, Modi AR, et al. Role of conventional and biotechnological approaches in genetic improvement of castor (*Ricinus communis* L.). Industrial Crops and Products. 2015;**74**:55-62

[74] Totaro G, Cruciani L, Vannini M, et al. Synthesis of castor oil-derived polyesters with antimicrobial activity. European Polymer Journal. 2014;**56**:174-184

[75] Malathi B, Ramesh S, Venkateswara Rao K, et al. Agrobacterium-mediated genetic transformation and production of semilooper resistant transgenic castor (*Ricinus communis* L.). Euphytica. 2006;**147**:441-449

[76] Sujatha M, Lakshminarayana M, Tarakeswari M, et al. Expression of the cry1EC gene in castor (*Ricinus communis* L.) confers field resistance to tobacco caterpillar (*Spodoptera litura* Fabr) and castor semilooper (*Achoea janata* L.). Plant Cell Reports. 2009;**28**:935-946

[77] Kumar AM, Sreevathsa R, Reddy KN, et al. Amenability of castor to an Agrobacterium-mediated in planta transformation strategy using a cry1AcF gene for insect tolerance. Journal of Crop Science and Biotechnology. 2011;**14**:125-132

[78] Auld DL, Rolfe RD, McKeon TA. Development of castor with reduced toxicity. Journal of New Seeds. 2001;**3**:61-69

[79] McKeon TA, Bo Shim K, He X. Reducing the toxicity of castor seed meal through processing treatments. Biocatalysis and Agricultural Biotechnology. 2013;**2**:159-161

[80] Liu Q, Singh SP, Green AG. High-stearic and high-oleic cottonseed oils produced by hairpin RNA-mediated post-transcriptional gene silencing. Plant Physiology. 2002;**129**:1732-1743

[81] Wang N, Ma J, Pei W, et al. A genome-wide analysis of the lysophosphatidate acyltransferase (LPAAT) gene family in cotton: organization, expression, sequence variation, and association with seed oil content and fiber quality. BMC Genomics. 2017;**18**:218

[82] Chapman KD, Austin-Brown S, Sparace SA, et al. Transgenic cotton plants with increased seed oleic acid content. Journal of the American Oil Chemists' Society. 2001;**78**: 941-947

[83] Baum SJ, Kris-Etherton PM, Willett WC, et al. Fatty acids in cardiovascular health and disease: A comprehensive update. Journal of Clinical Lipidology. 2012;**6**:216-234

[84] Liu F, Zhao Y-P, Zhu H-G, et al. Simultaneous silencing of GhFAD2-1 and GhFATB enhances the quality of cottonseed oil with high oleic acid. [Epub ahead of print]. 2017. DOI: 10.1016/j.jplph.2017.06.001

[85] Grosso NR, Julio AZ, Lamarque AL, et al. Proximate, fatty acid and sterol compositions of aboriginal peanut (*Arachis hypogaea* L) seeds from Bolivia. 1997. [Epub ahead of print]. DOI: 10.1002/(SICI)1097-0010(199703)73:3<349::AID-JSFA736>3.0.CO;2-E.

[86] Carrín ME, Carelli AA. Peanut oil: Compositional data. European Journal of Lipid Science and Technology. 2010;**112**:697-707

[87] Chenault KD, Burns JA, Melouk HA, et al. Hydrolase activity in transgenic peanut. Peanut Science. 2002;**29**:89-95

[88] Chenault KD, Payton ME, Melouk HA. Greenhouse testing of transgenic peanut for resistance to *Sclerotinia minor*. Peanut Science. 2003;**30**:116-120

[89] Chenault KD, Melouk HA, Payton ME. Field reaction to Sclerotinia blight among transgenic peanut lines containing antifungal genes. Crop Science. 2005;**45**:511-515

[90] Jonnala RS, Dunford NT, Chenault K. Nutritional composition of genetically modified peanut varieties. Journal of Food Science. 2005;**70**:S254-S256

[91] Ng EC, Dunford NT, Chenault K. Chemical characteristics and volatile profile of genetically modified peanut cultivars. Journal of Bioscience and Bioengineering. 2008;**106**:350-356

[92] FAOSTAT Database. Food and Agriculture Organization of the United Nations. Rome, 2013. http://faostat.fao.org/

[93] Uylaşer V, Yildiz G. The historical development and nutritional importance of olive and olive oil constituted an important part of the Mediterranean diet. Critical Reviews in Food Science and Nutrition. 2014;**54**:1092-1101

[94] de la Rosa R, Angiolillo A, Guerrero C, et al. A first linkage map of olive (*Olea europaea* L.) cultivars using RAPD, AFLP, RFLP and SSR markers. Theoretical and Applied Genetics. 2003;**106**:1273-1282

[95] Bartolini G, Prevost G, Messeri C, et al., editors. Olive Germplasm: Cultivars and Worldwide Collections.

[96] Agrimonti C, Vietina M, Pafundo S, et al. The use of food genomics to ensure the traceability of olive oil. Trends in Food Science and Technology. 2011;**22**:237-244

[97] Rugini E, Cristofori V, Silvestri C. Genetic improvement of olive (*Olea europaea* L.) by conventional and in vitro biotechnology methods. Biotechnology Advances. 2016;**34**:687-696

Chapter 16

# Application of Mechanics to Plant Seeds as a Granular or Particulate Material

Michal Petrů and Ivan Mašín

Additional information is available at the end of the chapter

http://dx.doi.org/10.5772/intechopen.70838

**Abstract**

The mechanical behavior of plant seeds as a granular or particulate material dramatically differs from the mechanical behavior of solid materials. This difference is caused by the possibility of partially autonomous movement and rotation of seeds, their mutual contacts, or due to the occurrence of the second fluid phase among the seeds at the stage of their moving or processing. For obtaining the economic effects from the seeds (energy, nutrients, livestock, etc.), seeds must often be subjected to mechanical treatment. In this context application of mechanics as science concerned with the behavior of physical bodies when subjected to forces or displacements is very important and needed. One of the goals of this chapter is therefore to provide an overview for readers who are not primarily concerned with mechanics but who are interested in the behavior of seeds in the context of biology, agriculture, and pharmacy or food industry. This chapter is therefore focused on both an overview of the principles of mechanics of granular or particulate materials and the presentation of experimental results particularly in the area of mechanical extraction of oil from seeds.

**Keywords:** seed, mechanics, granular material, particulate material, rheological model, FEM model, oil extraction

## 1. Introduction

Human life has become dependent on plants for the qualities and developments that they provide, which include agriculture, food production, and chemical industry. Plant seeds are one of the most important agricultural materials which affects billions of people. Seeds are the result of sexual reproduction in plants. Seeds are of immense biological and economic importance. They contain protein, carbohydrate, starch, and oil reserves that help in the early stages of growth and development in a plant. These reserves are what make many plant seeds important source for a large proportion of the world's inhabitants. For obtaining the economic effects from the seeds, they must often be subjected to mechanical treatment (manipulation, conveying, separation, hulling, pressing, purification, packaging, etc.). That is why studies of mechanical behavior of seeds are nowadays an important field of science and engineering. The research of how seeds as granular or particulate material are deformed, cracked, or how they flow has big importance for industries as biotechnology, pharmacy, agriculture, or food processing and various nanotechnologies. Results of that research are very important for both proper design of machines or equipment and for technological processes optimization. Current research has shown that seeds mechanical behavior is changing over time within the context of their moisture or oil content, acting forces, geometrical parameters, process of crack formation, porosity, visco-elastic properties, or in connection with the changes in the mutual arrangement or reorganization of the seed layers. The resulting nonlinear mechanical behavior and complex motion or spatial orientation of individual seeds or seed layers are very difficult to study, analyze, describe, and predict. For these scientific and engineering tasks, it is necessary to combine knowledge of basic physical seed characteristics, trivial calculations, modeling with the help of simplified or rheological models, or modeling using numerical methods such as FEM (finite element method) and DEM (discrete element method) and experimentation combined with relevant measurements. Also, special methods utilizing the principle of FEM with variable geometry like FDMs (fictitious domain methods) and IB-BCE (immersed boundary—body conformal enrichment) can be applied. Almost all the physical processes can be solved using numerical methods. The difficulty lies in the time-consuming calculation of highly complex and nonlinear problems. In this case, the calculation time can only be reduced by using supercomputers. The following chapter is therefore divided into several sections dealing with descriptive seed properties, seed mechanics principles, rheological models, and advanced modeling of seeds behavior.

## 2. Geometrical and mechanical properties of seeds

Seeds can generally be seen as granular or particular material (**Figure 1**). The actual state of the seeds as a material depends on the time from their harvest, moisture/oil content, and on mechanical, thermal, chemical, or other effects induced by the other subjects or by environment. Granular material is composed of small, discrete entities as opposed to being continuous. The granules

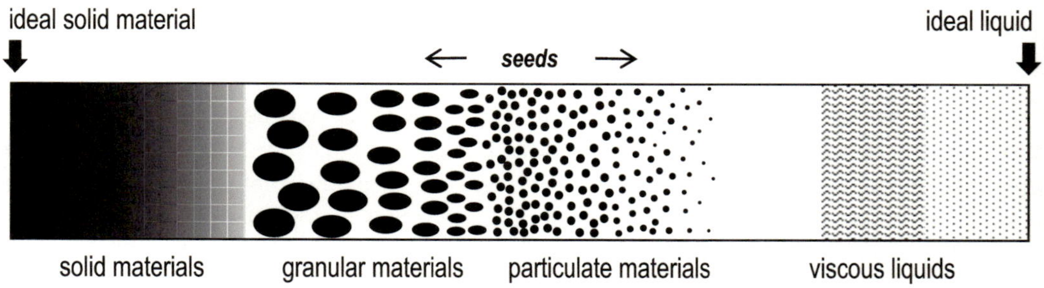


**Figure 1.** Plant seeds as granular or particulate material.

can be characterized as a mathematical set of macroscopic particles defined by their shape. Particulate material is substance composed of mutually contacting solid particles, or structural units, within the liquid and/or the gaseous phase [1].For various reasons, physical properties of seeds need to be studied. A physical property is any property of matter or energy that can be measured. It is an attribute of matter that can be observed or perceived. The physical properties of seeds can be divided into following categories: geometrical, mechanical, chemical, electrical, optical, and others. For mechanical engineering tasks, geometrical and mechanical properties are most important. Geometrical properties are those that can be derived from the geometry of a seed body. Seeds, however, come in a great variety of shapes and sizes. Seed size varies among species over a range of 10 orders of magnitude with extremes represented by orchids, e.g., *Goodyera repens* (weight 0.000002 g) and by the double coconut palm *Lodoicea maldivica* (weight 18,000–27,000 g) [2]. Seeds may be round, egg-shaped, triangular, long and slender, curved, coiled like a snail or irregular in shape (**Figure 2**). Some have a groove or depression with a ridge along the length; others may be flattened at one or both ends. Moreover, the seed shape may vary in immature or poorly developed seed. Since seed shapes are highly variable, their shapes are simplified (very often on an oval shape), and standardized geometrical parameters are used for the purpose of solving tasks from basic or advanced mechanics and simulation. This simplification is very important as a mean by which the size and shape of an irregular shaped seed can be easily described and quantified whether the seed is treated as an individual unit or as one that is a representative of many seeds in layer or bulk. Selected and often used geometrical properties are illustrated in **Figure 3** and described in **Table 1**. Size of seeds determines the efficiency of processing and storage and the quality of semi-product or final product. For this reason, it is also necessary during mechanical problems

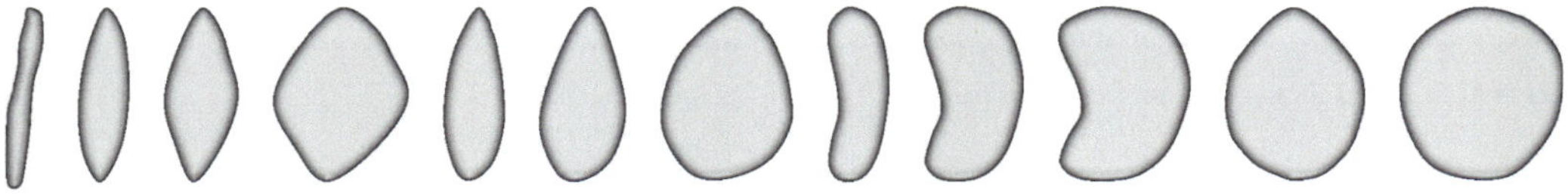

**Figure 2.** Different shapes of seeds.

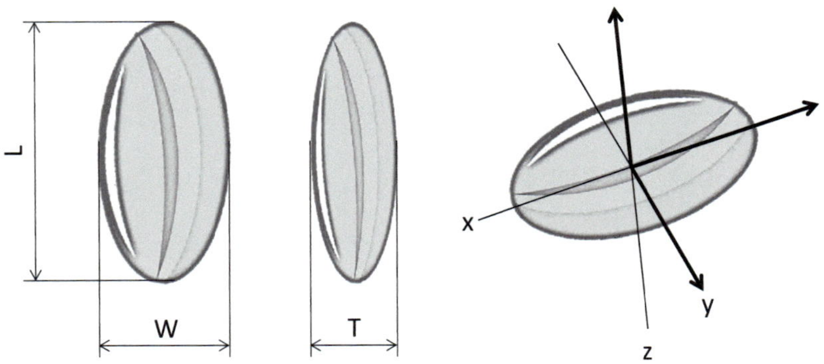


**Figure 3.** Basic dimensions of seed with oval shape (L—length, W—Width, and T—Thickness).

solving to take into account the stochastic nature of geometrical properties, for example, distinguish size fractions of one type plant seed (most commonly divided into groups of small, medium and large seeds), frequency distribution curves, or coefficient of variation of individual geometrical property. Mechanical properties are physical properties that a material exhibits upon the application of forces. Examples of mechanical properties are the modulus of elasticity, tensile strength, elongation, hardness, and fatigue limit. Mechanical properties occur as a result of the physical properties inherent to each material and are determined through a series of standardized mechanical tests. Some of important mechanical properties of seeds are listed in **Table 2**.

Research on geometrical and mechanical properties has been focused on various types of seeds, such as rapeseed (*Brassica napus* L.) [2], Jatropha (*Jatropha curcas* L.) [4, 5], sunflower

| Geometrical property of seed | Nomenclature | Unit | Formula | |
|---|---|---|---|---|
| Length of one seed | L | mm | | |
| Width of seed | W | mm | | |
| Thickness of one seed | T | mm | | |
| Arithmetic mean diameter | $D_a$ | mm | $D_a = (L + W + T)/3$ | (1) |
| Geometric mean diameter | $D_g$ | mm | $D_g = (L^* B^* T)^{1/3}$ | (2) |
| Equivalent diameter | $\varphi$ | mm | $(4 S_m/\pi)^{1/2}$ $S_m$ – *measured area* | (3) |
| Seed surface area | S | $mm^2$ | $S = \pi(D_g)^2$ | (4) |
| Sphericity | $S_\varphi$ | – | e.g., $S_\varphi = D_g/L$ | (5) |
| Volume of one seed | V | $mm^3$ | e.g., $V = 4/3\pi(L^* B^* T)$ | (6) |

**Table 1.** Selected geometrical properties of seeds.

| Mechanical property of seed | Nomenclature | Unit |
|---|---|---|
| Bulk modulus | K | MPa |
| Coefficient of friction | $\mu$ | – |
| Compression energy | $W_C$ | $J\,mm^{-3}$ |
| Consumed energy at rupture point | $W_R$ | J |
| Deformation coefficient of mechanical behavior | $B_D$ | $mm^{-1}$ |
| Elasticity modulus | E | MPa |
| Energy used for rupture | $E_R$ | MPa |
| Rupture force | $R_F$ | N |
| Seed oil dynamic viscosity | $\eta$ | Pa s |
| Shear modulus | G | MPa |
| Stress in the structure | $\sigma_S$ | MPa |

**Table 2.** Selected mechanical properties of seeds.

| Seed | L (mm) | W (mm) | T (mm) | $D_g$ (mm) | $E_R$ (Nmm) | $R_F$ (N) | μ (steel) (–) |
|---|---|---|---|---|---|---|---|
| *Brassica napus* (fraction: medium) [3] | 2.12 ± 0.108 | 1.91 ± 0.093 | | | | 13.53 ± 2.649 | |
| *Jatropha curcas* L. (var. Kanlueang, nut) [4] | 21.02 ± 1.03 | 11.97 ± 0.30 | 9.58 ± 0.28 | 13.40 ± 0.3 | 124.44 ± 19.95 | 146.63 ± 14.82 | |
| *Helianthus annuus* L. (var. Morden, moisture 6.2%) [6] | 9.27 ± 0.68 | 4.78 ± 0.34 | 3.32 ± 0.27 | 5.39 ± 0.416) | | | 0.45 |
| *Phaseolus vulgaris* (var. Hinis) [8] | 11.76 ± 0.77 | 8.85 ± 0.50 | 7.66 ± 0.58 | 9.26 ± 0.53 | 63.79 ± 23.25 (X ax) | 145.88 ± 33.54 (X ax) | 0.227 ± 0.0038 |
| *Dacryodes edulis* (moisture 10.3%) [9] | 19.00 ± 1.1 | 12.20 ± 0.8) | 10.10 ± 0.8) | 13.2 ± 1.4 | | | metal 0.25 ± 0.009 |

**Table 3.** Selected geometrical and mechanical properties of seeds.

(*Helianthus annuus* L.) [6, 7], bean (e.g., *Phaseolus vulgaris*) [8], arigo seeds (*Dacryodes edulis*) [9], pine (*Pinus pinea*) [10], chia seeds (*Salvia hispanica*l.) [11], wild legumes (e.g., *Canavalia cathartica*) [12], melon seeds (*Cucumis melo* L.) [13], tung seed (*Aleurites Fordii*) [14], sugarbeet seed (*Beta vulgaris* L.) [15], sesame (*Sesamum indicum* L.) [16], vetch seeds (*Vicia sativa* L.) [17], green soybean (*Glycine max*) [18], quinoa (*Chenopodium album* L.) [19], hemp (*Cannabis sativa* L.) [20], and so on. Geometrical and mechanical properties of selected seeds types are listed in **Table 3**.

## 3. Mechanical behavior of single seed

When analyzing the mechanical behavior of individual seed, we can identify states similar to the mechanical behavior of the elastic body. Therefore, a model of an idealized flexible body can be used to study and describe the behavior of individual seeds. In this case, the forces and moments of the forces acting on the seed cause its state of strain accompanied by its deformation and fracture (**Figures 4** and **5**)

Initially consider the compression of seed as a compression of linear elastic substance and constant side pressure coefficient $K_0|_{=const.}$. Then, according to Hook's law:

$$\varepsilon_S^E = \frac{1-2\mu}{E_S}(1+2K_0)\sigma_S^E \tag{7}$$

where $\varepsilon_S^E, \sigma_S^E$ are axial deformation and stress in the direction of compression during elastic deformation, $\mu$ is Poisson's ratio, and $E_S$ is initial stiffness modulus.

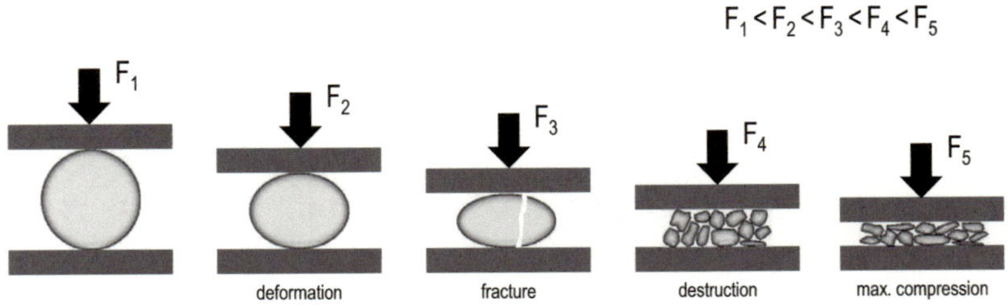


**Figure 4.** Seed deformation at different forces.

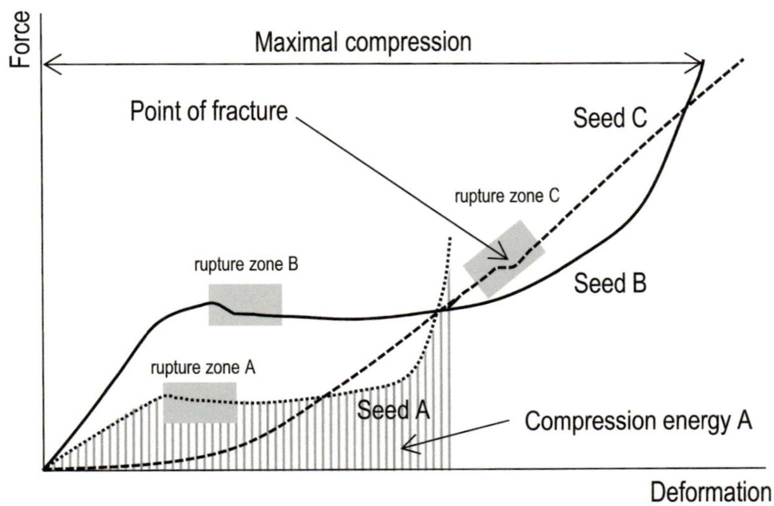


**Figure 5.** Selected types of deformation diagrams for different plant seeds.

During an elastic deformation, the structure of the substance does not change, and therefore, initial stiffness modulus will be constant $E_S|_{=const.}$. Deformation of single seed is a function of stress $\sigma_S^E$ which characterizes the intensity of structural changes. Then, Eq. (7) can be overwritten as:

$$d\left(\frac{w}{w_0}\right) = d\left(\frac{V}{V_0}\right) = \frac{d\sigma_S^E}{E_S} \tag{8}$$

where $w_0$ is the initial seed width.

The stress in the structure of the sample $\sigma_S$ (MPa) can be resolved into the sum of stresses consisting of the stress elastic component $\sigma_S^E$ and stress in the plastic component $\sigma_S^P$, and which is characterized by oil viscosity component $\sigma_S^\eta$. This is responsible for the damping of the structure. Friction component $\sigma_S^\mu$ of the *seed* sample causes significant permanent deformation with geometrical and physical damage from certain value of strain. By using numerical analysis of the isothermal compression T=*const.* for the seed sample, it has been observed that the elastic component of stress a function which is defined approximately by the Hooke's law [21]. This strain is characterized by a low oiliness point/low point of structure, where the isotropic properties of the structure can be considered. Values of low oiliness point for example tested samples of *J. curcas* L. are shown (**Table 4**). This area is characterized by a constant volume strain $\gamma|_{\varepsilon_{S\le 40\%\ \text{strain}}}$, which can be defined by Eq. (9), and in principle, it is same as Eq. (8).

$$\gamma = d\left(\frac{V}{V_0}\right) = -\psi \cdot \Delta C \tag{9}$$

where $V$ is the volume (in an initial position: compression time $t$ = 0), $C$ is the compression pressure, and $\psi$ the compressibility [22].

$S$ is the projected area of the seed, and $\varepsilon_{Slow}$ is deformation of the seed (low oiliness point).

This can be visualized as a reversible elastic compression of the Hooke member, which is defined by the initial elastic modulus $E|_{\varepsilon_{S\le\text{low point of structure}}}$ and which can be supplemented by the damping factor of the Newtonian viscous member $\eta|_{\varepsilon_{S\le\text{low point of structure}}}$. This behavior could be partly described by a generalized Maxwell rheological model (or generalized Maxwell model). The higher compression of the test structure from the point of oiliness or oil point passes the functional dependence of elastic stress in the elastic or plastic or visco-plastic components $\sigma_S^E = f(\varepsilon_S)|_{\le\text{low point of structure}} \rightarrow \sigma_S^P = f(\varepsilon_S)|_{\ge\text{low point of structure}}$. This is reflected in breach of the yield stress of the test sample structure (**Figure 5**). Here, a comprehensive change in the volumetric strain $\gamma|_{\ge\text{low point of structure}}$ was reflected and spread with the speed of plastic strain $\varepsilon_S^P$ which caused increased friction $\mu = \mu_S$ and creep structure due to extrusion of the oil component from the structure. Functional dependency was reflected in a significant increase in compression strength with a parabolic character which is different to the elastic-plastic deformation of elastic structures. There is a whole range of analytical and empirical relations in the literature that extend the seed deformation description by porosity. Define that the density of the undried seed is $\rho_S = m_S/V_S$ and the density of the dried seed at 105°C is $\rho_D = m_S/V_D$. The pore volume can be then defined by the porosity number:

$$e = \frac{\rho_S - \rho_D}{\rho_D} = e_S - 1 \tag{10}$$

where $e_S = \rho_S/\rho_D$ is the ratio by which the compression can be defined according to (9).

In principle, it is a semi-logarithmic Walker-Balsin's law applicable to, for example, powder metallurgy:

| Maturity stage | T (mm) | S (mm$^2$) | $\varepsilon_{SLow}$ (−) |
|---|---|---|---|
| Ripe | 10.0 ± 1.8 | 180.1 ± 19.3 | 0.5 |
| Over-ripe | 10.2 ± 1.4 | 181.8 ± 19.5 | 0.4 |
| Unripe | 9.2 ± 1.9 | 172.2 ± 18.3 | 0.4 |

**Table 4.** Dimensions of the seeds prior deformation tests and the coefficients characterizing the deformation of seeds (data in the table are mean ± SD).

$$e_S = (1 + e_0)(1 - C\log\sigma_S) \tag{11}$$

where $e_0$ is the initial porosity number at $\sigma_S = \sigma_0$.

Let us assume a single seed is comprised under pressure $\sigma_{S1}$ to width $w_1$ and subsequently comprised under pressure $\sigma_{S2}$ ($\sigma_{S1} < \sigma_{S2}$) to width $w_2 < w_1$. Then, we can introduce Eq. (12), which (in the case $\sigma_0$ is unit pressure) can be generalized to Eq. (13).

$$\frac{w_1}{w_0} - \frac{w_2}{w_0} = C\left[\left(\frac{\sigma_0}{\sigma_1}\right)^m - \left(\frac{\sigma_0}{\sigma_2}\right)^m\right] \tag{12}$$

$$\frac{w}{w_0} = C\sigma_S^{-m} \tag{13}$$

For FEM simulation of such anisotropic behavior, we can use equilibrium equations or applications of the constitutive equations of a rheology model with modified parameters like Maxwell model, Kelvin model, Saint-Venant model or Perzyna model. Acceptance of certain seed states as a particulate material leads to the use of rheology in describing its mechanical behavior. The science of rheology is important to the study of the flow behavior of solids suspended in fluids. Rheological models (visco-elastic models) used to characterize seeds are classified as non-Newtonian fluids. For these fluids, no constant of proportionality exists between shear stress and shear rate; their viscosity varies with changing shear rate. For seeds, it is advisable to use the Perzyna rheological model in particular (**Figure 6**).

The Perzyna model with visco-plastic parameters includes the strain rate, and its rheological behavior is applicable not only for the description of elastic–plastic material damage but also

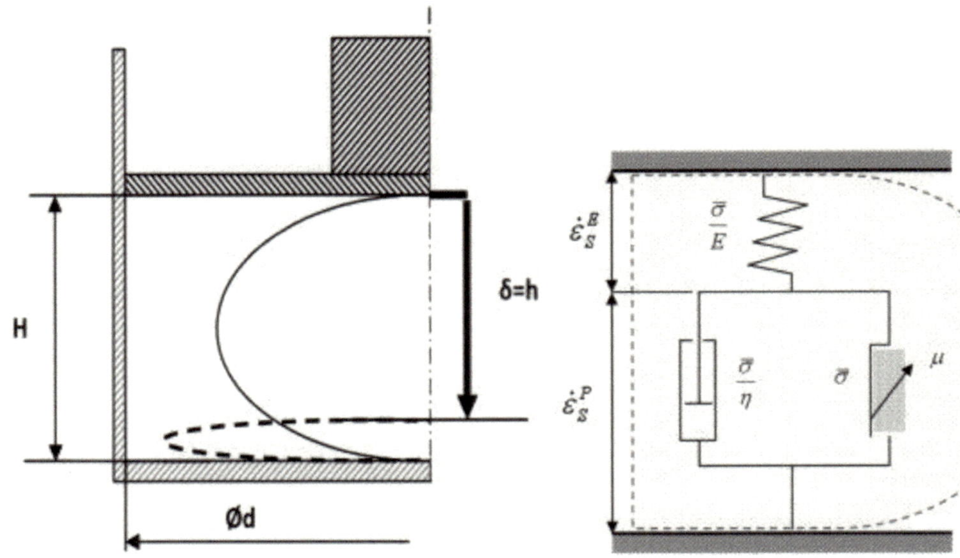


**Figure 6.** Perzyna rheology model for single *Jatropha curcas* L. seed compression [21].

for the behavior of anisotropic structures of compressed seed samples of different crops (**Figure 6**) in the plastic strain state $\varepsilon_S^P\big|_{\geq \text{low point of structure}}$. The Perzyna model derivation is based on studies of the stress limit description, which was defined by Bingham in 1920 to describe a plastic deformation of elastic materials. It is possible to use this model for quasi-static strain behind yield stress. Also, the model defines the time-dependent response of the structure cohesion, for example, the determination of oil extrusion initiation (so-called low oiliness point), immediate friction, and viscous and elastic parameters as functions or constants. Also, the Perzyna model can be used for modeling of the flow of soil such as clay. The velocity of stress $\dot{\sigma}_S$ propagation during plastic deformation can be expressed by the constitutive Eq. (14), where the progress of plastic deformations is given by changing of volumetric strain $\gamma$, which can assume various values:

$$\dot{\sigma}_S = E_S\left(\dot{\varepsilon}_S - \dot{\varepsilon}_S^P\right) \tag{14}$$

where $\dot{\sigma}_S$ is stress rate of structure, $E_S$ initial elastic modulus, $\dot{\varepsilon}_S$ is strain rate of structure, and $\dot{\varepsilon}_S^P$ is visco-plastic strain rate.

The plastic deformation of material structure denoted by the stress $\sigma_S$ can be resolved according to Eq. (15) to the sum of the initial yield stress $\overline{\sigma}_S\left(\overline{\sigma} \leq \sigma_S^P\right)$ and product of equivalent visco-plastic strain $\gamma_S^P$ and function of the strain $h_S^P$ that describes following condition: if $h_S^P < 0$, then a softening of the structure occurs $h_S^P > 0$, and then a hardening occurs:

$$\sigma_S = \overline{\sigma}_S + h_S^P \cdot \gamma_S^P \tag{15}$$

The numerical model can be subsequently designed for isothermal or thermal compression of different seed types. An example is the numerical isothermal study of three samples of Jatropha seeds (mature, immature, and precisely mature) with the same initial geometry $\Omega_{S(ripe)} = \Omega_{S(unripe)} = \Omega_{S(overripe)}\Big|_{t=0}^{T=const.}$. Differently mature seeds show different deformation of the original geometry during compression $t \neq t_0$, which has been described in detail in [21]. The degree of compression of magnitude $\delta$ is also the compressibility function $\psi$ according to Eq. (9). Seed dimensions for the CAD/FEM model for differently mature seeds are based on data listed in the **Table 4**. Modeled seeds had almost ellipsoidal shape with dimensions ~$17 \times 10 \pm 1.8 \times 9.3$ mm, as shown in **Figure 7**.

Compression experiments and simulations of *J. curcas* L. seeds were performed up to plastic deformation (**Figure 8**). The model showed the highest concentration of the main stress on the seed circumference for each level of maturity of tested seeds. This is because the greatest stress is concentrated in the center of the seed and consequently causes the extrusion of the structure from the center toward the periphery. This effect changed and damaged the smallest radius of the seed shape as seen from **Figures 8** and **9**. From this, it can be deduced that the seed is damaged by the internal pressure resulting from external compression forces.

The energetic behavior of individual seeds during compression is shown in the following figures. **Figure 10** shows the comparison of compression forces derived from mathematical

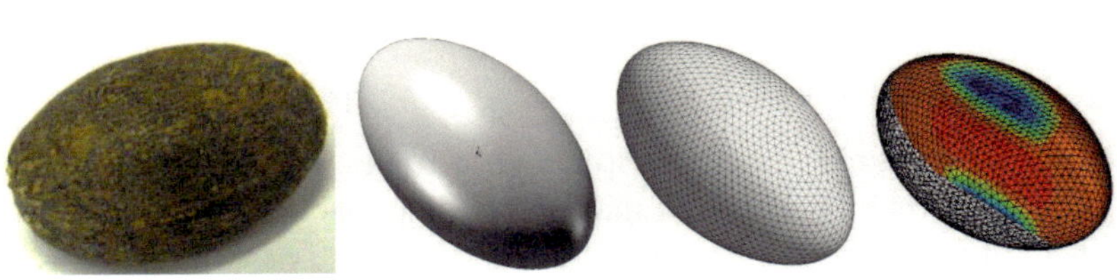

**Figure 7.** *Jatropha curcas* L. seed (CAD model, FEM model, and simulation model).

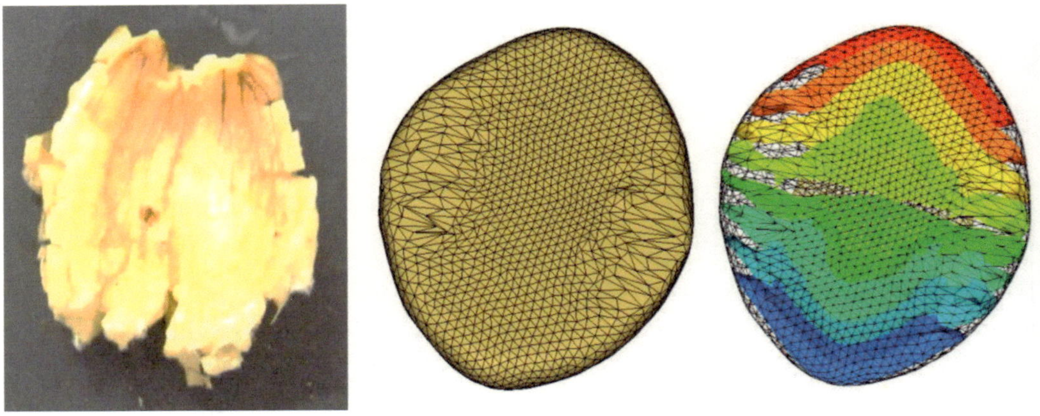

**Figure 8.** Deformation of *Jatropha curcas* L. seed (real sample, FEM model, vector plot of FEM model).

**Figure 9.** Time response of *Jatropha curcas* L. seed deformation.

equations and forces obtained by numerical models. It can be seen from **Figure 10** that mathematical relationships (7)–(13) do not allow to sufficiently describe the "point of fracture" as they are generally based on the empirical analysis and using graphs of standard mathematical functions (such as a parabola). Point of fracture can be better and more accurately determined from numerical models. Numerical and FEM models can be used to solve a whole range of engineering tasks. For example, using a FEM model, it is possible to determine the compressibility of mature, immature, and overmatured seeds (**Figure 11**) or to effectively assess the

http://dx.doi.org/10.5772/intechopen.70838

energy intensity of obtaining the oil from the seeds. The results of the experiment also showed differences in the values of the stress of individual seeds, which depend on their maturity.

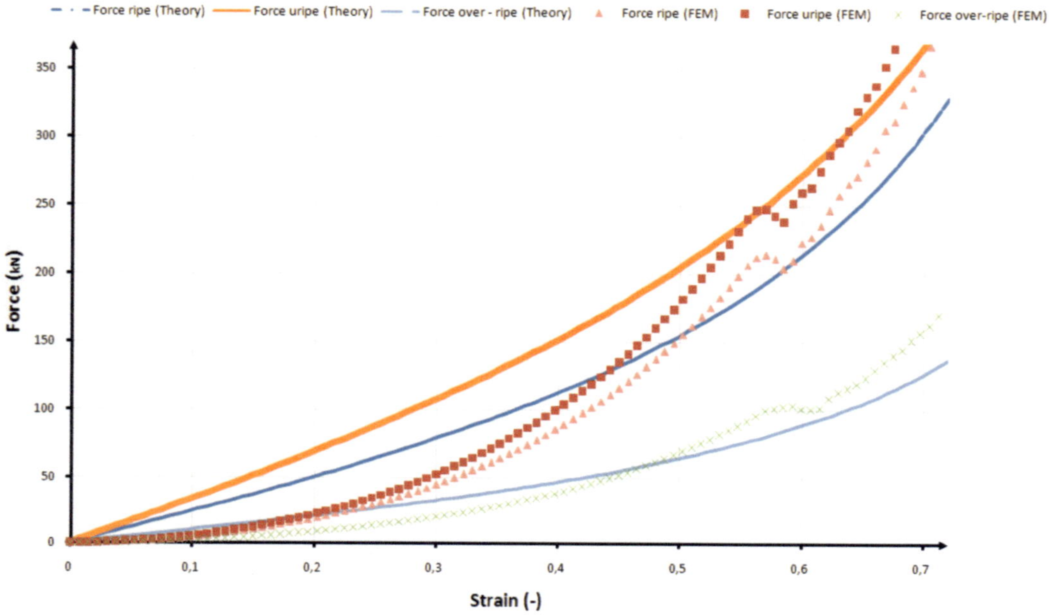


**Figure 10.** Comparison of theory and FEM model of *Jatropha curcas* L. seed for compressive force.

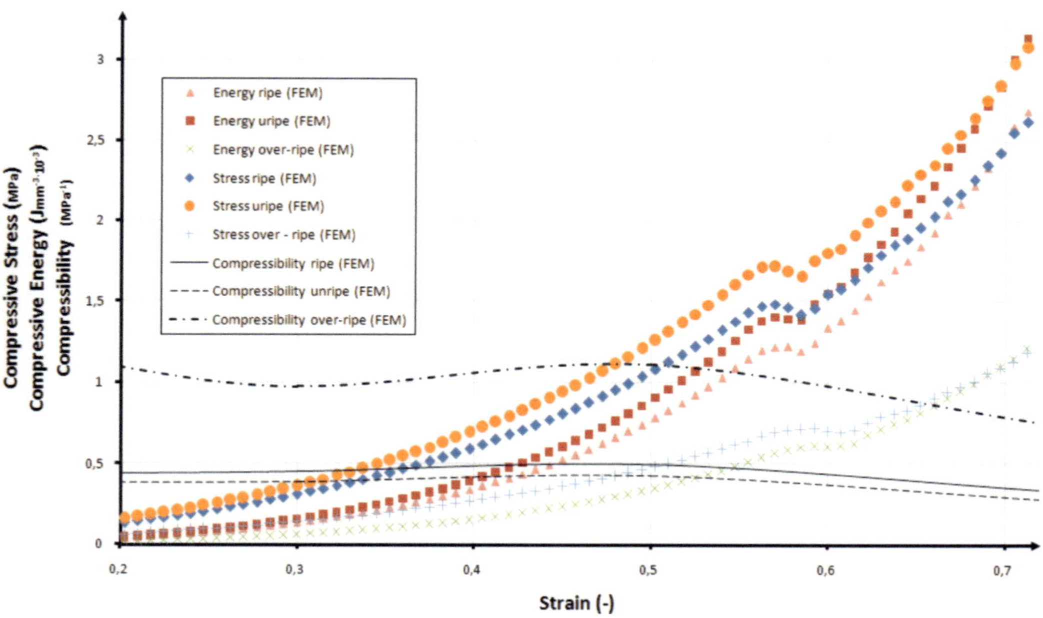


**Figure 11.** Strain response: compressive stress, compressive energy, compressibility (individual seed).

These findings can be used to find a proper seed position to control orientation of the seed in the press or to optimize the geometry of the press.

## 4. Mechanical behavior of seeds as a granular or particulate materials

Seeds as a granular or particulate material consists of individual seeds and adjacent voids filled with gas (air) or/and liquid (e.g., oil). Understanding the mechanical behavior of such kind of material, we must therefore take into account the geometrical presentation that describes spatial distribution of seeds and voids, orientation of seeds, and contacts of seeds, etc. Systems composed of seeds consist of mutually contacting phases, which can be solid (geometry, solid structure of seed, etc.), liquid phases (oil component of seed), and gas phases (gaseous environment). The liquid and gas phases fill the pores of the solid skeleton of the seed. In addition to these basic phases, there are mutual bonds between the seeds which, in terms of mechanical properties of the compressed seeds, are mainly frictional. When compressing seeds, we have to study frictional bonds seed-seed, seed-seeds, seeds-container wall, etc. To understand this complex process, we can start on the general theory of friction of solids. The two ideal solid particulate seeds, when moved relative to one another, can either slide or roll in an elastic state. Seeds in a plastic state can penetrate and break into each other. In solving mechanical problems, generally we have to distinguish dry seed surfaces, hydrodynamically lubricated surfaces, and limiting friction. The friction or contact bond is given by the normal stress $\sigma_n$ on the geometric contact area $a_g$ (**Figure 12**). The normal stress of the contact points $\sigma_n^c$ on the surface area $a_S$ is then given by Eq. (16). Similarly, sliding friction $\tau_n$ can be described by Eq. (17).where $\tau_n^c$ is the shear strength of contact point (assumed to be the same everywhere). From Eqs. (16) and (17) is Eq. (18):

$$\tau_n = \frac{\tau_n^c}{\sigma_n^c}\sigma_n = \nu\sigma_n \tag{16}$$

$$\sigma_n^c = \sigma_n \frac{a_g}{\sum a_S} \tag{17}$$

$$\tau_n = \tau_n^c \frac{\sum a_S}{a_g} \tag{18}$$

According to Eq. (18), we obtain the friction response between the seed surfaces. It follows that shear friction in the case of seeds also does not depend on the size of the surface and is directly proportional to the normal stress. The area of contact point plasticity allows a more accurate description of the plastic state of the stresses in it as well as for the emerging strain. The surface area $a_k$ is inversely proportional to the stress $\sigma_n^c$ according to Eq. (16) and at the same time increases the shear stress of the contact point [23]. This effect is particularly evident in elastic materials such as steel, in which case their friction coefficient $\nu$ increases considerably. In the case of structures such as seeds, this problem is limited by their skeletal fragility. In general, according to the friction adhesion theory, the intergranular friction coefficient is equal to the

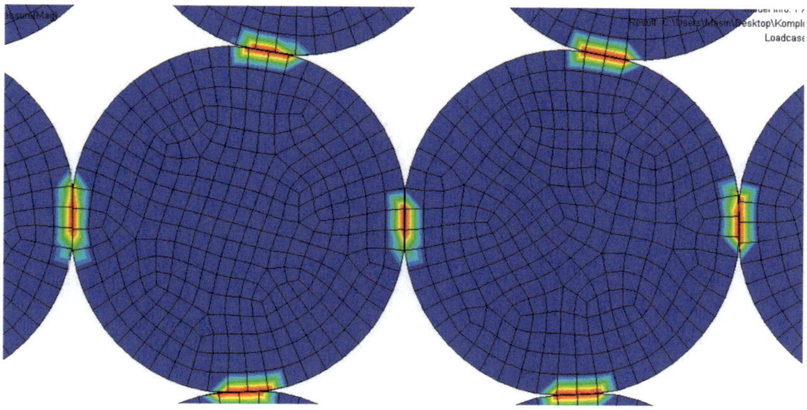

**Figure 12.** Contact area $a_S$ of the seeds on their outer boundary.

shear and normal strength of the contact point. Through this theory, we gain a function dependent on the amount of stress concentration at the boundaries between the seeds.

Therefore, the description of the mechanical behavior of the seeds as granular or particulate material must, for various reasons, be based on:

- initial seed configuration (seeds geometry, seeds orientation, shape regularity, and ripeness);
- the volume ratio of the individual phases in the press $V_{Si}|_{V_A \neq Vsi \neq Voi}$ ($V_A$ the volume of air, $V_{si}$ the volume of individual seeds, $V_{oi}$ the volume of the oil fraction of each seed);
- the ratio of the initial elastic modules $E_{Si}|E_{ripe} \neq E_{unripe} \neq E_{overripe}$ ($E_{ripe}$ the ripe seed module, $E_{unripe}$ the unripe seed module, $E_{overripe}$the overripe seed module);
- initial volumetric modules$G_{Si}|G_{ripe} \neq G_{unripe} \neq G_{overripe}$; and
- initial shear modules$K_{Si}|K_{ripe} \neq K_{unripe} \neq K_{overripe}$.

As an example, we can consider a seed-system that will be compressed in an experimental container. Such a system will be loaded with axial stress $\sigma_a$ and all-sided pressure $\sigma_r$. We will not consider a homogeneous system where the stress $\sigma_a, \sigma_r$ would be totally consistent with the continuum, but we will consider a particle system where each part of the filled container (seeds, air, walls, and piston) contributes to $\sigma_a, \sigma_r$. Terzaghi [24] introduced in 1923 the construct of effective stress, which is a function of the total stress and the particle stress. The study states that the change in chamber pressure does not affect the dependency of $(\sigma_a - \sigma_r)$ on axial strain of the system $\varepsilon_a$. Terzaghi also introduced the boundary condition that we can simplify the particular systems in the individual planes of symmetry (**Figure 13**). The greater the seed filling in the container, the more accurate the calculation will be, not only as a continuous system but also as a particle system. This claim can be demonstrated by the example of compressing a system of regularly arranged ellipsoidal shaped seeds (such as *J. curcas* L.). If we make a simple axial section of the seeds, we obtain a 2D arrangement of identical circles with a radius $r$, where adjacent circles touch each other (**Figure 13**).

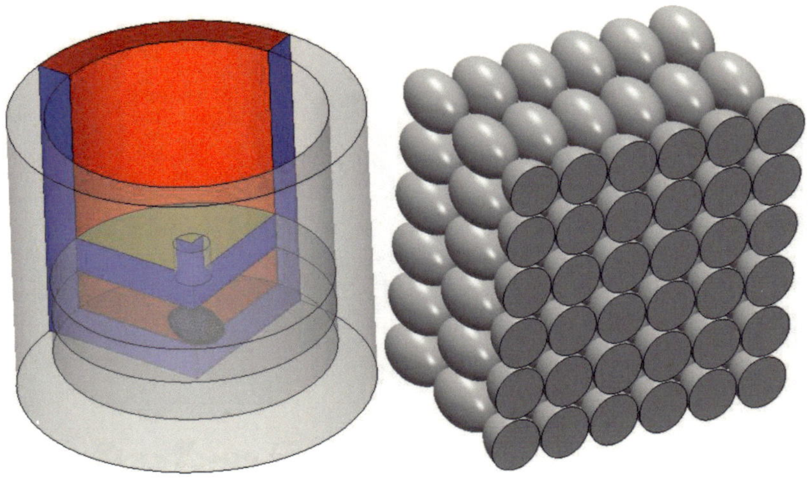

**Figure 13.** Symmetry planes for modeling (left) and axial section for 2D model (right).

During compression, the radius $\Delta r$ will change. Radius $\Delta r$ will be a function of contact forces that can be derived from the Hertz or Mindlin relationship. Basically, if the contact forces are known, deformation of the circles for axial strain and lateral strain can be calculated according to Eqs. (19) and (20).

$$\varepsilon_S^a = \left(\frac{3(1-\nu_S)\sqrt{2}}{G_S}\right)^{2/3} \cdot \left[2\left(\frac{1}{1+\nu_S}\frac{\sigma_a}{\sigma_r}\right)^{2/3} - \left(2\frac{1-\nu_S}{1+\nu_S}\frac{\sigma_a}{\sigma_r}\right)^{2/3} - 1\right] \cdot \sigma_{r^{2/3}} \quad (19)$$

$$\varepsilon_S^r = \left(\frac{3(1-\nu_S)\sqrt{2}}{G_S}\right)^{2/3} \cdot \left[\left(2-\frac{1-\nu_S}{1+\nu_S}\frac{\sigma_a}{\sigma_r}\right)^{2/3} - 1\right] \cdot \sigma_{r^{2/3}} \quad (20)$$

It can be seen from Eqs. (19) and (20) that the size of the strain depends on the ratio of the main stresses $\sigma_1/\sigma_3 \equiv \sigma_a/\sigma_r$. There is a special case, where the contact forces of the ideal circular arrangement are statically determined (in case of hydrostatic tension, respectively, when $\sigma_a = \sigma_r$). These mathematical derivations of stress and deformation give us insights about mechanical changes of the compressed seed system. The Terzaghi's principle is well described in [23], where a symmetrical numerical model was created to study the strain of the seed system during compression.

**Figure 14** shows the transformation of seeds stored in a regular hexagonal arrangement where the initial seed filling $\alpha_S$ will gradually increase during compression as it $\alpha_S < \alpha_{S1} < \alpha_{Sn}|_{t_0 < t1 < tn}$. **Figure 14** also shows the reorganization of the seeds and their subsequent clustering and joining in the case of a plastic failure. **Figure 15** is a comparison of the modeled individual seeds crack propagation with a real experiment. This gives us insight into the compressive behavior of individual seeds and thus the knowledge about the efficiency of the pressing process. By increasing the crack in the individual seeds, we also provide information on the movement and direction of the extruded oil.

**Figure 14.** Seed-system deformation in time.

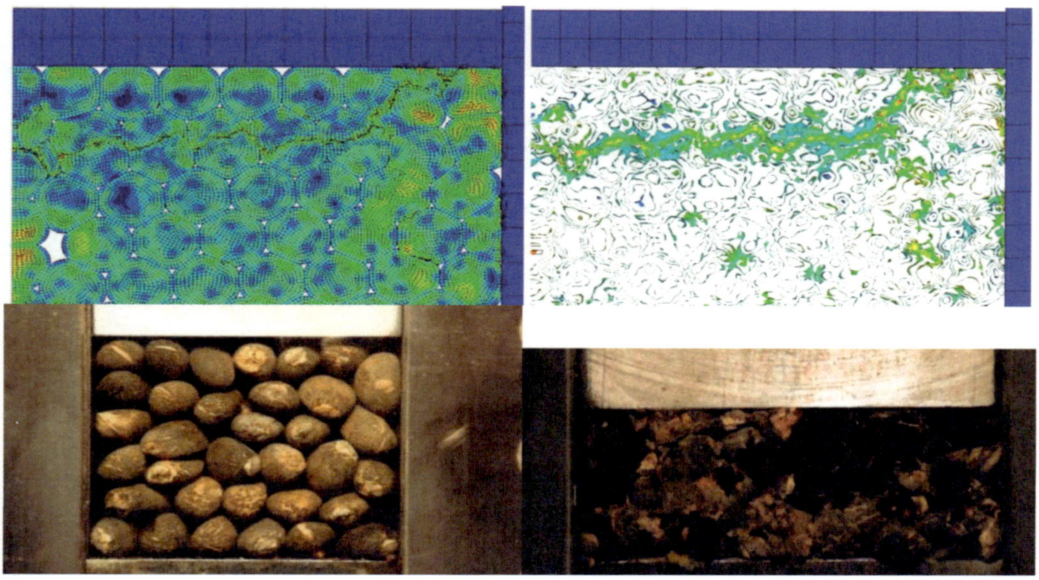

**Figure 15.** Growth of the cracks in seeds during pressing (FEM model—above, experiment—bottom).

Consequently, the energy behavior of the seed system during compression can be evaluated, as shown in **Figures 16** and **17**. Figures compare numerical models and the experimental data (compressive forces and compressibility) for the different cylinder diameters. The results show a very good match that can be attributed to the simplified 2D symmetric model, which approximates empirically established relationships (19) and (20). A 2D seed-system modeling not only simplifies calculation and reduces time for computations but also allows for more accurate results compared to 3D modeling of single seed behavior (**Figure 10**). Appropriate simplification of mathematical models describing the mechanical

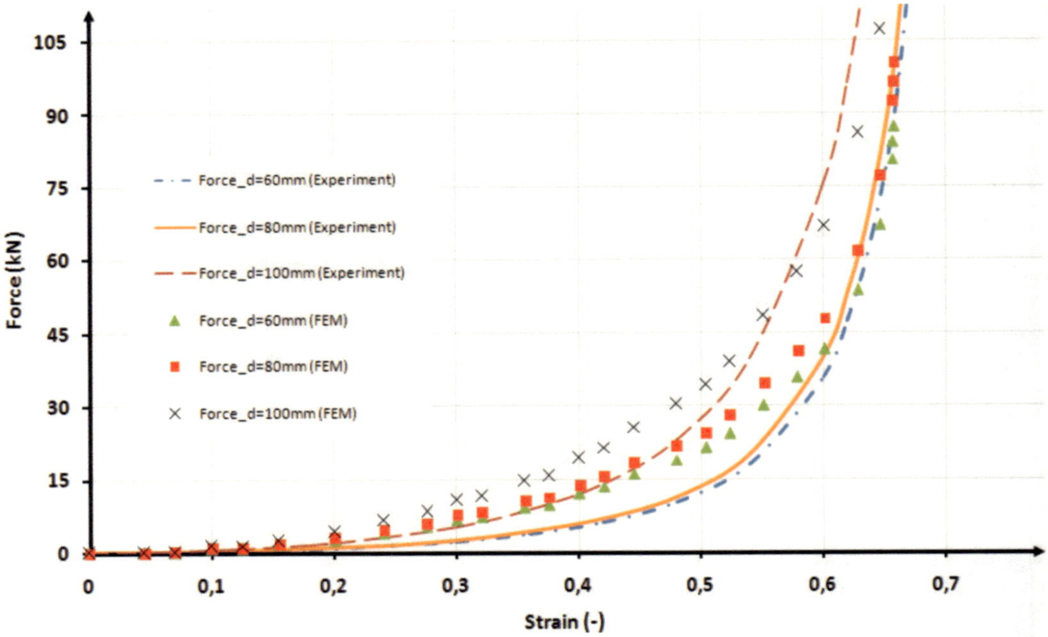


**Figure 16.** Comparison of experiment and FEM model for compressive force.

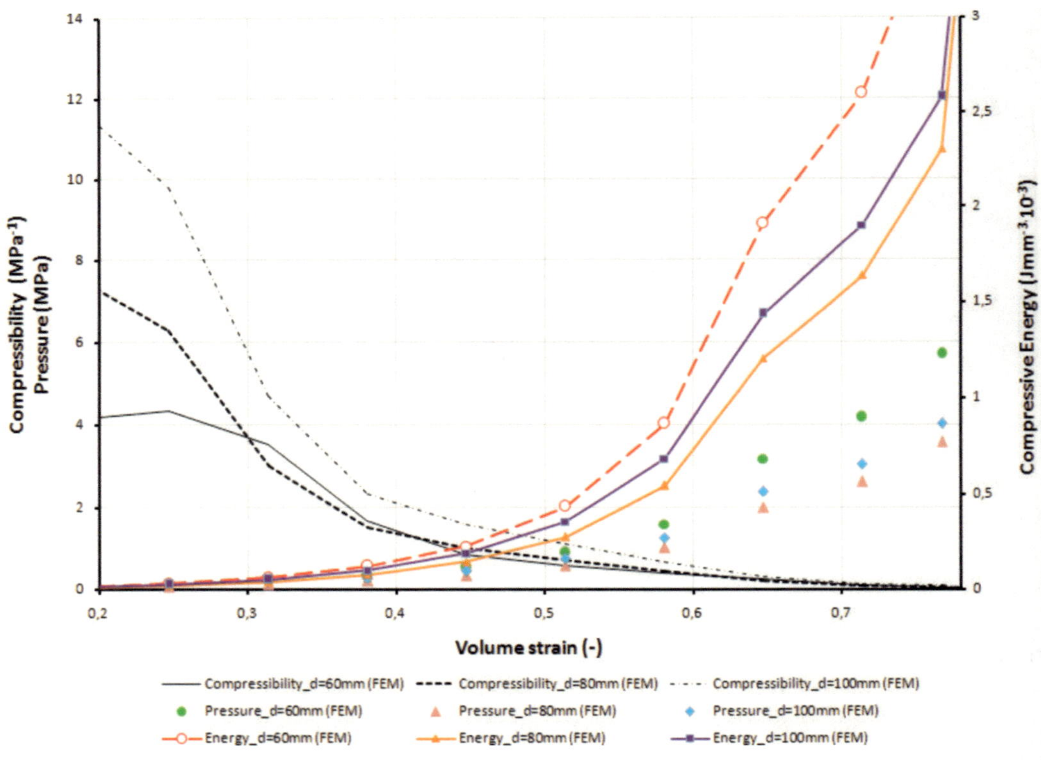


**Figure 17.** Strain response: compressive stress, compressive energy, compressibility (system of seeds).

behavior of seeds-system (edge conditions, symmetry, etc.) can provide us with results that are sufficient to comprehend complex processes and to optimize seed treatment technologies, compared to sophisticated models with disproportionate time-consuming calculation and interpretation. Mathematical models of seeds mechanical behavior are very important for different agricultural and engineering tasks (e.g., tillage mode, processing technology design, treatment control and optimization, procedures effectiveness increasing, seeds behavior prediction, etc.).

As this topic goes beyond the scope of this publication, we refer readers to various works focused on various aspects of mechanic behavior of seeds, using different theories, models, and modeling methods:

- modeling of seeds compression [21, 22, 23, 25, 31];
- modeling of stress or force relaxation of seeds [30];
- modeling of seeds movement and orientation [26];
- various contact and friction models utilization [1, 29];
- FEM application for seeds behavior modeling [21, 23]; and
- DEM application for seeds behavior modeling [27, 28].

## 5. Conclusion

A detailed understanding of the mechanical behavior of plant seeds is currently an important science and technology challenge. Post-harvest geometrical and mechanical properties of the seeds or seeds mechanical behavior are important from the point of view of treatment process optimization and machines design. Basic dimensions and geometrical properties (length, width, thickness, mean diameters, seed surface area, sphericity, etc.) were described in this overview type of paper. Selected mechanical properties (e.g., compression energy, stress in the structure, coefficient of friction, rupture force, etc.) were also described. Mechanical behavior of seeds has been described and discussed in terms of both single seed and seeds-system. Selected experimental results from the research of mechanical behavior of *J. curcas* L. seeds were used to bring seeds behavior as a granular and particulate material closer to readers.

## Author details

Michal Petrů* and Ivan Mašín

*Address all correspondence to: michal.petru@tul.cz

Institute for Nanomaterials, Advanced Technologies and Innovation, Technical University in Liberec, Liberec, Czech Republic

## References


[1] Feda J. Mechanics of Particulate Materials: The Principles. Amsterdam, Oxford, New York: Elsevier Scientific Publishing Company; 1982

[2] Harper JL, Lovell PH, Moore K. The shapes and sizes of seeds. Annual Review of Ecology, Evolution, and Systematics. 1970;**1**:327-356

[3] Tanska M, Konopka I, Rotkiewicz D. Relationship of rapeseed strength properties to seed size, colour and coat fibre composition. Journal of the Science of Food and Agriculture. 2008;**88**:2186-2193

[4] Sirisomboon P, Kitchaiya P, Pholpho T, Mahuttanyavanitch W. Physical and mechanical properties of *Jatropha curcas* L. fruits, nuts and kernels. Biosystems Engineering. 2007; **97**:201-207

[5] Karaj S, Müller J. Determination of physical, mechanical and chemical properties of seeds and kernels of *Jatropha curcas* L. Industrial Crops and Products. 2010;**32**:129-138

[6] Gupta RK, Das SK. Physical properties of sunflower seeds. Journal of Agricultural Engineering Research. 1997;**66**:1-8

[7] Khodabakhshian R, Emadi B, Abbaspour Fard MH. Some engineering properties of sunflower seed and its kernel. Journal of Agriculture, Science and Technology. 2010;**4**:37-46

[8] Ozturk I, Kara M, Yildiz C, Ercisli S. Physico-mechanical seed properties of the common Turkish bean (Phaseolus vulgaris) cultivars 'Hinis' and 'Ispir'. New Zealand Journal of Crop and Horticultural Science. 2009;**37**:41-50

[9] Davies RM. Some physical properties of arigo seeds. International Agrophysics. 2010; **24**:89-92

[10] Özgüven F, Vursavus K. Some physical, mechanical and aerodynamic properties of pine (*Pinus pinea*) nuts. Journal of Food Engineering. 2005;**68**:191-196

[11] Ixtainaa VY, Nolascoa SM, Tomás MC. Physical properties of chia (*Salvia hispanica* L.) seeds. Industrial Crops and Products. 2008;**28**:286-293

[12] Niveditha VR, Sridhar KR, Balasubramanian D. Physical and mechanical properties of seeds and kernels of Canavalia of coastal sand dunes. International Food Research Journal. 2013;**20**:1547-1554

[13] Mansouri A, Mirzabe AH, Ráufi A. Physical properties and mathematical modeling of melon (*Cucumis melo* L.) seeds and kernels. Journal of the Saudi Society of Agricultural Sciences. 2017;**16**:218-226

[14] Sharma V, Das L, Pradhan RC, Naik SN, Bhatnagar N, Kureel RS. Physical properties of tung seed: An industrial oil yielding crop. Industrial Crops and Products. 2011;**33**: 440-444

[15] Dursun I, Tugrul KM, Dursun E. Some physical properties of sugarbeet seed. Journal of Stored Products Research. 2007;**43**:149-155

[16] Tunde-Akintunde TY, Akintunde BO. Some physical properties of sesame seed. Biosystems Engineering. 2004;**88**:127-129

[17] Yalcın I, Özarslan C. Physical properties of vetch seed. Biosystems Engineering. 2004; **88**:507-512

[18] Sirisomboon P, Pornchaloempong P, Teeranud Romphophak T. Physical properties of green soybean: Criteria for sorting. Journal of Food Engineering. 2007;**79**:18-22

[19] Vilche C, Gely M, Santalla E. Physical properties of quinoa seeds. Biosystems Engineering. 2003;**86**:59-65

[20] Sacilik K, Öztürk R, Keskin R. Some physical properties of hemp seed. Biosystems Engineering. 2003;**86**:191-198

[21] Petrů M, Novák O, Herák D, Simanjuntak S. Finite element method model of the mechanical behaviour of *Jatropha curcas* L. seed under compression loading. Biosystems Engineering. 2012;**111**:412-421

[22] Raji AO, Favier JF. Model for the deformation in agricultural and food particulate materials under bulk compressive loading using discrete element method II: theory, compression of oilseeds. Journal of Food Engineering. 2004;**64**:373-380

[23] Petru M, Novak O, Herak D, Masin I, Lepsik P, Hrabe P. Finite element method model of the mechanical behaviour of *Jatropha curcas* L. bulk seeds under compression loading: Study and 2D modelling of the damage to seeds. Biosystems Engineering. 2014;**127**:50-66

[24] Terzaghi K. Die Berechnung der Durchlassigkeitsziffer des Tones aus dem Verlauf der hydrodynamischen Spannungserscheinungen. Sitzungsber, Ak.Wiss., Math. - Naturwiss. Klasse, 132,3–4 (1923), 125–138.

[25] Hasseldine BPJ, Gao C, Collins JM, Jung HD, Jang TS, Song J, Li Y. Mechanical response of common millet (*Panicum miliaceum*) seeds under quasi-static compression: Experiments and modeling. Journal of the Mechanical Behavior of Biomedical Materials. 2017;**73**:102-113

[26] Koller AA, Taylor RK, Raun WB, Weckler PR, Buser MD. Modelling and validation of maize seed orientation by pushing. Biosystems Engineering. 2016;**151**:338-349

[27] Horabik J, Beczek M, Mazur R, Parafiniuk P, Ryzak M, Molenda M. Determination of the restitution coefficient of seeds and coefficients of visco-elastic Hertz contact models for DEM simulations. Biosystems Engineering. 2017;**161**:106-119

[28] Boac JM, Ambrose RPK, Casada ME, Maghirang RG, Maier DE. Applications of discrete element method in modeling of grain postharvest operations. Food Engineering Reviews. 2014;**6**:128-149

[29] Mohsenin NN. Physical Properties of Plant and Animal Materials. London: Gordon and Breach Science Publishers; 1986

[30] Herak D, Kabutey A, Choteborsky R, Petru M, Sigalingging R. Mathematical models describing the relaxation behaviour of *Jatropha curcas* L. bulk seeds under axial compression. Biosystems Engineering. 2015;**131**:77-83

[31] Masin I, Petru M. Complex approach to conceptual design of machine mechanically extracting oil from *Jatropha curcas* L. seeds for biomass-based fuel production. BioMed Research International. 2016:77-83